7th International Conference on Nuclear and Renewable Energy Resources (NURER2024)

Antalya, Turkiye
27-30 October 2024

Editors:

H. Mehmet Sahin
Erdem Ciftci
Sulenur Asal
Coskun Yuksel

ISBN: 979-8-3313-2087-4

Printed from e-media with permission by:

Curran Associates, Inc.
57 Morehouse Lane
Red Hook, NY 12571

Printed with permission by Curran Associates, Inc. (2025)

For permission requests, please contact Association for Scientific and Academic Research (ASAR) at the address below.

Association for Scientific and Academic Research (ASAR)

Phone: +91 8754929172

www.asar.net.in

Additional copies of this publication are available from:

Curran Associates, Inc.
57 Morehouse Lane
Red Hook, NY 12571 USA
Phone: 845-758-0400
Fax: 845-758-2633
Email: curran@proceedings.com
Web: www.proceedings.com

7th International Conference on

Nuclear and Renewable Energy Resources

October 27-30, 2024

Antalya, Türkiye

NURER 2024

nurer2024.org

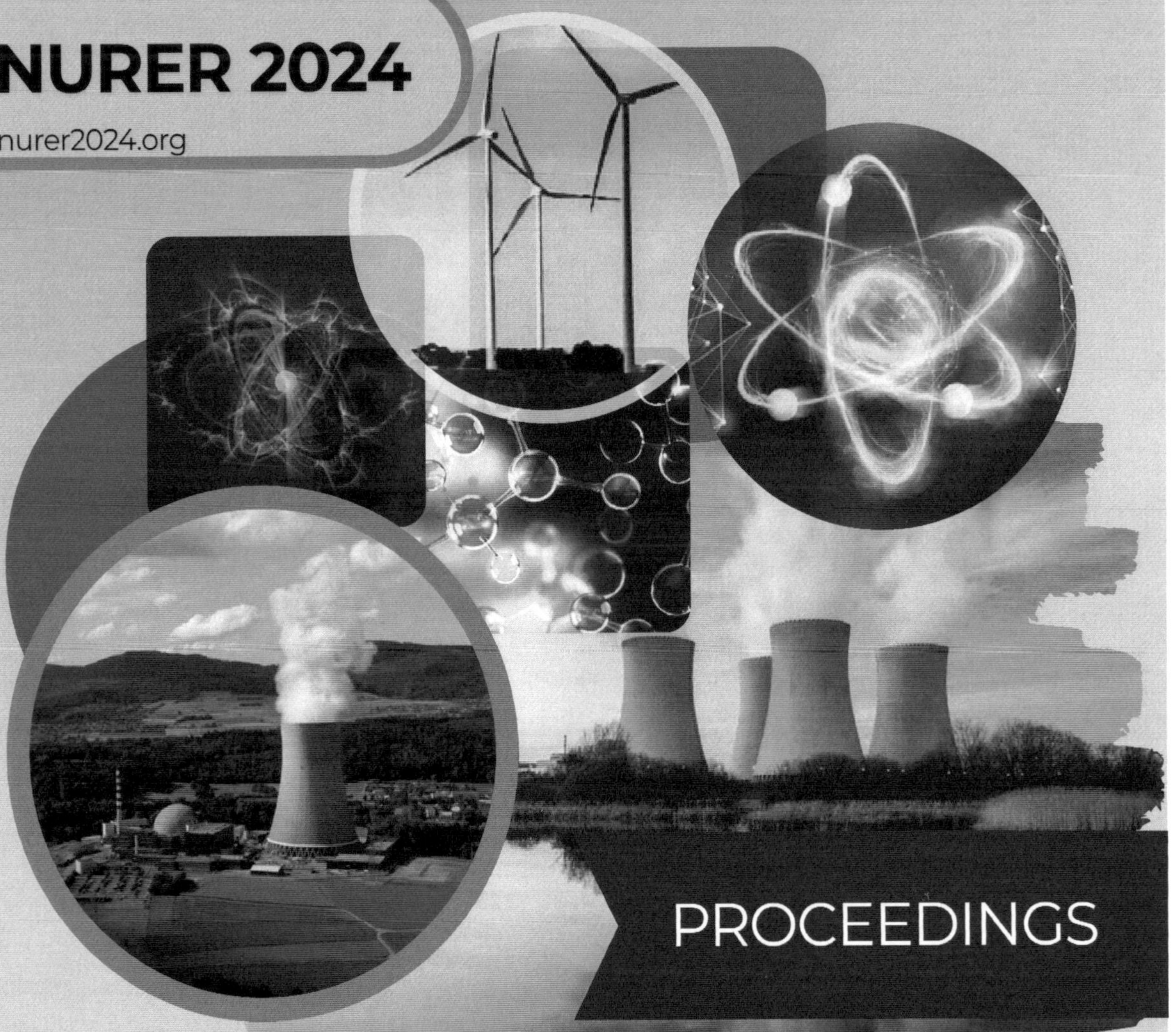

PROCEEDINGS

IAEA
International Atomic Energy Agency

American Nuclear Society

İstanbul Nişantaşı Üniversitesi

BAU
Bahçeşehir University

Sinop Üniversitesi

7th International Conference on

Nuclear and Renewable Energy Resources

(NURER 2024)

October 27-30, 2024

Antalya, Türkiye

CONFERENCE PROCEEDINGS

Edited By

Prof. Dr. H. Mehmet SAHIN

Assoc. Prof. Dr. Erdem CIFTCI

Res. Assist. Sulenur ASAL

Res. Assist. Coskun YUKSEL

INDEX

Foreword

Dear Colleagues,

The main objective of NURER conference series provide an international scientific and technical forum for scientists, engineers, industry leaders, policy makers, decision makers and young scientists/professionals who will shape future energy supply and technology, for a broad review and discussion of various advanced, innovative and non-conventional nuclear energy and renewable energy systems with UNTHINKABLE IDEAS on sound scientific-technical basis.

This conference underlines the importance of integrating two different concepts such as nuclear and renewable energies. The idea behind this is that both of these energy disciplines are specified as 'clean' energy sources. All of humanity stands to benefit when we apply our highest intellectual abilities to produce cutting-edge technologies in nuclear and renewable energy disciplines. We believe this is important because both energy types are clean, sustainable, safe, reliable and more abundant energy solutions. The idea is that nuclear and renewable energy sources are not competitors but complements. Nuclear power provides continuous and stable energy production, balancing the seasonal and weather changes of renewables. Nuclear energy is the base energy when these two energy sources are considered together.

The conference covers a variety of topics that are of great importance for all countries in terms of energy. The issues discussed here not only indicate the present situation of energy systems and technologies, but also concern future perspectives and innovative solutions that can shape the future. Therefore, the advancement of nuclear and renewable energy technologies will contribute to a better environment and living conditions for humanity in a peaceful manner.

The Organization Committee is proud to announce the NURER2024 that 65 papers have been accepted from 20 countries and have been presented in 15 sessions (5 plenary sessions). These papers are on the following subjects: 6 SMR, 3 ADS, 2 Thorium, 5 MSR, 15 Neutronic and reactor simulations, 5 CFD-thermal-hydraulics, 3 Combustion, 4 VVER, and 20 renewable energies, hydrogen production and applications.

Yours sincerely,

Prof. Dr. Sumer SAHIN
Honorary Chairman

Prof. Dr. Haci Mehmet SAHIN
General Chairman

Committees

Honorary Chair

Prof. Dr. Sumer SAHIN

General Chair

Prof. Dr. H. Mehmet SAHIN

Organizing Committee

Prof. Dr. Serhat KARYEYEN
Assoc. Prof. Dr. Erdem CIFTCI
Assis. Prof. Dr. Guven TUNC
Assis. Prof. Dr. Huseyin SAHINER

Standing Committee

Sumer SAHIN
Hacı Mehmet SAHIN
Yican WU
Yonghee KIM

Scientific Committee

Carlo R. C. RUBBIA	Xue-Nong CHEN
T. Nejat VEZIROGLU	V. JAGANNATHAN
Ibrahim DINCER	Minghuang WANG
Mohamed ABDOU	Ugur ATIKOL
José M. MARTINEZ VAL	Erol KURT
Senay YALÇIN	Necmettin SAHIN
Yunus A. CENGEL	Ruzhu WANG
Ashwani K. GUPTA	Sidik PERMANA
Abdul WARIS	José Rubens MAIORINO
Zaki SU'UD	Cláubia Pereira Bezerra LIMA
Ertugrul BALTACIOGLU	Aleksander GEORGIEV
Adnan MIDILLI	Bulent YANIKTEPE
Hiroshi SEKIMOTO	Wejdan ABU ElIHAIJA
José Manuel PERLADO	Francesco Paola ORSİTTO
Emilio MINGUEZ TORRES	Rizwan UDDIN
Ilhami COLAK	Xiaojing LIU
Mustafa ILBAS	Yuksel KAPLAN
Nader KARIMI	Ilker YILMAZ
Seref SAGIROGLU	Huseyin Turan ARAT
Mohammad Reza NEMATOLLAHI	Shoaib USMAN
Adem ACIR	Bora TIMURKUTLUK
Kamal HADDAD	Mustafa K. BALKI
Waclaw GUDOWSKI	Yunus UZUN

Technical Support

Res. Assist. Sulenur ASAL

Res. Assist. Coskun YUKSEL

7th International Conference on

Nuclear and Renewable Energy Resources

(NURER 2024)

SCIENTIFIC PROGRAM

NURER 2024 –PROGRAMME
7TH INTERNATIONAL CONFERENCE ON NUCLEAR AND RENEWABLE ENERGY RESOURCES

PG:			
	27 Oct 2024 Sunday		
	15:30-20:30 Registration		
	28 Oct 2024 Monday		
	08:00-09:00 Registration		
	Opening Session		
	09:00 - 09:15	Prof. Dr. H. Mehmet Şahin (Chairman)	
	09:00 - 09:15	Prof. Dr. Sümer Şahin (Honorary Chairman)	
	Plenary Session 1 (Hall A) **Chairman: Rizwan Uddin**		
181	**09:15-10:00**	**Introduction to accelerator-driven systems: the power and promise of renew**	**Jean-Pierre Revol**
N/A	**10:00-10:40**	**Problems and solutions in the field of geothermal energy sector**	**Yunus Çengel**
	10:40:11:20	**AEA platform on SMRs and their applications**	**Dohee Hahn**
	11:20 – 11:40 Coffe Break		
	Session 1 (Hall A)	**Small Modular Reactors**	Chairman: **Virginie Cartier**
77	**11:40-12:00**	Analysis of SBO sequences in a Small Modular Reactor using the TRACE code	**Jorge Arrabal** **David Canal** **Cesar Queral**
84	**12:00-12:20**	Probabilistic risk assessment of Molten Salt Reactors: Evaluating safety systems and reliability	**Ümmühan Barlak** **Huseyin Sahiner**
83	**12:20-12:40**	Radiological consequences of the SMRs - first steps towards deployment	**Adam Kecek**
37	**12:40-13:00**	LOCA sequence analysis in NuScale SMR with TRACE code	**Yago Martinez-Gonzalez** **Jorge Sanchez-Torrijos** **Cesar Queral**
	13:00-14:00 Conference Photo and Lunch		
	Plenary Session 2 (Hall A)	**Chairman: Victor Hugo Sanchez-Espinoza**	
183	**14:00-15:00**	**Status of SMR and advanced reactors supply-chain, economic impact and licensing requirement for participation**	**Shoaib Usman**
	Session 2 (Hall A)	**Simulations of Nuclear Reactors**	Chairman: **Jean-Christophe de Mestral**
29	15:00-15:20	Simulation of MOTEL experimental tests MS-SG01 and MS-SG02 using TRACE code	**Yago Martinez-Gonzalez** **Cesar Queral** **Elena Redondo-Valero** **Jorge Sanchez-Torrijos**
132	15:20-15:40	Mitigating recirculation zones in molten salt fast reactor models	**Ismail Yagci** **Zekeriya Parlak** **Huseyin Sahiner**
	15:40-16:00 Coffee Break		
	Session 3 (Hall A)	**Renewable Energies and Applications**	Chairman: **Shoaib Usman**
78	16:00-16:20	Influence of reduced graphene oxide on anode functional layer performance in Solid oxide fuel cells	**Furkan Toruntay**
124	16:20-16:40	Characterization of microstructure and corrosion properties of powder metallurgy Ti-Al-Sn-Zr-Mo alloy for nuclear applications	**Eda Ergün Songül** **İlven Mutlu**
200	16:40-17:00	An Investigation on large scale green hydrogen production through electrolysis: Case study of AKKUYU nuclear power plant	**Fethi Mustafa Çimen** **Saltuk Buğrahan Yeşilgöl**
156	17:00-17:20	Design and assessment of the ARC fusion reactor integrated energy system to produce power and hydrogen	**Sulenur Asal** **Adem Acır**

PG:	29 Oct 2024 Tuesday		
	Planary Session 3 (Hall A) Chairman: **Adam Kecek**		
182	**09:00-10:00**	**Transmutex**	**Jean-Christophe de Mestral**
	Session 4 (Hall A)	**Analysis of VVERs**	Chairman: **Cesar Queral**
86	10:00-10:20	KIT Investigations of the VVER-1000 plant behavior under design basis accidents	**Victor Hugo Sanchez-Espinoza** **Nuri Beydogan** **Gianfranco Huaccho Zavala**
5	10:20-10:40	Analysis of the impact of passive safety systems on accidental sequences of VVER-1000 reactors.	**Elena Redondo-Valero** **Cesar Queral** **Kevin Fernandez-Cosials**
19	10:40-11:00	Comparative analysis of accidental sequences with conventional and Advanced Tolerant Fuels (ATF) in VVER-1000 reactors	**Elena Redondo-Valero** **Cesar Queral** **Jorge Sanchez-Torrijos** **Emilio Castro-Gonzalez**
	11:00-11:20 Coffee Break		
	Session 5 (Hall A)	**Neutronic Analysis and Reactor Physics**	Chairman: **Khurrum Saleem Chaudri**
45	11:20-11:40	Optimized fuel cycle and burnup analysis for pebble-bed reactors	**Sefa Bektaş** **Volkan Seker** **Üner Çolak** **Thomas Downar**
205	11:40-12:00	Neutronic Study of a Molten Salt Uranium Mixture in a Fusion Reactor Blanket	**H. Mehmet Şahin** **Güven Tunç** **Alper Karakoç**
113	12:00-12:20	Reactor physics analysis of accident tolerant fuel cladding system for 24-month cycle of APR-1400	**Khurrum Saleem Chaudri** **Mohammad Alrwashdeh** **Saeed A. Alameri**
	13:00-14:00 Conference Photo-2 and Lunch		
	Planary Session 4 (Hall A) Chairman: **Jean-Pierre Revol**		
N/A	**14:00-14:40**	**Integration of multiple energy sources (including nuclear) and multiple uses in an optimally designed and optimally operated energy network**	**Rizwan Uddin**
	Session 6 (Hall A)	**Nuclear Management**	Chairman: **Hidemasa Yamano**
1	14:40-15:00	International cooperation for nuclear management: an institutional framework	**Mehmet Şahin** **Ömer Faruk Kocatepe**
85	15:00-15:20	Environmental impacts of energy production and alternatives	**Huseyin Sahiner**
131	15:20-15:40	Tsunami flood protection - system	**Jens Wieneke**
81	15:40-16:00	Role of nuclear-renewable hybrid energy systems – a case study of natriumtm energy system for power and hydrogen production	**Hailei Wang**
82	16:00-16:40	Synthetic time series modeling and sensitivity study for robust techno-economic analysis of advanced nuclear energy systems	**Hailei Wang**
	16:40-17:00 Coffee Break		
	Session 7 (Hall A)	**Renewable-Hydrogen Technologies**	Chairman: **Hailei Wang**
54	17:00-17:20	Hydrogen technology development during a revolutionary era in Europe	**Virginie Cartier** **Elisabeth A. Cazier**
161	17:20-17:40	Long-term planning of power generation volumes for the Akkuyu nuclear power plant	**Saltuk Buğrahan Yeşilgöl** **Fethi Mustafa Çimen**
122	17:40-18:00	Prediction of fuel variability impacts on the combustion characteristics of $NH_3/H_2/N_2$ fuel mixtures using machine learning techniques	**Israfil Soyler** **Cihat Emre Ustun** **Amin Paykani** **Nader Karimi**
125	18:00-18:20	Numerical analysis of air-channel photovoltaic thermal collectors and evaluation of hydrogen production	**Erdem Çiftçi** **Murat Öztürk**
180	18:20-18:40	Ultra-low NOx emissions and increased power in a	**Berre Kumuk**

PG:			
		combustor-SOFC System under CDC conditions	
	19:30 GALA DINNER		
	30 Oct 2024 Wednesday		
	Session 8 (Hall A)	**Thermal-Hydraulic and Energy Storage**	Chairman: **Gulnur Tyulepberdinova**
76	09:40-10:00	Safety design technologies for sodium-cooled fast reactor coupled to thermal energy storage system with sodium-molten salt heat exchanger	**Hidemasa Yamano** **Kazuya Takano** **Kenichi Kurisaka** **Shin Kikuchi** **Toshiki Kondo** **Masaaki Hayashi**
100	10:00-10:20	Numerical simulation of underground hydrogen storage in a depleted gas reservoir: Effect of well configurations	**Emrah Sarı** **Erdem Çiftçi**
123	10:20-10:40	Numerical simulation of thermal-hydraulic performance of staggered tube bank with longitudinal triangular fins	**Mustafa Söyler**
	10:40-11:00 Coffee Break		
	Session 9 (Hall A)	**Renewable Energy and Applications**	Chairman: **Farimehr Ghanizadeh Geraili**
145	11:00-11:20	Utilizing gamma-irradiated Iranian propolis (bee glue) extract for the synthesis of green silver nanoparticles	**Farimehr Ghanizadeh Geraili** **Seyed Mahmoud Reza Aghamiri** **Marzieh Heidarieh**
166	11:20-11:40	Heat pump system of renewable energy sources utilization as a basis for the development of a digital twin	**Gulshat Amirkhanova** **Murat Kunelbayev** **Gulnur Tyulepberdinova** **Madina Mansurova** **Baurzhan Amirkhanov**
133	11:40-12:00	Experimental studies of a double-circuit solar installation with thermosiphon circulation with a heat pump	**Gulnur Tyulepberdinova** **Murat Kunelbayev** **Gulshat Amirkhanova** **Ardak Tolepberdinova**
	12:00- 12:20 Planery Session 6 Closing Talk (*Prof. Dr. H. Mehmet Şahin*)		
	13:00-14:00 LUNCH		

7th International Conference on

Nuclear and Renewable Energy Resources

(NURER 2024)

PROCEEDINGS

International Cooperation for Nuclear Management: An Institutional Framework

Mehmet Şahin
National Defence University, Ankara, Türkiye, mesahin@kho.msu.edu.tr,
ORCID: 0000-0002-0142-6666

Ömer Faruk Kocatepe
National Defence University, Ankara, Türkiye, omerfarukkocatepe@hotmail.com,
ORCID: 0000-0001-8647-4634

Cite this paper as: *Şahin, M, Kocatepe, OF. International Cooperation for Nuclear Management: An Institutional Framework. 7th Int. Conf. on Nucl. and Ren. Energy Res. 27-30 October 2024, Antalya, Turkey*

Abstract: Energy deficiency has been exacerbated by the world's growing urban population. This highlights three major issues: underdevelopment, energy insecurity, and pollution. Although nations are inclined to accept these difficulties as national security concerns, none of the aforementioned issues can be addressed alone. Instead, energy security, environmental difficulties, and underdevelopment are global concerns. Thus, contemporary security studies, such as the Critical School or the Copenhagen School, focus on the securitization of non-traditional security sectors, such as environmental security or energy security, which has a growing body of literature. Nonetheless, energy security is not adequately addressed in security studies. While resource-rich countries approach it from a national standpoint, developing or resource-poor countries are tempted to see it as an international issue. As a result, an international framework for energy management should be strengthened to address global issues and foster energy cooperation. Cooperation in nuclear energy complicates the situation. In this regard, comprehensive norms and systems for worldwide nuclear energy management are required to guarantee security while also providing for long-term development. Thus, multilateral collaboration in energy management is critical for nuclear security. However, nuclear energy management is heavily reliant on bilateral partnerships. In this regard, this article will look at the challenges that come with international nuclear cooperation. It will be argued that global collaboration is required in nuclear management to address underdevelopment and environmental concerns. Furthermore, multinational collaboration is required to improve safety and secure the peaceful use of nuclear energy. From this perspective, the article will investigate the institutional structure of international nuclear cooperation. That is, it will be claimed that the institutional development of a viable international nuclear energy regime is limited because nuclear energy is overshadowed by power imbalance in an oligopolistic market. In other words, nuclear energy is viewed from a power perspective rather than a developmentalist one, preventing cooperation and the building of a competent international regime.

Keywords: *International Energy Management, Nuclear Cooperation, Energy Security*

1. INTRODUCTION

The transfer of nuclear technology is one of the most debated topics in the nuclear proliferation literature and there are many research questions in this regard. Ultimately, the transfer of nuclear technology can initiate a process that leads to relative gains for the recipient state, which is unacceptable for the donor country from the realist perspective. As Fuhrmann argues, 78% of the countries that produced the nuclear bomb received some assistance, and no country acquired weapons without receiving aid from 1953 to 2000 [4] but still states share knowledge of nuclear technology. Given this reality, why do states engage in nuclear cooperation is one of the research question in security studies. In the end, a security hierarchy exists, with military security being the primary concern of states, and nuclear threats being paramount. Thus, states are expected to be reluctant to share nuclear technology. However, especially among Global North countries, there is significant nuclear cooperation. This question, which is a significant research problem from a realist perspective, has been addressed in the literature with various explanations. These include supply-side factors [4, 7], diplomatic influence [6], economic competition [3, 8], forming alliances [5] and geopolitical considerations [12].

However, it is also a fact that states are selective in their dissemination of nuclear technology. While economic competition and diplomatic influence can explain cooperation to a certain extent, they fall short in fully accounting for the phenomenon. If this were merely a marketing effort, one would expect to see a significant expansion of the market regardless the receiving country. Similarly, while states seek to expand their political influence, cooperation in this area remains limited. Thus, while existing approaches provide partial explanations, they still leave gaps. Similarly, alliance formation and geopolitical considerations explain another part of the story but leave other aside. In general, realist explanations miss to explain cooperation among states and liberal explanations miss the selectivity part of the story.

2. CONTRADICTIONS IN NUCLEAR MANAGEMENT

It is natural for such gaps to exist given the nature of nuclear policies. Indeed, the transfer of nuclear technology is built upon this contradiction. While it is a crucial tool for economic growth and gaining influence, it is also a risky process that can lead to relative losses for provider states. From a constructivist perspective, it can be argued that nuclear technology is more freely transferred among liberal states, while authoritarian governments are also more open to sharing knowledge with each other. However, this claim needs to be empirically tested. Still, it can be said that Global North countries are more generous in transferring technology to liberal countries due to the desecuritisization processes among each other. On the other hand, it is clear that lacking nuclear technology of third parties has become a security issue in the modern world. Technological backwardness, and more importantly, underdevelopment, significantly contributes to global inequality. This, especially in the liberal world, leads to processes of societal and political securitization.

This contradiction and complexity necessitates institutional cooperation and governance. While nuclear security is the paramount threat in the security hierarchy, underdevelopment also stands as a significant security challenge. Therefore, peaceful technology transfer is subject to international regulations under the IAEA. However, global energy governance has been relatively unexplored in the literature, partly because it is difficult to conceptualize as a coherent field [3]. That is to say, energy security is a broad and contemporary field in international relations but it has competing and contrasting issues in each sub-field. When it comes to nuclear governance, the limited literature mostly criticizes IAEA. It has been heavily criticized for being non-independent [1] or practical difficulties in detecting violations of the NPT [11].

Institutional failure is another perspective that some scholars emphasize. The ineffectiveness of the IAEA is often attributed to factors such as security considerations of nation states [10], limited data sharing [9], and political resistance [2]. While these factors highlight different aspects of the problem, they all stem from the underlying fear of states that joining the organization, sharing information, or granting the IAEA more authority would infringe upon their sovereignty. This poses a significant challenge for the IAEA, especially in an area as sensitive as nuclear non-proliferation. In essence, there is a tension between the international system, which seeks to regulate the spread of nuclear technology, and states, which prioritize their autonomy.

3. BILATERALISM AND MULTILATERALISM

Furthermore, or may be for this reason, nuclear cooperation and management is heavily relying on bilateral partnerships instead of multilateral negotiations and management. This bilateralism causes one of the main reasons for lack of institutionalism. These agreements can have certain advantages, such flexibility and customized solutions, but they also greatly add to the absence of a strong international institutional framework.

On the other side, multilateral management and talks offer a forum for tackling nuclear matters in a more thorough and cooperative way. Multilateral techniques can aid in establishing shared standards and procedures, promoting transparency, and establishing trust by encouraging communication among a broader range of stakeholders.

There are various ramifications when there is no robust international organization for nuclear cooperation. First of all, it may result in inconsistent safety regulations and standards, which could raise the possibility of mishaps or spread. Second, it can make it challenging to handle global issues that call for coordinated international action, such nuclear waste management and nonproliferation. Lastly, it can lead to an imbalance of power as, in bilateral discussions, larger countries may have more sway over smaller ones.

A more international approach is urgently needed to handle the challenges of the twenty-first century and fortify the global nuclear transfer. This would entail creating or fortifying international organizations tasked with advancing nuclear cooperation, guaranteeing safety, and addressing nonproliferation issues. These organizations, particularly IAEA might be extremely important in creating and enforcing global standards, promoting information exchange, and offering a platform for conflict resolution. However, states generally do not prefer multilateralism because it ultimately limits the sovereignty of nation-states. As a result, states primarily prefer bilateralism due to security concerns and sovereignty issues.

Therefore, the transfer of nuclear technology encompasses two significant paradoxes. The first is the paradox of security and underdevelopment, and the second is the paradox of control and autonomy. While the first paradox relates to international relations, the second is associated with international institutionalization. This study will focus on these paradoxes. The main argument here is that the first paradox is the primary cause of the second one. In other words, international rules and regulations are shaped by the hierarchy of security. Moreover, the limited cooperation on technology transfer causes demand asymmetry, where technology owners insist more on control, while developing countries are more in favor of autonomy.

RESEARCH QUESTION

Why has this gap broadened? This can be explained by either a constructivist approach or realism, emphasizing the power asymmetry of nuclear countries. Constructivists may explain this through the lens of liberal identity. Accordingly, liberal states do not fear sharing technology with other liberal states. In contrast, cooperation between liberal and illiberal states in nuclear technology transfer is limited. Realists, on the other hand, may point to power asymmetry within the IAEA. Accordingly, the interests of developing countries are less considered by international institutions due to the use of power asymmetry as a trump card by nuclear owners in the IAEA and similar institutions. In other words, the United States and other Western Bloc countries are reluctant to share nuclear knowledge with their non-allies and use the IAEA as a means to impose strict rules to limit proliferation. However, as has been asserted, since nuclear sharing means a possible relative loss, the realist approach has a shortcoming in explaining nuclear policies in general. Therefore, this study will seek to explain it from a constructivist perspective.

REFERENCES

[1] 2011. A watchdog with bite. *Nature* 472, 7344, 389.
[2] Boureston, J. and Ogilvie-White, T. 2010. Expanding the IAEA's nuclear security mandate. *Bulletin of the Atomic Scientists* 66, 5, 55–64.
[3] Dubash, N. K. and Florini, A. 2011. Mapping Global Energy Governance. *Global Policy* 2, s1, 6–18.
[4] Fuhrmann, M. 2009. Spreading Temptation: Proliferation and Peaceful Nuclear Cooperation Agreements. *International Security* 34, 1, 7–41.
[5] Fuhrmann, M. 2009. Taking a Walk on the Supply Side. *Journal of Conflict Resolution* 53, 2, 181–208.
[6] Gartzke, E. and Kroenig, M. 2009. A Strategic Approach to Nuclear Proliferation. *The Journal of Conflict Resolution* 53, 2, 151–160.
[7] Kroenig, M. 2009. Importing the Bomb: Sensitive Nuclear Assistance and Nuclear Proliferation. *The Journal of Conflict Resolution* 53, 2, 161–180.
[8] Lantis, J. S. 2014. Economic Competition and Nuclear Cooperation. *The Nonproliferation Review* 21, 1, 21–41.
[9] Ogilvie-White, T. 2014. The IAEA and the International Politics of Nuclear Intelligence. *Intelligence and National Security* 29, 3, 323–340.

[10] Rettig, E. 2016. Limits to cooperation: why Israel does not want to become a member of the International Energy Agency. *Israel Affairs* 22, 2, 512–527.

[11] Rockwood, L. 2018. How the IAEA verifies if a country's nuclear program is peaceful or not: The legal basis. *Bulletin of the Atomic Scientists* 74, 5, 317–325.

[12] Tandler, J. 2014. French Nuclear Diplomacy. *The Nonproliferation Review* 21, 2, 125–148.

NURER 2024
Nuclear and Renewable Energy Resources

Analysis of the impact of passive safety systems on accidental sequences of VVER-1000 reactors

Elena Redondo-Valero
NFQ Advisory Services S.L., Madrid, Spain, elena.redondo@nfq.es,
ORCID: 0000-0002-9842-0892

Cesar Queral
Universidad Politécnica de Madrid (UPM), Madrid, Spain, cesar.queral@upm.es,
ORCID: 0000-0001-9384-2474

Kevin Fernandez-Cosials
Universidad Politécnica de Madrid (UPM), Madrid, Spain, kevin.fcosials@upm.es,
ORCID: 0000-0001-6722-1084

Victor Hugo Sanchez-Espinoza
Karlsruhe Institute of Technology (KIT), Karlsruhe, Germany, victor.sanchez@kit.edu
ORCID: 0000-0001-9397-7600

Cite this paper as: *Redondo-Valero, E, Queral, C, Fernandez-Cosials, K. Analysis of the impact of passive safety systems on accidental sequences of VVER-1000 reactors. 7th Int. Conf. on Nucl. and Ren. Energy Res. 27-30 October 2024, Antalya, Türkiye*

Abstract: VVER are one of the most common reactor types in the world. Moreover, a significant percentage of the Gen-III /III+ reactors that are currently being built or have recently come into operation are VVER. Therefore, there is a growing interest in studying their behavior under both anticipated and accidental transients. These Gen-III /III+ reactors have improved their safety margins by implementing passive safety systems that do not require human actions or external power sources. Some advanced VVER designs have incorporated a passive emergency core cooling system, the second stage hydroaccumulator (HA-2), and a Passive Heat Removal System (PHRS). In the present study, the impact of the HA-2 and the PHRS has been analyzed by means of a TRACE model for a VVER-1000 reactor. For this purpose, a LOCA along with SBO sequence have been simulated. The results show that the combined performance of HA-2 and PHRS in LOCA under SBO sequences allows avoiding fuel damage for 24 hours.

Keywords: *VVER-1000, Passive Heat Removal System (PHRS), Second Stage Hydroaccumulators (HA-2), SBO, LOCA*

1. INTRODUCTION

This paper presents a detailed discussion of the impact of two passive safety systems, the second-state hydroaccumulator (HA-2) and the air-cooled Passive Heat Removal System (PHRS), on the advanced VVER-1000. For this purpose, a thermal-hydraulic model has been developed for the TRACE code of the air-cooled PHRS system. Then it has been incorporated into a full plant model that already included the HA-2. The accidental sequence selected to study the impact of these passive safety systems is the Double Ended Guillotine Break (DEGB) Loss Of Coolant Accident (LOCA)

Previous studies have developed isolated thermal-hydraulic models of the air-cooled PHRS, see [1], [2], [3]. There are also several references on experimental facilities for these passive safety systems, see [4], [5], [6], [7], [8], [9], [10], [11]. However, only a few references analyse in detail the effect of the air-cooled PHRS and the HA-2 in a LOCA under SBO conditions with full plant thermal-hydraulic models, see [12], [13].

The present work has been carried out in the framework of the "Integrated Safety Analysis of Modular and Evolutive Reactors" (ISASMORE). Within this project, accidental sequences (SBLOCA, MBLOCA, LBLOCA, SBO, ...) have been analyzed with the thermal-hydraulic TRACE code, focusing on the conventional safety systems of GEN II VVER reactors, see [14], [15]. This includes the verification of the Probabilistic Safety Analysis (PSA) Level 1 Event Trees (ET) and the proposal of new ones. In addition, these accidental sequences have been re-examined taking into account the actuation of HA-2, see [16].

The previous experience of the Universidad Politécnica de Madrid (UPM) with TRACE (also some analyses with MELCOR) regarding the passive safety systems is related to the following reactor designs and experimental facilities:

- MOTEL facility (natural circulation, helical-coil SGs, H2020 McSafer project) [17],
- ACME facility (PHRS, CMT, ACC, IRWST),
- Sirio facility (PHRS, H2020 PIACE Project) [18],
- Plant model: NuScale (1D, 3D models, ECCS and PHRS, H2020 McSafer project),
- Plant model: AP1000 (full plant model, PHRS, CMT, ACC, IRWST) [19], [20],
- Plant model: CAREM-like (PHRS, 1D and 3D models).

The paper is structured as follows: first, section 2 discusses the PHRS design selected for analysis. Then, in section 3, the modelling for the TRACE code of the air-cooled PHRS and its implementation in a full plant model including the HA-2 is shown. Subsequently, section 4 presents the DEGB LBLOCA under SBO conditions simulated considering the HA-2 and the air-cooled PHRS. A comparison of this sequence without consideration of both passive safety systems is given in section 5. Finally, the conclusions are presented in section 6.

2. DESCRIPTION OF THE AIR-COOLED PHRS

The PHRS thermal-hydraulic model developed at the UPM corresponds to the air-cooled PHRS present in the VVER-1000/V412, see [21]. This choice was made for two reasons:

1. In addition to simulating SBO sequences, the aim was also to simulate LOCA sequences with SBO, and the advanced VVER reactors with a long-term passive ECCS, the HA-2, are those which incorporate the air-cooled PHRS.
2. Among the VVER plants with air-cooled PHRS, the one for which the most public information is available is the VVER-100/V412, commissioned at the Kudankulam Nuclear Power Plant (NPP).

The PHRS incorporated in the VVER-1000/V412 consists of four trains, one per SG, see Figure 1. Each of the PHRS trains contains 3 HXs, which are located in an elevated position, right where the containment dome begins, [3]. It is important to note that in the VVER-1200/V392M/V509/V523 and the VVER-TOI, the PHRS also consists of four trains, but each has two HXs instead of three, [22]. The design criterion for the air-cooled PHRS in all these reactor designs is that a maximum of 3 out of 4 trains are required to fulfil their safety function, [1].

Each HX is placed in a shell inside an air duct. In the air duct, upstream and downstream of the HX, there are air gates with the capacity to isolate the HX from the atmosphere, [21]. Moreover, between the HX and the upstream air gate valve, there is a regulator with the capacity to control the air flow through the HX, see Figure 2 and [23]. It is important to note that air gate valves have only two positions, closed and open, i.e. no intermediate positions.

Each HX contains a steam collector and a condensate collector, being the steam collector above the condensate collector, see Figure 2. On the one hand, the steam collector is connected to the condensate collector by a total of 630 horizontal U-shaped tubes with a slight slope, [24]. On the other hand, the steam collectors and condensate collectors are connected to the SGs thought the steam lines and the condensate lines.

The PHRS is designed to remove decay heat for extended periods of time in the event of unavailability of the active heat removal system through the SGs or total loss of AC power, i.e. SBO conditions, [13], [21]. Therefore, two actuation signals have been found:

- AC power loss, i.e., SBO conditions.
- High SG pressure, i.e., SGs pressure above 8.44 MPa.

The actuation signals of the air gate valves have a delay of 30 s, see [25]. Moreover, the PHRS HXs take 90 s to reach full power, [13]. Besides, the PHRS has two operating modes: SGs pressure maintenance mode and RCS cool-down mode, [12], [13]:

- SGs pressure maintenance mode: The aim is to remove the decay heat from the core through the SGs by maintaining the pressure of the SGs between 5.35 MPa and 6.05 MPa, [3]. This is achieved by controlling the regulator area in the air duct and therefore the air flow through the HX tubes. When, the PHRS starts to operate, the regulators are fully open. As the pressure in the SGs drops below 5.35 MPa, the regulators begin to close, drastically reducing the power that the PHRS can dissipate compared to the power it could remove if the regulators were fully open, see the red line in Figure 3.

- RCS cool-down mode: If the SBO sequence or the loss of the active heat removal systems is followed by a LOCA, the PHRS operates in RCS cool-down mode. When the subcooling margin in the hot legs is below 8°C, [13], the control that regulates the regulators area is inhibited, forcing them to remain fully open. In this model, the PHRS behavior is characterized by a power vs. SG pressure curve, see Figure 3.

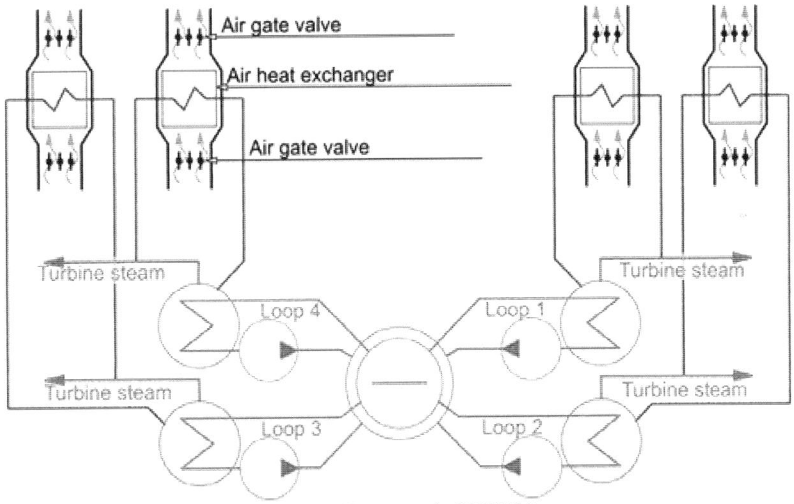

Figure 1. VVER air-cooled PHRS layout

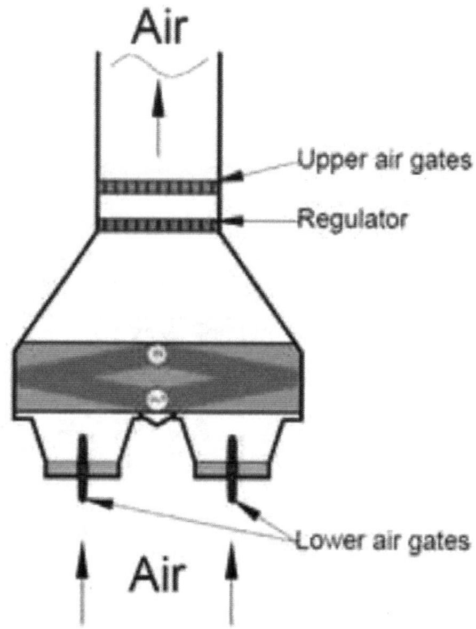

Figure 2. PHRS heat exchanger, regulator and air gates

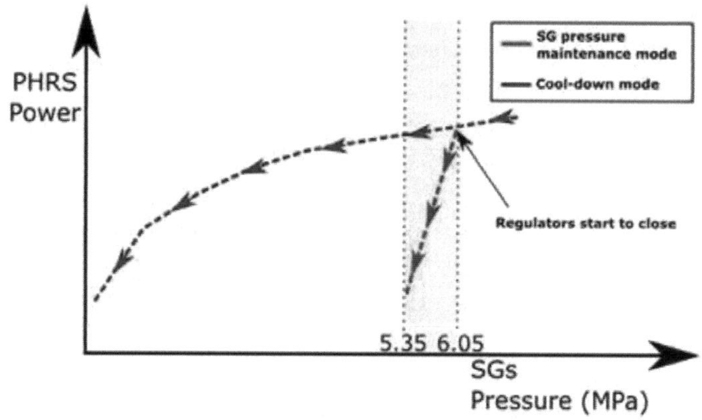

Figure 3. PHRS power vs. SGs pressure curve in both operating models

3. PHRS MODEL FOR THE TRACE CODE

A thermal-hydraulic model of the air-cooled PHRS has been developed for the TRACE code. As the four PHRS trains are symmetrical, only one of them has been developed as an isolated model with boundary conditions in the steam line inlet and the condensate line outlet. The thermal-hydraulic model includes the air ducts from the air inlet to the outlet, the U-tube of the HXs, the steam lines and the condensate lines, see Figure 4 (left).

In the TRACE model, the air gate valves are controlled by two signals which cause them to open: SBO conditions and SG pressure above 8.44 MPa, [13]. The opening of the air gates allows the air to pass through the air duct, which leads the PHRS to start operation. On the other hand, the area of the regulator is determined by a control that maintains the SG pressure between 5.35 MPa and 6.05 MPa. In addition, a hot lets subcooling signal has also been implemented on the thermal-hydraulic model, which forces the regulators to remain fully open, overriding the regulator area control. In this way it is possible to switch from SG pressure maintenance mode to RCS cool-down mode.

To verify the performance of the air-cooled PHRS model developed at the UPM, the power that a train can remove at different pressures has been determined. The results obtained has been then compared with the results from the FSAR thermal-hydraulic code in [3] for an isolated PHRS model of the Kudankulam NPP. In order to have a comprehensive view of the behavior of the PHRS under different outdoor conditions, the PHRS power vs. pressure curves have been obtained for various air temperatures: 10 °C, 20 °C, 30 °C and 40 °C, see Figure 4 (right). Each curve has been obtained for a total of nine pressure values, from 8 MPa to 0.1 MPa. It can be seen as the curve is very close to the Kudankulam curve from 8 MPa to 1 MPa. The value for the atmospheric pressure of the Kudankulam NPP is not available in [3], but it can be found in [7] that the power for a pressure of 0.15 MPa is about 15 MW for an outside temperature of 38 °C at the Novovoronezh NPP, which is about 3.75 MW per train, see Figure 4 (right).

The PHRS model, has been implemented in a thermal-hydraulic model for the TRACEV5P5 code [26] of a VVER-1000 reactor, see Figure 5 and Figure 6. The nodalization of the full plant model was built based on a VVER-1000 RELAP5 model, [27], and has been applied for MB/LBLOCA and SBLOCA analyses in previous studies, see [14], [15], [16].

The thermal-hydraulic model of the VVER-1000 reactor includes an advanced PSS, the HA-2, a passive low-pressure ECCS connected downstream to the First Stage Hydroaccumulator (HA-1) discharge lines and upstream to the cold legs. This system consists of four trains of 33% capacity each and two 120 m^3 tanks which inject borated water into the RCS in four stages, see Figure 7 and [13], when the RCS pressure is below 1.5 MPa. The HA-2 actuation signal has a delay of 100 s.

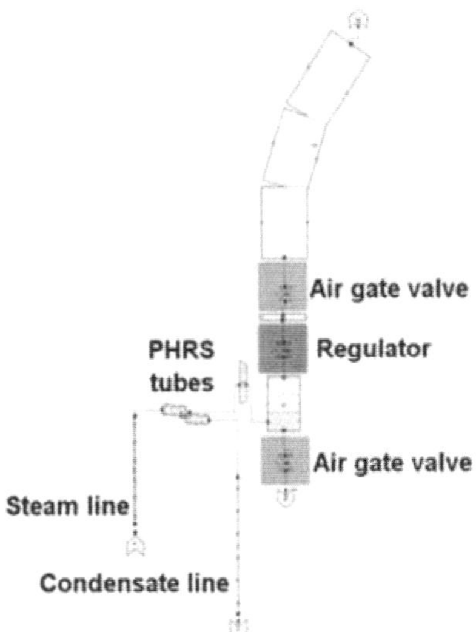

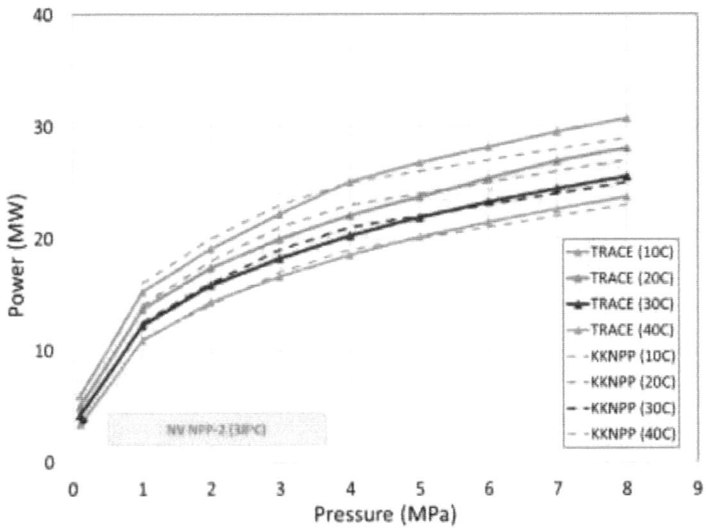

Figure 4. PHRS (one train) isolated thermal-hydraulic model for TRACE (left), PHRS power per trains vs. pressure (right)

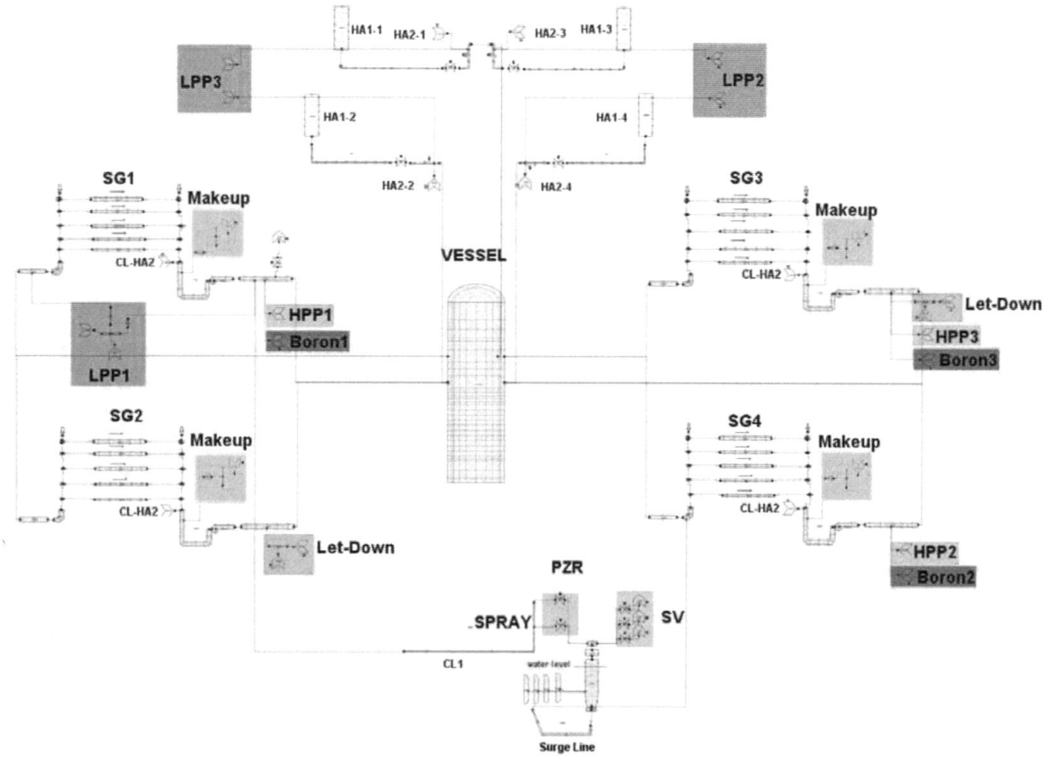

Figure 5. Reactor Coolant System with the HA-2 view of the VVER-1000 reactor TRACE model

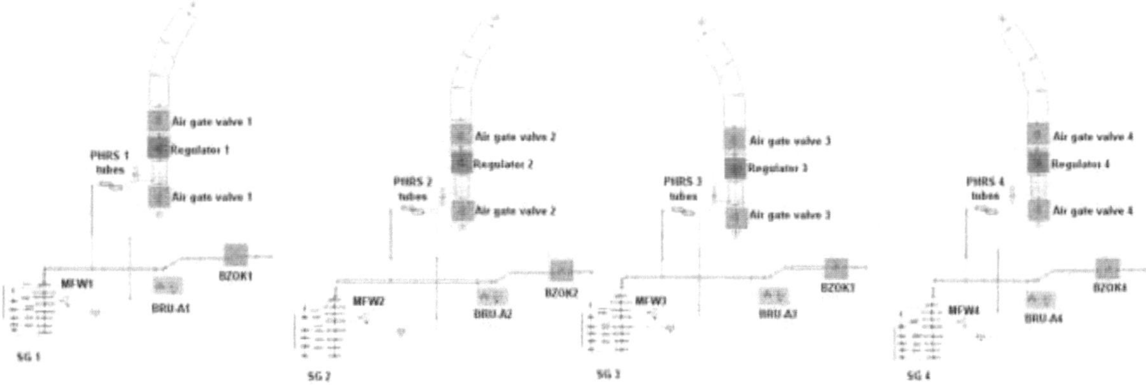

Figure 6. Steam Generators secondary side with the PHRS of the VVER-1000 reactor TRACE mode

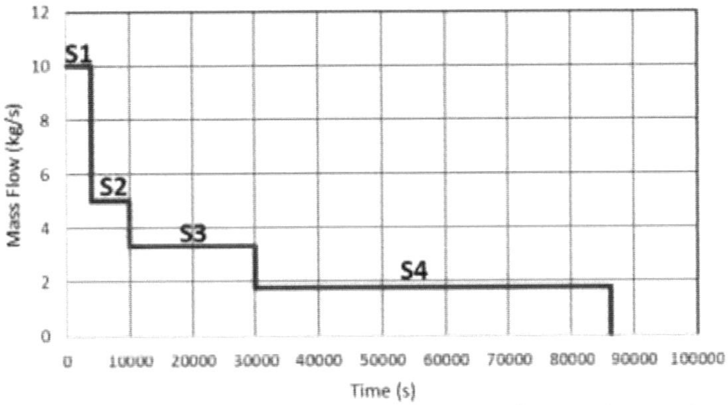

Figure 7. Second Stage Hydro-accumulators mass flow rate (per train)

4. PERFORMACE OF THE PHRS AND THE HA-2 IN A LBLOCA UNDER SBO CONDITIONS

The aim of this section is to determine the performance of the PHRS in RCS cool-down mode along with the HA-2 in an LOCA sequence under SBO conditions. The LBLOCA chosen corresponds to a DEGB, with an equivalent break diameter of 47 inches, located in the C L1. The atmospheric temperature chosen for this analysis is 303.15 K (30°C). In addition, the success criteria considered for the passive safety systems involved are those of the design criteria:

- HA-1: 2 out of 4 trains
- HA-2: 3 out of 4 trains
- PHRS: 3 out of 4 trains

It has been found in the literature that the containment pressure in LOCA sequences between 0.15 MPa and 0.25 MPa, see [13], [28]. This is due to the RCS inventory discharge through the break. Therefore, in the present analysis, the containment is considered to be pressurized at 0.2 MPa.

The LBLOCA starts 300 seconds from the beginning of the simulation. At that instant there is a total loss of AC power, without starting the diesel generators, which causes SCRAM, tripping of the RCPs, loss of the CVCS, tripping of the MFW pumps, tripping of the Turbine and loss of the condenser. The evolution of the events is shown in Table 1. Due to the large size of the break, the RCS depressurizes rapidly and reaches atmospheric pressure within a few seconds, see Figure 8. The HA-1 therefore injects within 15 s of the beginning of the sequence and the HA-2 set point is reached 10 s later, which causes the increase in the collapsed liquid level, see Figure 10.

The second stage of the HA-2 injection starts at 4325 s, the third stage at 10325 s and finally the fourth stage at 30325 s. On the other hand, due to the emptying of the HA-2, there is a small inventory from the CLs sucked up by the HA-2 upstream, see Figure 11.

The PHRS actuation signal is generated by the SBO conditions. Despite the PHRS actuation signal delay of 30 s, the SG pressure drops from the beginning of the sequence because the RCS cools the SGs by its abrupt pressure decrease. The subcooling signal (Tsat –THL < 8 oC) is reached within 5 seconds, which means that the air duct dampers do not close at any time during the sequence. The signal for the MSIV occurs at 495 seconds. Subsequently, as a result of the PHRS performance, the SGs pressure drops below the RCS pressure at 3500 s. From this moment, the SGs start to cool-down the RCS, see Figure 8.

In the air ducts, it is observed that the air flow is about 200 kg/s per train at the beginning of the sequence and decreases to about 125 kg/s after 24 hours. The decrease in the air flow is due to a cooling of the inventory passing through the PHRS tubes, Figure 12. The temperature difference between the air which is at 303.15 K and the SG inventory which is in saturation results in a natural circulation in the PHRS steam lines of about 2.26 kg/s after 24 hours of the sequence.

Besides, the saturated SG inventory is at a lower pressure than the RCS, which is also in saturation conditions, so there is a temperature difference between the two, see Figure 13, which leads to a slight natural circulation in the 3 RCS loops with the PHRS available. Finally, it can be concluded that no CD occurs during the first 24 hours of the SBLOCA along with SBO sequence, see Figure 14.

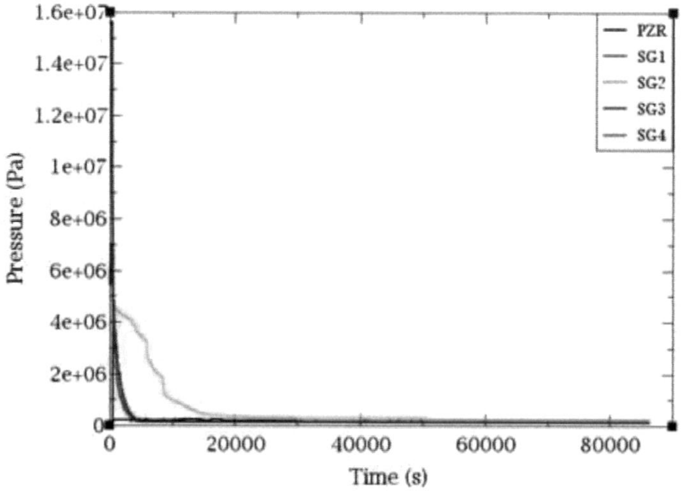

Figure 8. RCS and SGs pressure, DEGB LBLOCA under SBO conditions

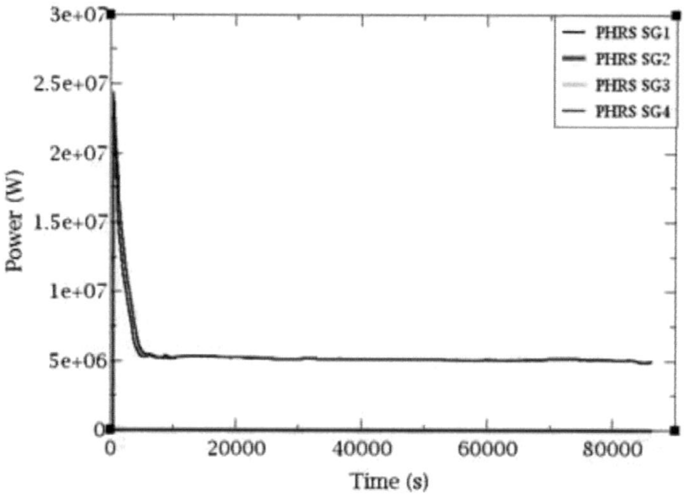

Figure 9. PHRS power per train, DEGB LBLOCA under SBO conditions

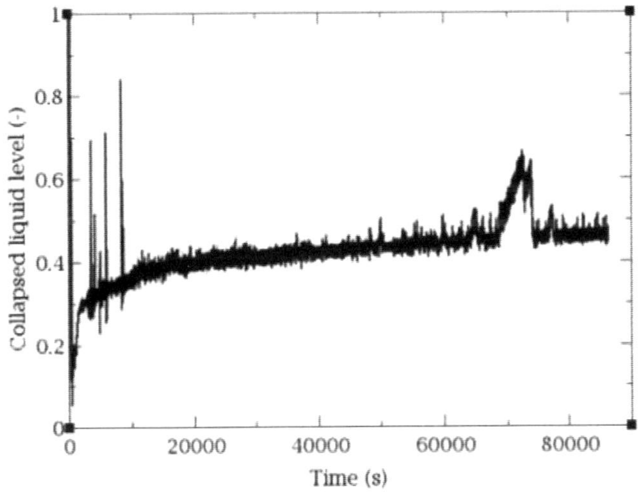

Figure 10. Collapsed liquid level, DEGB LBLOCA under SBO conditions

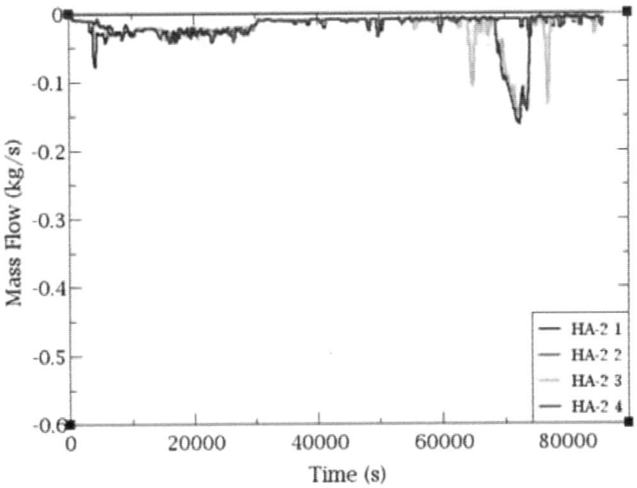

Figure 11. Cold legs to HA-2 mass flow rate, DEGB LBLOCA under SBO conditions

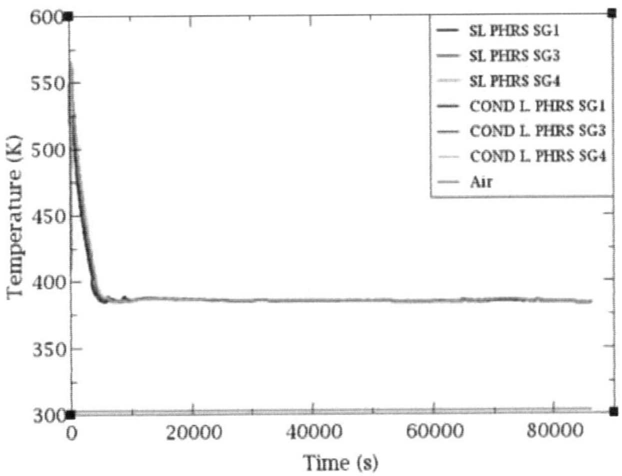

Figure 12. PHRS steam lines and condensate lines temperature, DEGB LBLOCA under SBO conditions

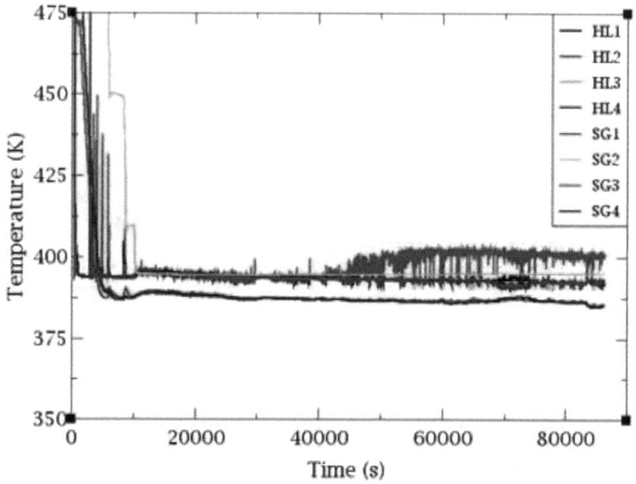

Figure 13. Hot legs and outlet steam generators temperatures, DEGB LBLOCA under SBO conditions

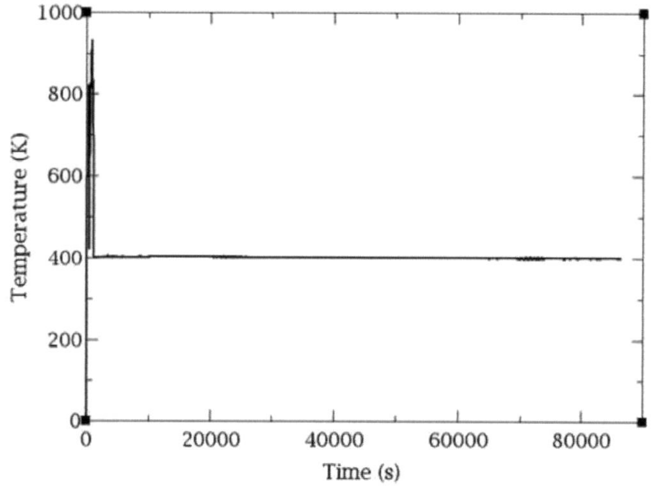

Figure 14. Peak Cladding Temperature, DEGB LBLOCA under SBO conditions

Table 1. DEGB LBLOCA under SBO conditions

Time (s)	Event
300	LBLOCA along with SBO (SCRAM, MFW pumps and RCPs trip, TT, loss of the condenser, CVCS off)
305	PHRS in cooldown mode (Tsat –THL < 8 °C)
315	HA-1 injection begins (RCS pressure < 6 MPa)
325	HA-2 injection set-point (RCS pressure < 1.5 MPa)
330	Air gate valves opening (SBO + 30 s delay)
420	HA-1 injection ends (HA-1 empty)
420	PHRS full capacity (90 s opening)
425	First stage HA-2 injection begins (HA-2 set-point + 100 s)
495	MSIV/BZOK close (SG pressure < 4.69 MPa)
4325	Second stage HA-2 injection begins (HA-2 set-point + 4000 s)
10325	Third stage HA-2 injection begins (HA-2 set-point + 10000 s)
30325	Fourth stage HA-2 injection begins (HA-2 set-point + 30000 s)
86400	End simulation (24 hours)

5. LBLOCA UNDER SBO CONDITIONSWITHOUT PHRS OR HA-2 AVAILABILITY

In this section, both cases are studied, the LBLOCA with SBO sequence without considering the PHRS actuation and without considering the HA-2 actuation. The aim is to know when the CD is reached if one of the two critical safety functions, the heat removal and the RCS coolant supply, is not fulfilled.

If only 3 out of 4 HA-2 trains (success criteria) are considered, and not the PHRS, it is found that the PHRS is not necessary for the HA-2 to start injecting water into the RCS, as the break is sufficient to allow the RCS pressure to reach atmospheric values in a few minutes, see Figure 16. However, from 40000 s onwards, the HA-2 alone is not sufficient to remove the decay heat, so the core is uncovered and subsequently the CD is reached, see Figure 17. This was found in a previous analysis, where it was also obtained that if 4 out of 4 HA-2 trains are considered instead of the success criteria number of trains, the HA-2 removal power is enough to avoid the CD for 24 hours, see [16].

On the other hand, if only 3 out of 4 trains (success criteria) of the PHRS are available, the CD is reached within a few seconds after the beginning of the accidental sequence, see Figure 17, since the inventory of the HA-1 is not sufficient to prevent the core uncovering. To conclude, it has been shown that for the LBLOCA sequence under SBO conditions the combined actuation of the HA-2 and the PHRS, considering the design criteria of both, is necessary to ensure core cooling for at least 24 hours.

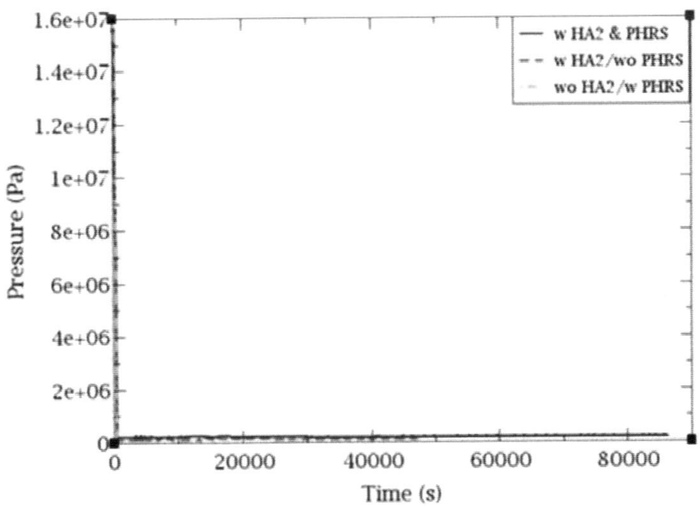

Figure 16. Reactor Coolant System pressure, DEGB LBLOCA sequence with HA-2 and PHRS, with HA-2 and without PHRS, without PHRS and with HA-2

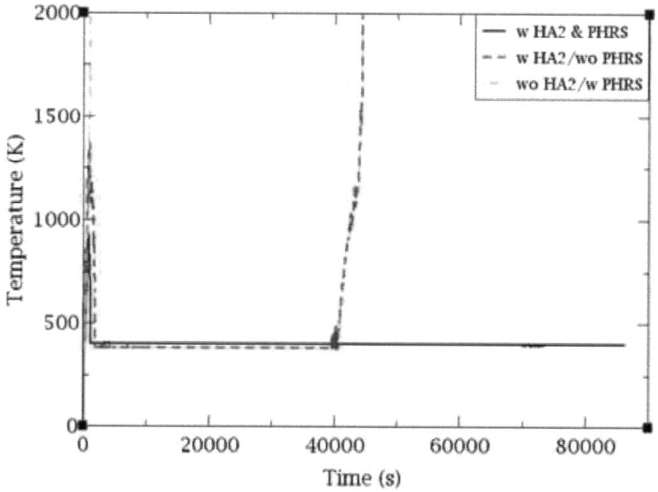

Figure 17. Peak Cladding Temperature, DEGB LBLOCA sequence with HA-2 and PHRS, with HA-2 and without PHRS, without PHRS and with HA-2

6. CONCLUSIONS

In this study, a comprehensive literature review of the air-cooled PHRS has been carried out to determine its configuration and operating modes. An isolated air-cooled PHRS model has then been developed for the thermal-hydraulic system code TRACE, which has been integrated into a thermal-hydraulic model of a VVER-1000 reactor. This has allowed a detailed analysis of the impact of the air-cooled PHRS performance on a DEGB LBLOCA under SBO conditions. The simulation of the LBLOCA sequence show that the performance of the PHRS, considering its design criteria, in RCS cooling mode, allows to keep the RCS cooled for 24 hours, while operating the HA-2 with its design criteria.

ACKNOWLEDGEMENT

This work in the framework of the Grant PID2019-108755RB-I00 (ISASMORE) funded by MCIN/AEI/ 10.13039/501100011033.

REFERENCES

[1] H. Ayhan and C. N. Sokmen, "Design and modeling of the passive residual hear removal system for VVERs," *Ann Nucl Energy*, vol. 95, pp. 109–115, 2016.

[2] H. Ayhan and C. N. Sokmen, "Investigation of passive residual hear removal system for VVERs: effect of finned type heat exchanger tubes," *Appl Therm Eng*, vol. 108, pp. 466–474, 2016.

[3] P. Khubchandani, A. Srivastava, and H. G. Lele, "Reliability assessment of the passive heat removal system of the VVER-1000 reactor at Kudankulam NPP," *Kerntechnik*, vol. 78, pp. 489–495, 2013.

[4] V. M. Berkovich, A. S. Korshunov, G. S. Taranov, S. G. Kalyakin, A. V. Morozov, and O. V Remizov, "Development and Validation of a technology for removal of noncondensing gases to ensure the operability of a passive heat removal system," *Atomic Energy*, vol. 100, no. 1, 2006.

[5] I. V. Elkin, O. I. Melikhov, V. I. Melikhov, S. M. Nikonov, A. V. Kapustin, and S. S. Selkin, "Experimental studies in the PSB-VVER bench of thermal hydraulics of emergency modes at NPPs with VVER-TOI," *Technologies for supporting the life cycle of NPP*, vol. 4, no. 14, 2018.

[6] I. I. Kopytov, M. B. Maltsev, G. S. Taranov, S. G. Kalyakin, O. V Remizov, and A. V Morozov, "Experimental study of NV NPP-2 steam generator model operation in condensation mode by steady states technique," in *19th International Conference on Nuclear Engineering (ICONE19)*, Chiba, Japan, May 2011.

[7] I. I. Kopytov, S. G. Kalyakin, V. M. Berkovich, A. V Morozov, and O. V Remizov, "Experimental investigation of non-condensable gases effect on Novovoronezh NPP-2 steam generator condensation power under the condition of passive safety systems operation," in *Proceedings of the 17th International Conference on Nuclear Engineering (ICONE17)*, Brussels, Belgium, Jun. 2009.

[8] A. V Morozov, O. V Remizov, and D. S. Kalyakin, "Experimental investigations of thermal–hydraulic processes arising during Operation of the passive safety systems used in new projects of nuclear power plants equipped with VVER reactors," *ISSN 00406015, Thermal Engineering*, vol. 61, no. 5, pp. 357–363, 2014.

[9] A. V Morozov and A. R. Sakhipgareev, "Experimental estimation of the effect of contact condensation of steam–gas mixture on VVER passive safety systems operation," *Nuclear Energy and Technology*, vol. 3, pp. 98–104, 2017.

[10] A. S. Shlepkin, A. V Morozov, and A. R. Sakhipgareev, "Experimental study of heat transfer processes during the operation of WWER steam generator in emergency mode on a single-tube model," *Journal of Physics: Conference Series. XXXVI Siberian Thermophysical Seminar (STS 36)*, 2020.

[11] A. S. Shlepkin, A. R. Sakhipgareev, and A. V Morozov, "Analysis of the influence of heat and mass transfer processes in the WWER equipment on the duration of effective operation of passive safety systems," *Journal of Physics: Conference Series. XXXVI Siberian Thermophysical Seminar (STS 36)*, 2020.

[12] IAEA, "Natural circulation phenomena and modelling for advanced water cooled reactors. IAEA-TECDOC-1677," 2012.

[13] ROSATOM, "Course on technological aspects of AES-2006 (VVER-1200) development of nuclear curricula on VVER technology"," Virtually, Dec. 19, 2022.

[14] E. Redondo-Valero, C. Queral, K. Fernandez-Cosials, and V. Sanchez-Espinoza, "Analysis of MBLOCA and LBLOCA success criteria in VVER-1000/V320 reactors: New proposals for PSA Level 1," *Nuclear engineering and technology*, vol. 55, pp. 623–639, 2023.

[15] E. Redondo-Valero, C. Queral, K. Fernandez-Cosials, V. H. Sanchez-Espinoza, M. Sanchez-Perea, and P. Groudev, "Management of the SBLOCA sequences with HPIS failure in VVER-1000/V320 reactors; comparison with Westinghouse PWR strategies (preprint)," *Progress in Nuclear Energy*, 2024.

[16] E. Redondo-Valero, C. Queral, K. Fernandez-Cosials, and V. Sanchez-Espinoza, "Safety margins improvement by means of the passive second stage hydroaccumulators in a VVER-1000/V320 reactor," *Nuclear Engineering and Design*, vol. 414, no. 112644, 2023.

[17] V. Sanchez-Espinoza *et al.*, "The H2020 McSAFER Project: Main Goals, Technical Work Program, and Status," *Energies (Basel)*, vol. 14, p. 6348, 2021.

[18] S. De Grandis, M. Apostol, and F. Saverio Nitt, "D1: Project Presentation. Passive Isolation Condenser. PIACE project," 2019.

[19] K. Fernández-Cosials, Z. Goñi, G. Jimenez, C. Queral, and J. Montero, "Three-dimensional simulation of a LBLOCA in an AP1000® containment building," *Energy Procedia*, vol. 127, pp. 234–241, 2017.

[20] C. Queral, J. Montero-Mayorga, J. Rivas-Lewicky, and M. Rebollo, "Verification of AP1000 low-margin PRA sequences based on best-estimate calculations," *Ann Nucl Energy*, vol. 104, pp. 9–27, 2017.

[21] S. Agrawal, A. Chauhan, and A. Mishra, "The VVERs at KudanKulam," *Nuclear Engineering and Design*, vol. 236, pp. 812–835, 2006.

[22] K. F. Galiev, S. V Yaurov, Ye. V Goncharov, and A. S. Volnov, "Experience of commissioning of the V-392M reactor plant passive heat removal system," *Nuclear Energy and Technology*, vol. 3, pp. 291–196, 2017.

[23] V. I. Polunichev, G. P. Shumailov, P. A. Gorbunov, M. M. Grigor'ev, A. A. Plakseev, and G. S. Taranov, "Development and investigation of the regulator of the passive heat-removal system for the Kudankulam nuclear power plant (India). UDC 621.039.562," *Atomic Energy*, vol. 102, no. 1, 2007.

[24] S. Chaudhary, P. Krishna Kumar, and H. P. Rammohan, "Study of performance of passive air cooled heat exchangers as a safety system for nuclear power plants," in *Proceedings of the 24th National and 2nd International ISHMT-ASTFE. Heat and Mass Transfer Conference (IHMTC-2017)*, BITS Pilani, Hyderabad, India, Dec. 2017.

[25] H. Ayhan and C. N. Sokmen, "Performance Optimization and Validation for the PRHRS of VVER with RELAP5 Code," *BgNS TRANSACTIONS*, vol. 21, no. 1, pp. 20–24, 2016.

[26] NRC, "TRACE V5.840 User Manual: Input Specification," Input Specification, User's Manual, 2017.

[27] V. Sanchez-Espinoza and M. Bottcher, "Investigations of the VVER-1000 coolant transient benchmark phase with the coupled system code RELAP5/PARCS," *Progress in Nuclear Energy*, vol. 48, pp. 865–879, 2006.

[28] A. A. Lebezov, A. V Morozov, A. S. Shlepkin, A. R. Sakhipgareev, and A. S. Soshkina, "Experimental study of heat and mass transfer processes affecting the duration of operation of VVER passive core cooling systems," *Nuclear Engineering and Desing*, vol. 419, no. 112963, 2024.

NURER 2024
Nuclear and Renewable Energy Resources

Comparative analysis of accidental sequences with conventional and Advanced Tolerant Fuels (ATF) in VVER-1000 reactors

Elena Redondo-Valero
NFQ Advisory Services S.L., Madrid, Spain, elena.redondo.valero@upm.es,
ORCID: ORCID: 0000-0002-9842-0892

Cesar Queral
Universidad Politécnica de Madrid (UPM), Madrid, Spain, cesar.queral@upm.es,
ORCID: 0000-0001-9384-2474

Jorge Sanchez-Torrijos
NFQ Advisory Services S.L. (NFQ), Madrid, Spain, jorge.sanchez@nfq.es,
ORCID: 0000-0001-6561-3551

Emilio Castro-Gonzalez
Universidad Politécnica de Madrid (UPM), Madrid, Spain, emilio.castro@upm.es,
ORCID: 0000-0002-6810-5031

Victor Hugo Sanchez-Espinoza
Karlsruhe Institute of Technology (KIT), Karlsruhe, Germany, victor.sanchez@kit.edu
ORCID: 0000-0001-9397-7600

Cite this paper as: *Redondo-Valero, E., Queral, C., Sanchez-Torrijos, J., Castro-Gonzalez, E.., Sanchez-Espinoza, V. Comparative analysis of accidental sequences with conventional and Advanced Technology Fuels (ATF) in VVER-1000 reactors. 7th Int. Conf. on Nucl. and Ren. Energy Res. 27-30 October 2024, Antalya, Türkiye*

Abstract: A significant percentage of reactors in operation, under construction or recently commissioned are VVER reactors. In parallel, there is a growing interest in analyzing the behavior of the Advanced Technology Fuels (ATF) under development in this kind of nuclear reactors. Among the new ATF designs, one of the most promising is the FeCrAl cladding. In addition to having a relatively high technology readiness level, this evolutionary cladding material offers an enhanced resistance against oxidation and hydriding, and an improved mechanical strength. All these properties are essential in accident sequences where high core temperatures are reached. In the present study, core damage sequences have been analyzed with a thermohydraulic model of a VVER-1000 reactor for the system code TRACE. The results show that the core damage temperature for the Zircaloy cladding cases is reached well before than for the FeCrAl cladding cases. The performance of this new cladding material provides an additional time for the start-up of the safety systems responsible for core cooling and replenishment of the inventory in the Reactor Coolant System.

***Keywords**:* *VVER-1000, ATF. FeCrAl, SBO, SBLOCA*

1. INTRODUCTION

VVER reactors are the most common reactor type in the world. Moreover, much of the Gen-III reactors that are being built or have recently come into operation are advanced VVER-1000 and VVER-1200. Therefore, there is a growing interest in studying their behavior under both anticipated and accidental transients. A peculiarity of the VVER reactors is the hexagonal fuel assembly design and the horizontal Steam Generators (SG) with a larger amount of coolant inventory compared to the Western PWR.

On the other hand, there is increasing interest in Accident Tolerant Fuels (ATFs) or Advanced Technology Fuels for light water reactors. These fuels have the potential to improve the nuclear safety and operational flexibility of commercial nuclear reactors. [1]. The ATFs incorporate advanced cladding materials and fuel compositions specifically designed to withstand higher temperatures, improve corrosion resistance and enhance the fuel performance under both normal operating conditions and accident scenarios [2].

Among the various ATF designs, FeCrAl cladding is one of the most promising. Made from an iron-chromium-aluminum alloy, it offers exceptional resistance to oxidation and corrosion under any circumstances, such as at high temperatures as in LOCA sequences [3]. This makes it particularly useful in the event of accidents, where conventional zirconium-based cladding can oxidase rapidly and the hydrogen generation can be relevant. The metal-water reaction under high temperature conditions generates large amounts of heat and hydrogen which, in turn, can create a risk of explosion when mixed with oxygen [1]. In addition, their superior thermal and mechanical properties make them a leading alternative to replace traditional cladding in current and future nuclear reactors.

It should be noted that the acceptance criterion considered for the Zircaloy (E110 for the VVER-1000 reactors) cladding is 1477 K, but for the FeCrAl cladding there is still no scientific consensus on which acceptance criterion should be used in safety analyses. Two values are under discussion, 1477 K as for the E110 cladding or 1773 K, which is the melting point of the FeCrAl alloy [4].

The present work has been developed within the project “Increasing safety margins in LWR through accident tolerant fuel (ATF)”, founded by the Spanish Nuclear Safety Council. Under that framework, several accidental sequences have been simulated in a VVER-1000 reactor considering the FeCrAl material for the claddings. The simulated accidental sequences correspond to transients in which one of the Critical Safety Functions (CSF) related to the removal of decay heat or the supply of coolant to the Reactor Coolant System (RCS) is not fulfilled. The simulations have been performed on a thermal-hydraulic (TH) model for the TRACE5P6 system code. An in-house version of the TRACE5P6 system code developed between UPM and NFQ has been used for this task, as TRACE5P6 is not designed to simulate ATF materials [5].

The paper is structured as follows. First, in Section 2, the modified version of TRACE5P6 is discussed, which includes the capability to model the FeCrAl cladding. Then, in Section 3, the TH model for the TRACE5P6 system code of the VVER-1000 reactor used in this work is described. Then, in Sections 4 to 6, the simulations of Station Blackouts (SBO), Large Break Loss of Coolant Accidents (LBLOCA) and Small Loss of Coolant Accidents (SBLOCA) are presented, followed by a comparison of the results obtained in Section 7. Finally, the conclusions of this work are presented in Section 8.

2. TRACE5P6 SYSTEM CODE AND ATF

UPM research group and NFQ consulting company are collaborating as a partner of the IAEA CRP ATF-TS project, which is, one of the most relevant international frameworks focused on ATF research activities nowadays [1]. This project is focused on improving the ATF-modelling capabilities of the current computational tools by means of experimental testing of ATF claddings and fuels and benchmarking activities, but also on the development of the methodology for LOCA fuel performance assessment and the ATF data basis with a view for its application to commercial nuclear power plants [2].

One of the key tasks carried out by the UPM-NFQ working group within the IAEA CRP ATF-TS project has to do with the integration of new capabilities in TRACE5P6 for simulating FeCrAl material [3]. To do so, the correlations for the material's physical properties along with the mechanical and corrosion models were implemented in TRACE5P6 to create the in-house version of TRACE used in this study. To achieve this, a thorough review of publicly available data on FeCrAl-related correlations and models was necessary to identify the appropriate information for incorporation into the in-house version of TRACE5P6 [6], [7], [8]. For further details on this executable, refer to [3].

3. THERMAL-HYDRAULIC MODEL OF A VVER-1000 WITH ATF

The TRACE5P6 TH model for the primary and secondary side including the safety and some control systems is built from the VVER-1000 RELAP5 model used in [9], see Figure 1. This TH model has been applied for LOCA and SBO analyses in previous studies, see [10], [11], [12]. The resulting integral plant model consists of following elements:

- **Primary loops**: Cold Legs (CLs), Hot Legs (HLs) and Reactor Coolant Pumps (RCPs).
- **Pressurizer (PZR)**: connected to HL 4 by the surge line. It contains spray lines, attached to the CL 1, three safety valves and four groups of heaters.
- **Reactor Pressure Vessel (RPV)**: it has been modelled by a 3D VESSEL component which is divided into 50 axial nodes, 6 azimuthal sectors and 6 radial sectors. In the core region, the three internal rings model the core whereas the fourth ring is dedicated to bypass. The core barrel is modelled by the fifth ring and the downcomer by the sixth. The VESSEL also contains the core lower and upper grid plates as well as the RPV upper head.
- **SGs**: In the primary side, the SG-tubes are modelled by three horizontal PIPEs. In the secondary side, the heat transfer zone is simulated with three horizontal levels. In the upper part is placed the liquid/steam SEPARATOR component.
- **Steam lines (SL)**: including one steam dump valve to the containment (BRU-A), two safety valves, one main isolation valve (BZOK) and a check valve in each SL.
- **Steam header**: containing steam dump valves to the condenser (BRU-K), steam dump valves to the atmosphere (BRU-SN) and the turbine connection.

The Emergency Core Cooling Systems (ECCS) are represented by four first Stage Hydro-accumulators (HA-1), two of them are connected to the reactor upper plenum and two others are connected to the downcomer, three Low Pressure Injection System (LPIS) trains and three HPIS trains. The model also includes the Emergency Boron Injection System (EBIS) and the makeup and letdown mass flows of the Control Volume and Chemical System (CVCS). The Main Feed Water (MFW) and the Emergency Feed Water (EFW) are also included in the model as boundary conditions. This model has been validated in steady state conditions against data from a VVER-1000/V320 NPP. The values obtained with the TRACEV5P6 model are very close to the reference plant data, see Table 1.

As the neutron capture cross-section of the FeCrAl material is much larger than that of the E110 material, a cladding thickness reduction of 0.185 mm is considered, as has been done in other studies on 3 loops Westinghouse PWR, see [3]. In addition, the fuel pellet radius has been increased to maintain the same gap width.

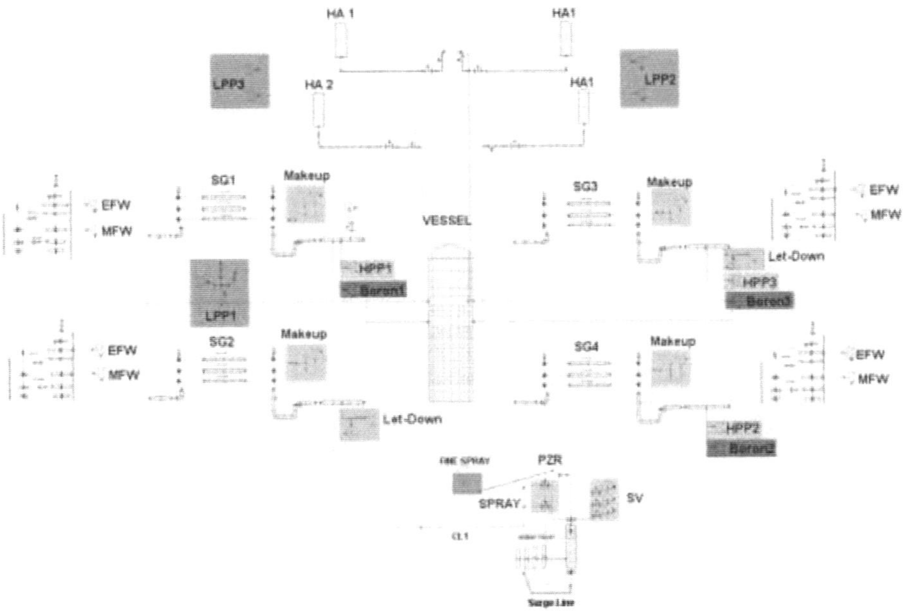

Figure 1. VVER-1000 RCS TRACE5P6 model

Table 1. Steady State parameters of the VVER-1000 TRACE5P6 model

Parameters	Reference NPP [13]	TRACE5P6
Core power [MW]	3010	3010
Core outlet pressure [MPa]	15.70	15.74
CLs temperature [K]	560.85	560.96
HLs temperature [K]	591.55	591.13
Average loop mass flow rate [kg/s]	4456	4457
SGs pressure [MPa]	6.27	6.27
MFW mass flow rate [kg/s]	409	408
SGs level [m]	2.5	2.45

4. STATION BLACKOUT SEQUENCE

The first scenario analyzed is a SBO taking into account the unavailability of the MFW, the AFW and EFW. During the first seconds of the sequence, the RCS inventory is cooled by the heat exchange with the SGs. Consequently, the level on the secondary side of the SGs drops as the water inventory boils off and no system replenishes it. Once the secondary side of the SGs is empty, the pressure in the RCS begins to rise until it reaches the set point for the PZR safety valves. At this point, the RCS inventory begins to deplete as the pressure is too high for the HA-1 to replenish the inventory lost through the safety valves.

Two simulations of the sequence have been performed using the TRACE5P6 system code. The first simulation considered E110 cladding and the second considered FeCrAl cladding. In both cases, core damage conditions are reached because two CSF, decay heat removal and RCS inventory replenishment, are not met.

The results show that the core damage temperature of 1477 K is reached after 13,562 s of simulation for the E110 case, whereas 1477 K is reached at 16,177 s and 1773 K at 17,415 s in the FeCrAl calculation. Thus, if the core damage is considered to be reached at 1477 K for FeCrAl, it takes approximately 3,000 s longer than for E110 to reach it. On the other hand, if the core damage conditions are considered at 1773 K for FeCrAl, it takes almost 4,000 s longer, see Figure 2.

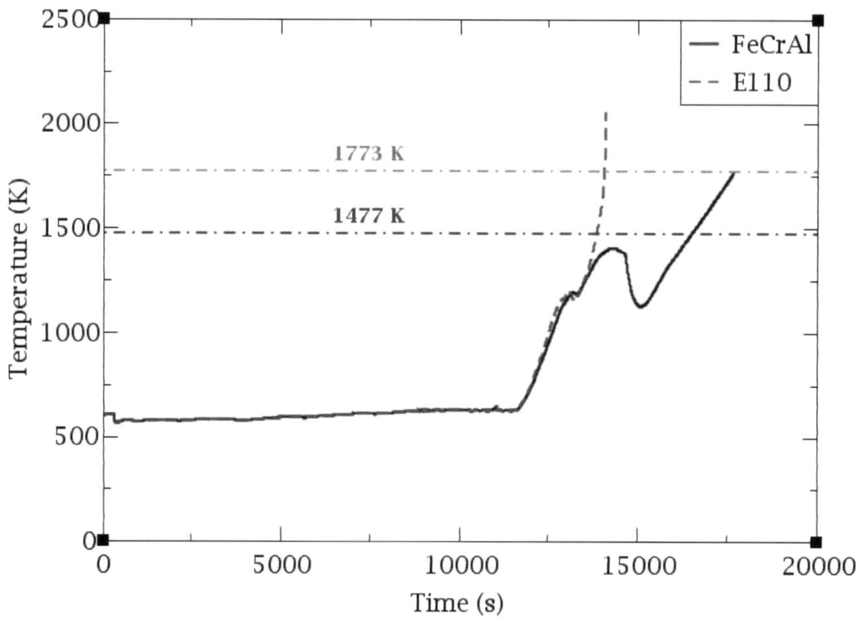

Figure 2. Peak Cladding Temperature, SBO sequence

5. LBLOCA SEQUENCES

The potential safety margin provided by the FeCrAl cladding has also been studied in the Double Ended Guillotine Break (DEGB) LOCA sequence. Three cases with different configurations of the ECCSs, previously analyzed in [10], have been considered:

- 1st case: LBLOCA Success Criteria (SC) [14], i.e., 2 out of 3 HPIS trains and 2 out of 4 HA-1 trains.
- 2nd case: LBLOCA with full availability of the HPIS without the performance of the LPIS and the HA-1.
- 3rd case: LBLOCA along with SBO, i.e., only the HA-1 are available.

The following results have been obtained for the three cases: In the 1st case, it has been found that the Peak Cladding Temperature (PCT) does not reach 1477 K for the E110 case, but it reaches values close to the acceptance criterion. However, with FeCrAl cladding, the PCT is almost 300 K lower than that obtained with E110 cladding, see Figure 3.

In the 2nd case, it can be seen that in the case where the cladding material is E110, the core damage is reached in the first seconds of the simulations. However, in the FeCrAl case, the PCT reaches 1477 K but does not reach 1773 K, see Figure 4. Therefore, if the acceptance criterion is considered to be 1773 K, the DEGB LOCA considering only the actuation of the HPIS does not reach the core damage criterion.

In the 3rd case, core damage is reached in both simulations with E110 and FeCrAl cladding. This is due to the fact that the CSF related to the long term replenishment of the RCS inventory is not fulfilled. However, the results show that the core damage for the E110 case is reached approximately 100 s before the FeCrAl case, see Figure 5.

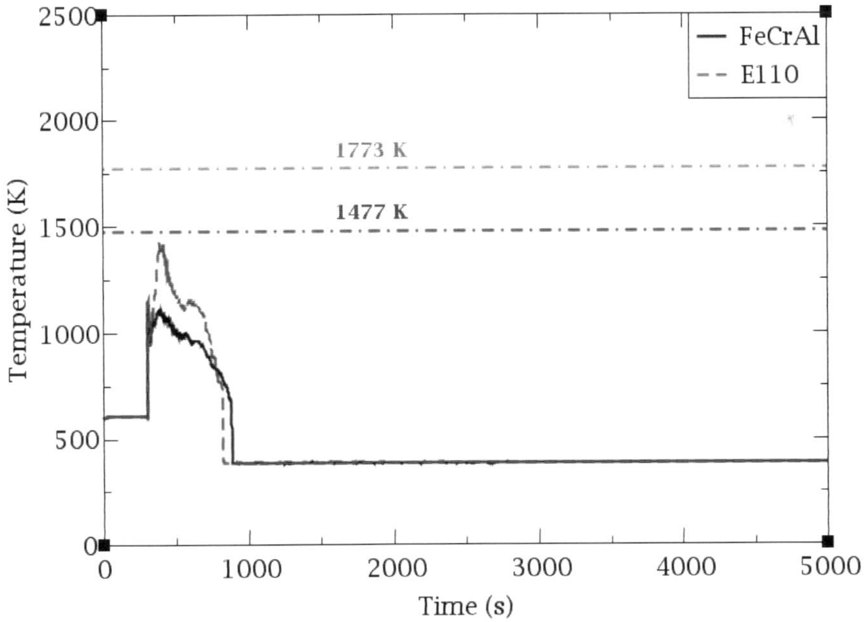

Figure 3. Peak Cladding Temperature, DEGB LBLOCA (1-2A-2H) sequence

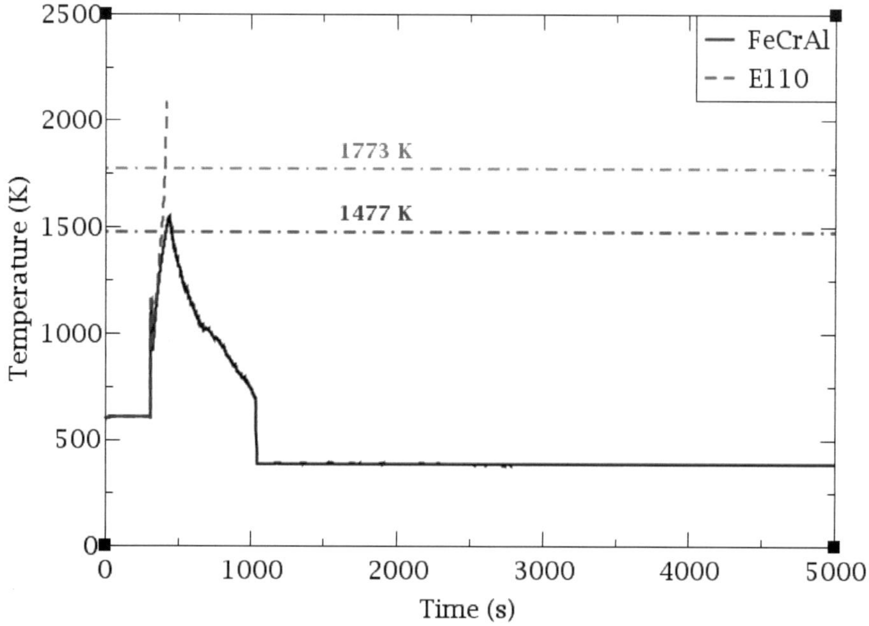

Figure 4. Peak Cladding Temperature, DEGB LBLOCA (l-a-3H) sequence

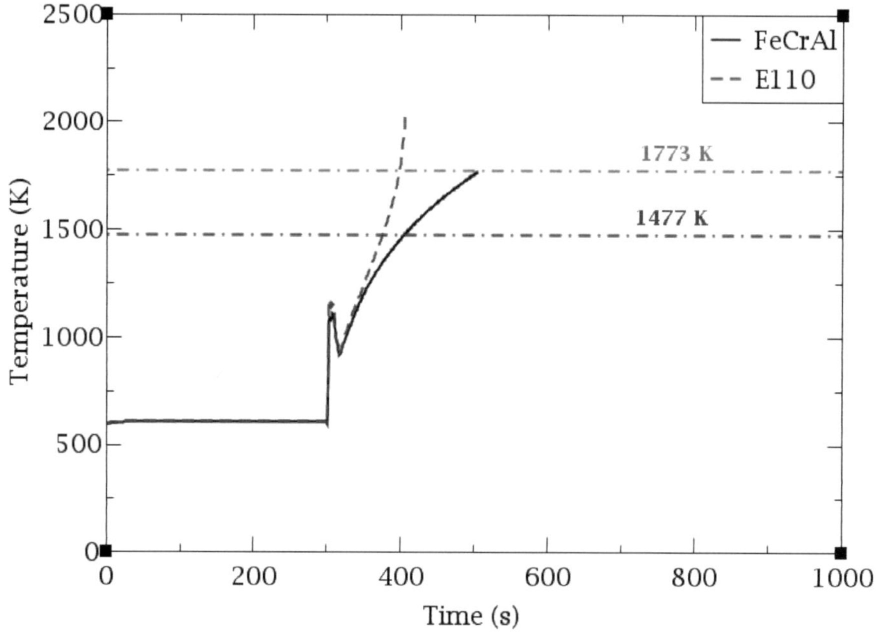

Figure 5. Peak Cladding Temperature, DEGB LBLOCA (l-2A-h) sequence

6. SBLOCA SEQUENCES

The enhanced performance of FeCrAl cladding has also been examined in the SBLOCA sequences. In previous analyses, see [15], it was observed that the SBLOCA sequences reach successful end states with the only actuation of 1 out of 3 HPIS trains, achieving very low values for the PCT. However, if the HPIS is not available, the sequences reach core damage despite the full availability of the HA-1 and the LPIS, see [11]. This is due to the fact that the pressure in the RCS remains above the set point of the HA-1 and the LPIS, therefore the systems do not inject water into the RCS before the PCT reaches 1477 K. For that reason, two SBLOCA sequences with different break sizes, 1 and 2 inches, have been selected. The results indicate that for the 1 inch SBLOCA, the sequence with FeCrAl cladding follows a similar trend to the one with E110 but with a favorable delay to FeCrAl case of around 1600 s in reaching the considered acceptance criterion (E110-1477 K; FeCrAl-1773 K), see Figure 6. On the contrary, for the 2 inches SBLOCA, if the FeCrAl material is considered for the cladding, the sequence does not reach either the 1477 K or the 1773 K, see Figure 7. It should be noted that for this sequence the SC for HA-1 (2 out of 4 trains) and for LPIS (1 out of 3 trains) have been taken into account [14].

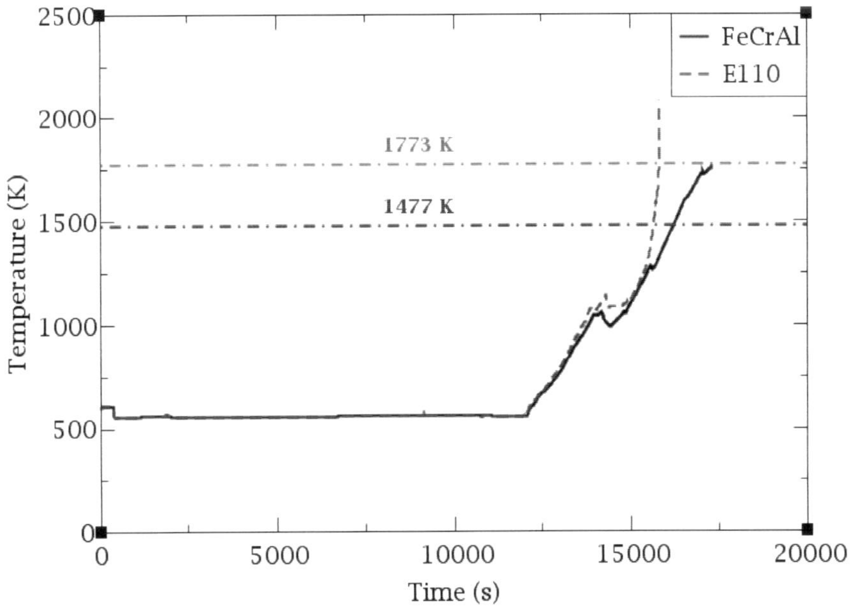

Figure 6. Peak Cladding Temperature, SBLOCA (1 inch) sequence

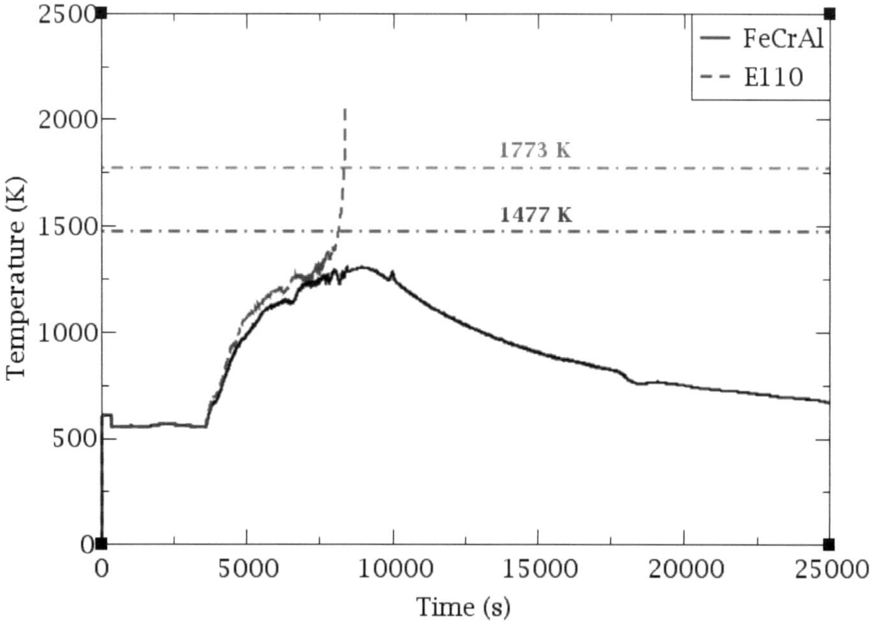

Figure 7. Peak Cladding Temperature, SBLOCA (2 inches) sequence

7. E110 VS. FeCrAl CLADDINGS CORE DAMAGE TIME COMPARISON

The objective of this section is to compare the different times at which core damage is reached for the sequences previously analyzed in Sections 4 to 6. In order to see how closed the different core damage times are to each other, the difference between the damage time for the E110 cladding and that for the two damage temperatures considered for the FeCrAl cladding has been obtained, see Table 2.

It can be seen that the major differences between the core damage times are found for the SBO sequences. Indeed, there are some sequences where core damage is reached with E110 cladding but not with FeCrAl cladding. It should be noted that the times given in Table 2 are from the beginning of the sequences, since a steady state of 300 s has been considered for all the sequences analyzed in this work.

Table 2. Accidental sequences core damage time comparison

Sequence	E110 (1477 K) [s]	FeCrAl (1477 K) [s] **(Δt) [min]**	FeCrAl (1773 K) [s] **(Δt) [min]**
DEGB LBLOCA (3/3 HPIS)	69	121 **(1)**	-
DEGB LBLOCA (4/4 HA-1)	77	107 **(0.5)**	205 **(2)**
2 inches SBLOCA (2/4 HA-1 and 2/3 LPIS)	8896	-	-
1 inches SBLOCA (2/4 HA-1 and 2/3 LPIS)	15425	16013 **(10)**	17092 **(28)**
SBO	13562	16177 **(43.5)**	17415 **(64)**

9. CONCLUSIONS

In the present work, a series of accidental scenarios, including SBLOCA, LBLOCA and SBO sequences, have been simulated with the aim of analyzing the safety margin provided by the implementation of the FeCrAl cladding material in a VVER-1000 reactor. For this purpose, an in-house TRACE5P6 version with the capability to model ATF, developed by NFQ and the UPM, has been used. The main conclusions obtained are as follows:

- In the SBO sequence, it has been observed that the time margin obtained when using FeCrAl cladding is 43 or 64 minutes depending on the acceptance criterion considered for FeCrAl, 1447 K or 1773 K.
- In the LBLOCA sequences, it is observed that the time saved by considering FeCrAl instead of E110 is very low, in the order of 2 minutes.
- In the SBLOCA sequences, the time margin obtained is between the margins for SBO and LBLOCA.

As a final conclusion, it is worth mentioning that the major improvement observed in this work when using ATF fuels is in the SBO sequence. Moreover, it is crucial to establish which is the acceptance criterion for FeCrAl, as the time margin differences obtained between the two criteria under discussion, 1477 K and 1773 K, are significant.

ACKNOWLEDGMENT

UPM Nuclear Safety group acknowledges technical and financial support granted in the framework of the project *"SUBV-16/2021 Safety Margins increase in LWR by means of Accident Tolerant Fuel (ATF)"* funded by the Spanish Nuclear Safety Council (CSN).

REFERENCES

[1] N. Doncel *et al.*, "R&D in advanced technology fuels (ATFs) in Spain," *Nuclear Engineering and Design*, vol. 424, no. 113246, 2024, doi: https://doi.org/10.1016/j.nucengdes.2024.113246.

[2] J. Zhang, P. Xu, M. Sevecek, K.-S. Sim, and A. Khaperskaia, "Contribution of IAEA coordinated research projects to light water reactors advanced technology fuel testing and simulation," *Nuclear Engineering and Design*, vol. 418, no. 112910, 2024.

[3] J. Sanchez-Torrijos, "Assesment of safety issues in NuScale and Gen II PWR under accident conditions using TRACE. PhD Thesis, Universidad Politécnica de Madrid," 2024.

[4] U. Baker *et al.*, "Source term analysis of FeCrAl accident tolerant fuel using MELCOR," *Ann Nucl Energy*, vol. 202, no. 110482, 2024.

[5] J. Sanchez-Torrijos *et al.*, "Assessment of ATF claddings performance under LOCA conditions using TRACE and TRANSURANUS codes," in *Workshop on Core and Plant Simulation with an Emphasis on Fuel Behaviour in Light Water Reactor Based Small Modular Reactors*, 2024.

[6] IAEA, "Analysis of Options and Experimental Examination of Fuels for Water Cooled Reactors with Increased Accident Tolerance (ACTOF)," Vienna, Austria, 2020.

[7] S. Wu and K. Shirvan, "System code evaluation of near-term accident tolerant claddings during boiling water reactor short-term and long-term station blackout accidents," *Nuclear Engineering and Design*, vol. 356, no. 110362, 2020.

[8] K. G. Field, M. A. Snead, Y. Yamamoto, and K. A. Terrani, *Handbook on the Material Properties of FeCrAl Alloys for Nuclear Power Production Applications*. 2018.

[9] V. Sanchez-Espinoza and M. Bottcher, "Investigations of the VVER-1000 coolant transient benchmark phase with the coupled system code RELAP5/PARCS," *Progress in Nuclear Energy*, vol. 48, pp. 865–879, 2006.

[10] E. Redondo-Valero, C. Queral, K. Fernandez-Cosials, and V. Sanchez-Espinoza, "Analysis of MBLOCA and LBLOCA success criteria in VVER-1000/V320 reactors: New proposals for PSA Level 1," *Nuclear engineering and technology*, vol. 55, pp. 623–639, 2023, doi: https://doi.org/10.1016/j.net.2022.10.006.

[11] E. Redondo-Valero, C. Queral, K. Fernandez-Cosials, V. H. Sanchez-Espinoza, M. Sanchez-Perea, and P. Groudev, "Management of the SBLOCA sequences with HPIS failure in VVER-1000/V320 reactors; comparison with Westinghouse PWR strategies, 177, 105414," *Progress in Nuclear Energy*, 2024, doi: https://doi.org/10.1016/j.pnucene.2024.105414.

[12] E. Redondo-Valero, C. Queral, K. Fernandez-Cosials, and V. Sanchez-Espinoza, "Safety margins improvement by means of the passive second stage hydroaccumulators in a VVER-1000/V320 reactor," *Nuclear Engineering and Design*, vol. 414, no. 112644, 2023, doi: https://doi.org/10.1016/j.nucengdes.2023.112644.

[13] N. Kolev *et al.*, "VVER-1000 Coolant Transient Benchmark - Phase 2 (V1000CT-2). Vol. 11: MSLB Problem - Final Specifications. NEA/NSC/DOC(2006)6," MSLB Problem - Final Specifications, May 2006.

[14] B. Skalozubov, A. Klyuchnikov, and B. Kolykhanov, *Fundamentals of management of design accidents with loss of coolant at the power plant (in Russian). National Academy of Sciences of Ukraine, Institute of NPP Safety Problems. - Chernobyl (Kiev region): Institute of NPP Safety Issues, 2010. ISBN 978-966-02-5203-5.* 2010. Accessed: Jul. 04, 2024. [Online]. Available: https://inis.iaea.org/collection/NCLCollectionStore/_Public/41/124/41124864.pdf.

[15] E. Redondo-Valero, C. Queral, and V. Sanchez-Espinoza, "Thermal-hydraulic analysis of accidental sequences in a VVER-1000/V320 reactor with TRACEv5p5 code (in Spanish)," in *Annual meeting of the Spanish Nuclear Society (RASNE 2021)*, Granada, Oct. 2021.

[16] L. Ibánez *et al.*, "Application of the Integrated Safety Assessment methodology to safety margins. Dynamic Event Trees, Damage Domains and Risk Assessment," *Reliab Eng Syst Saf*, vol. 147, pp. 170–193, 2016, doi: https://doi.org/10.1016/j.ress.2015.05.016.

THERMAL-HYDRAULIC ANALYSIS OF THE MOTEL EXPERIMENTAL FACILITY USING TRACE

Yago Martínez González
Universidad Politécnica de Madrid, Madrid, Spain, yago.mgonzalez@alumnos.upm.es
ORCID: 0009-0006-5485-9891

César Queral
Universidad Politécnica de Madrid, Madrid, Spain, cesar.queral@upm.es
ORCID: 0000-0001-9384-2474

Elena Redondo-Valero
Universidad Politécnica de Madrid, Madrid, Spain, elena.redondo.valero@upm.es
ORCID: 0000-0002-9842-0892

Jorge Sánchez-Torrijos
NFQ Advisory Services S.L., Madrid, Spain, jorge.sanchez@nfq.es
ORCID: 0000-0001-6561-3551

Cite this paper as: *Martínez-González, Y, Queral, C, Redondo-Valero, E, Sánchez-Torrijos, J. Thermal-Hydraulic Analysis of the Motel Experimental Facility Using TRACE. 7th Int. Conf. on Nucl. and Ren. Energy Res. 27-30 October 2024, Antaya, Turkey*

Abstract: Small modular reactors (SMRs) are gaining relevance in the nuclear industry. In these reactors, the steam generators play a crucial role due to its more compact design compared to the steam generators of large power plants. The steam generators of an integral SMR are generally composed of multiple tubes arranged in a helical coil configuration. Understanding the thermal-hydraulic behavior of this type of steam generator is essential to ensure safe and efficient operation of SMRs.
The objective of this study is to validate the TRACE system code by comparing its results with the experimental data obtained at the MOTEL facility, located in LUT (Finland), with a steam generators configuration similar to the NuScale reactor design. The work focuses on the study of helical steam generators, deepening the understanding of their behavior in both stationary and transient conditions. Experimental tests have been carried out in both medium/high power (250 kW, 500 kW, 750 kW and 1000 kW) and low power (75 kW, 100 kW, 125 kW and 150 kW) conditions.
The simulation results show that the TRACE code accurately reproduces the main parameters of interest recorded during the experimental tests.

Keywords: *TRACE, SMR, MOTEL, EXPERIMENTAL FACILITY.*

1. INTRODUCTION

MOTEL (MOdular Test Loop) is a modular integral test facility including replaceable modules, making it flexible to be adapted for investigations of different reactor configurations and future research needs. The first MOTEL version models a prototypical SMR with a helical coil steam generator, closely resembling the NuScale SMR design. The facility comprises a pressure vessel, which includes the SMR model, and auxiliary systems. The pressure vessel houses a core element, riser, downcomer, and pressurizer that form the primary side for the facility. The core of MOTEL is a general representation of a nuclear fuel rod bundle in a rectangular grid configuration. The secondary side consists of a feedwater injection system, the helical coil steam generator with 16 steam generator tubes positioned inside the annular downcomer, and a steam outlet system.

The functioning of the facility primary side is based on natural circulation: the primary water heats up in the core, then flows up inside the above riser, after which turns down into the annular downcomer, and finally flows back to the core The heat sink for this natural circulation system is at the top volumes of the downcomer, where the hot primary water

flows on the shell side of the steam generator tubes releasing heat to the secondary side water inside the tubes. The secondary side functions by the means of forced flow: feedwater is pumped via a pipeline to a common feedwater manifold, from where flow enters four pipelines and cold collectors, each of which are connected to four individual steam generator tubes, where boiling occurs. Then, the resulting secondary steam is directed out from the tubes through the following four hot collectors and pipelines that are finally joined to a common steam manifold and pipeline. From there steam is directed out of the secondary system.
The instrumentation of MOTEL is mostly based on traditional point-form temperature, pressure, and differential pressure measurements as well as traditional flow rate measurements. In addition, ultrasonic flow meters are used to measure the primary flow rate in the downcomer. The extensive instrumentation of MOTEL enables a diverse detection of thermal hydraulic phenomena especially in the core and steam generator. The low and wide core structure enables the investigation of 2D/3D core phenomena.
In terms of technical specifications, both on the primary and secondary sides, the design pressure and temperature are 40 bar and 250 °C.

The experiment MS-SG01 consists of three sub-experiments: MS-SG01-R, MS-SG01-A and MS-SG01-B. During this experiment, four quasi steady-state periods were observed, with heating power ranging from 250 to 1000 kW, and the primary pressure was consistently maintained at 3.5 MPa.
Similarly, in the MS-SG02 experiment, another set of four steady states was examined, with core heating power ranging from 75 to 150 kW. The primary pressure was also kept constant at 3.5 MPa during these periods.

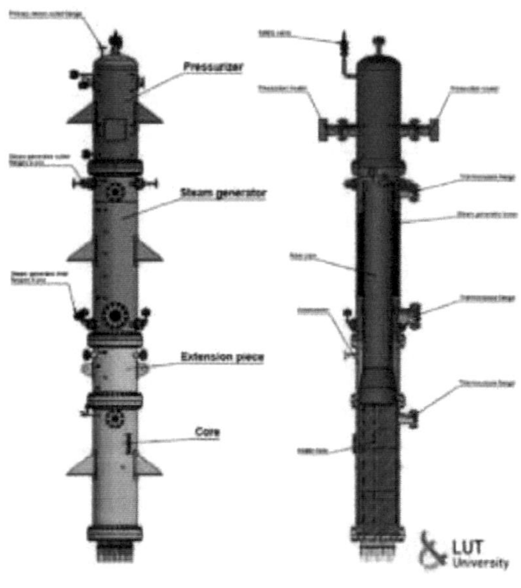

***Figure 1**. Overall scheme of the MOTEL facility. [1]*

Table I. Averaged boundary conditions of MOTEL facility in the MS-SG01R, MS-SG01A and MS-SG01B experiments [1]

Test	Core Power (kW)	Feedwater Mass Flow Rate (kg/s)	Secondary Pressure (MPa)	Feedwater Temperature (°C)
MS-SG01R	250	0.12	1	17.0
	500	0.20	1	15.0
MS-SG01-A	500	0.30	1	15.0
MS-SG01-B	1000	0.38	1	17.5

Table II. Averaged boundary conditions of MOTEL in the MS-SG02 experiment [1]

Test	Core Power (kW)	Feedwater Mass Flow Rate (kg/s)	Secondary Pressure (MPa)	Feedwater Temperature (°C)
MS-SG02	75	0.03	1	17.5
	100	0.04	1	17.5
	125	0.05	1	17.5
	150	0.06	1	17.5

2. TRACE THERMAL-HYDRAULIC MODEL AND SIMULATION RESULTS

The TRACE model for the MOTEL facility consists of 24 components, Figure 2. The primary side, representing the reactor coolant system of an SMR, is a closed loop of natural circulation, as there are no pumps. The core is modelled with a PIPE component divided into 10 cells, with a total height of 1.83 m. Also, the core PIPE is connected to a HEAT STRUCTURE and a POWER component to set the power boundary conditions.

The TRACE model for the helical Steam Generators (SGs) of MOTEL uses the 'Tube Bank Crossflow' PIPE and 'Curved pipe' PIPE components to model the primary and secondary sides. The secondary side in MOTEL contains four tube bundles, each containing four steam generator tubes of different lengths and angles. In the TRACE model, the SGs tubes of similar length are modelled together as one PIPE component. In addition, each of the PIPE components modelling a bundle of tubes includes a FILL component to set the inlet flow and temperature boundary conditions, and a BREAK component to set the pressure boundary condition. It has been assumed that the total mass flow rate is divided equally into each SG tube bundle.

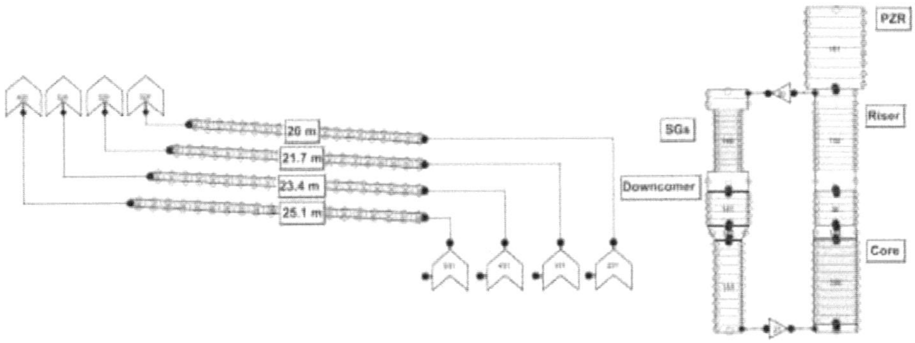

***Figure 2**. TRACE model developed by UPM research group. [2]*

2.1 MS-SG01 EXPERIMENT

In the MS-SG01 experiment, core power was increased in steps of 250 kW, 500 kW, 750 kW, and 1000 kW, with each step lasting 3600 seconds to reach quasi-steady state conditions. Notable fluctuations were observed in one of the mass flow rate sensors, especially in the highest power cases . The measurement error for mass flow rate is ±0.91 kg/s.

TRACE simulations, Figures 3-5, show no mass flow rate fluctuations and align more with the lower measured values (MOTEL 2) but tend to overestimate at higher power levels. Figures 6-8 compare the temperature trends, revealing that TRACE fits well at 250 kW but initially overpredicts temperatures. At 750 kW and 1000 kW, TRACE results are comparable to experimental data, though stable primary side conditions were not achieved during the higher power periods. Besides, the coolant temperature profiles along the upper downcomer, Figure 9, where the helical coil steam generator tubes are located, show a good agreement between TRACE simulations and experimental results.

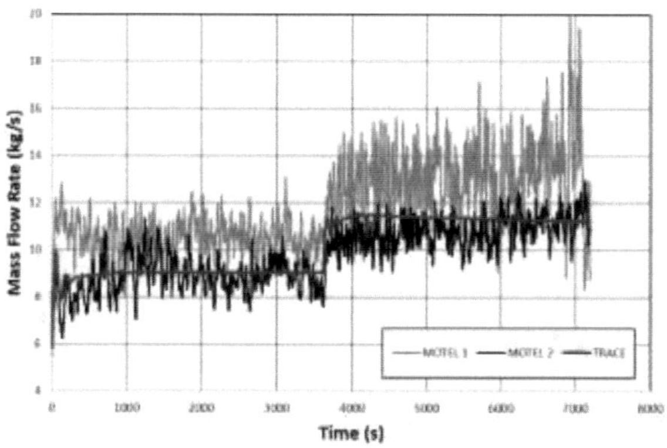

Figure 3. *Downcomer mass flow rate (MOTEL test, MS-SG01R, 250-500 kW) [2]*

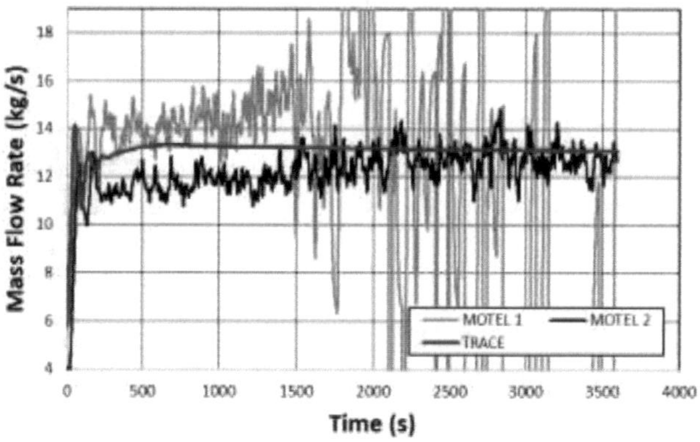

Figure 4. *Downcomer mass flow rate (MOTEL test, MS-SG01-A, 750 kW) [2]*

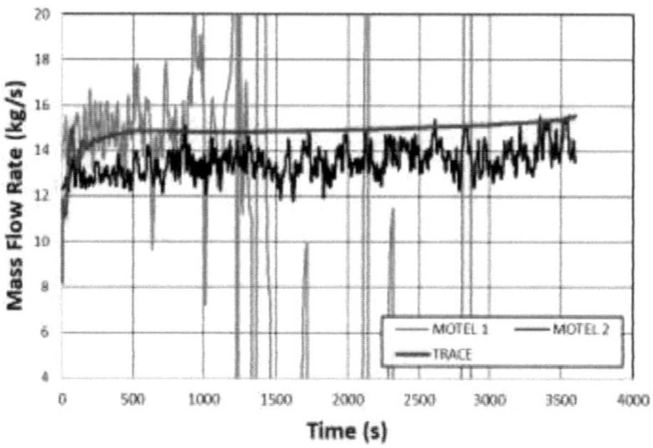

Figure 5. *Downcomer mass flow rate (MOTEL test, MS-SG01-B, 1000 kW) [2]*

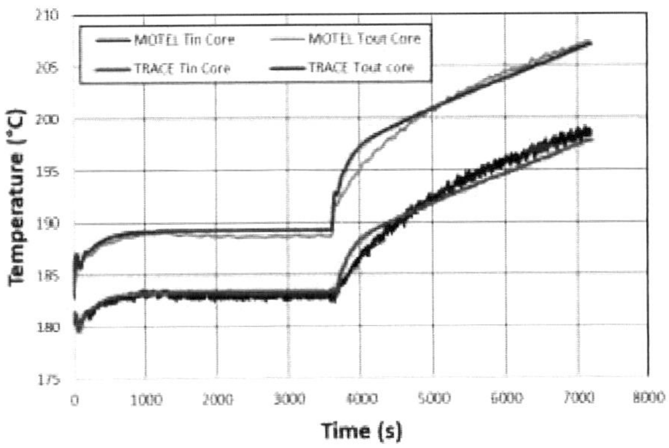

Figure 6. *Core inlet and outlet temperature (MOTEL test MS-SG01-R, 250-500 kW) [2]*

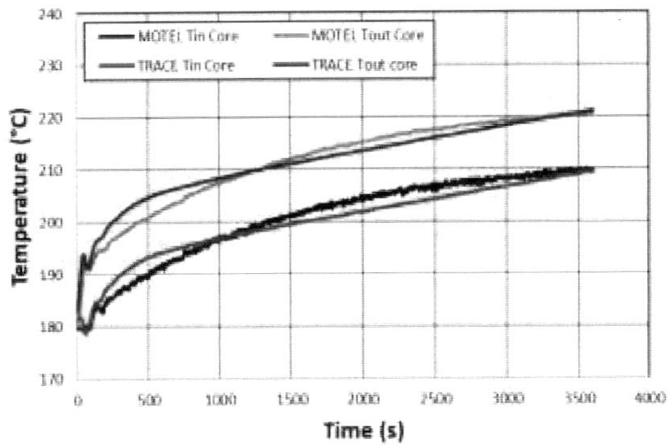

Figure 7. *Core inlet and outlet temperature (MOTEL test MS-SG01-A, 750 kW) [2]*

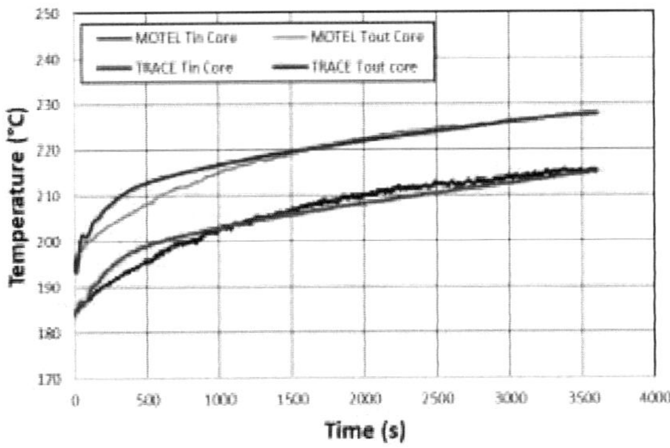

Figure 8. Core inlet and outlet temperature (MOTEL test MS-SG01-B, 1000 kW) [2]

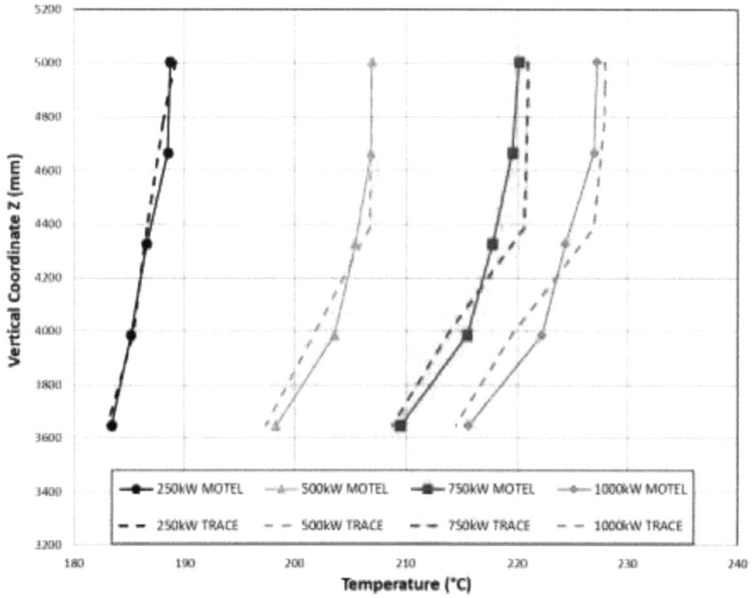

Figure 9. Axial temperature profiles on the primary side of the SG (MOTEL tests MS-SG01) [2]

2.1 MS-SG02 EXPERIMENT

In the MS-SG02 experiment, core power was increased in steps of 75 kW, 100 kW, 125 kW, and 150 kW, with each step lasting 3600 seconds, except the 150 kW step, which lasted longer to test the effect of a reduced secondary side feedwater mass flow rate. The primary side mass flow rate in the annular downcomer, measured in MOTEL and simulated with TRACE, are shown in Figure 10. Notable fluctuations were observed in the two measured values, especially at the beginning of the experiment, with a measurement error of ±0.91 kg/s. Simulated values stabilized efficiently, aligning more with the lower measured value trend and showing a clear increasing trend with rising core power and feedwater flow, except during the final period when the feedwater flow was reduced.

The measured and simulated temperature trends at the core inlet and outlet during the 75 kW to 150 kW power steps, Figure 11, shows differences between the inlet and outlet during each step were 3-4 °C, indicating the heat-up of the primary flow. While the measured values showed minor fluctuations, the simulations were more stable, generally staying within the error band of ±2 °C. TRACE model results fit well with the experimental data, remaining within signal noise at all power levels, except at 100 kW, where the simulations were slightly higher than the experimental values. Besides, the coolant temperature profiles along the upper downcomer, Figure 12, show a good agreement between TRACE simulations and experimental results.

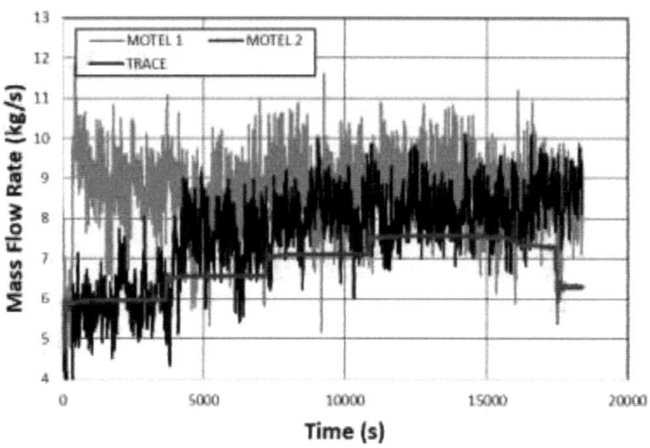

Figure 10. *Downcomer mass flow rate (MOTEL test MS-SG02) [2]*

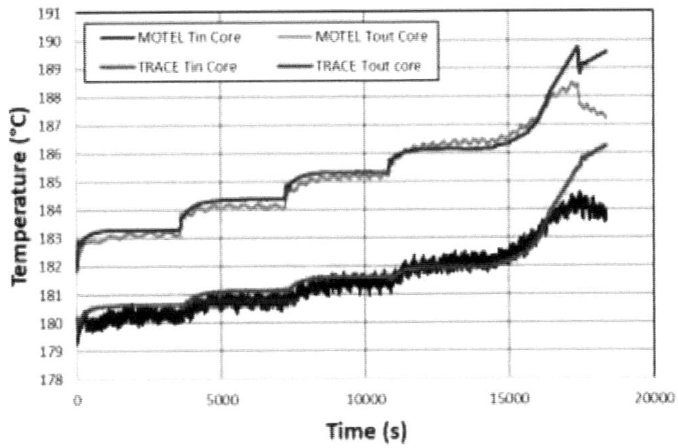

Figure 11. *Core inlet and outlet temperature (MOTEL test MS-SG02) [2]*

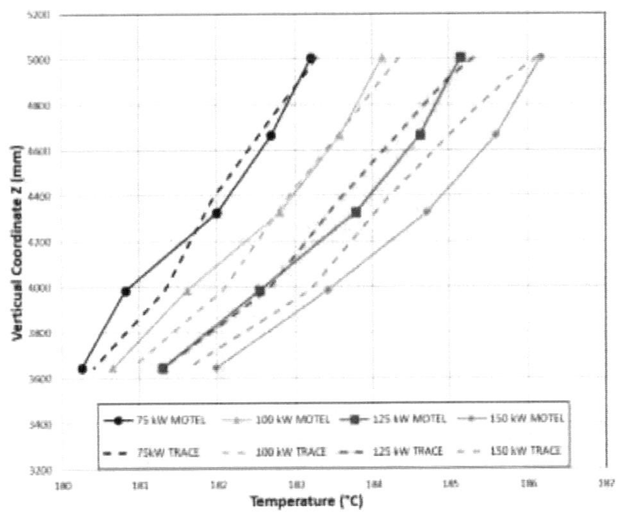

Figure 12. Axial temperature profiles on the primary side of the SG (MOTEL tests MS-SG02) [2]

3. CONCLUSIONS

Both experiments, MS-SG01 and MS-SG02, carried out in the MOTEL facility, have been simulated with the TRACE Thermal-hydraulic system code by the UPM research group. The following conclusions can be drawn from this study:

- In the MS-SG01 experiment, involving power steps from 250 kW to 1000 kW, TRACE simulations closely matched experimental core inlet and outlet temperatures, particularly at higher power levels. Moreover, the simulated mass flow rates were also consistent with the experimental data.

- The MS-SG02 experiment, with lower power steps from 75 kW to 150 kW, achieved quasi-steady state conditions in all periods. TRACE accurately reproduced the core inlet and outlet temperatures within the experimental error bands, and the simulated mass flow rates were also consistent with the experimental results.

- Globally, TRACE simulations reproduced the axial temperature profiles on the primary side of the steam generators, with temperature differences between the top and bottom locations aligning well with experimental values, demonstrating the model capability to reproduce the MOTEL experiments accurately.

Finally, it can be concluded that the model developed by the UPM for the TRACE system code is able to reproduce accurately the MS-SG01 and MS-SG02 experiments carried out in the MOTEL experimental facility.

ACKNOWLEDGMENT

McSafer project has received funding from the Euratom research and training program 2023-2024 under the grant agreement No 945063.

The authors would acknowledge LUT university for providing the experimental series used in this work.

The content of this presentation reflects only the author's views, and the European Commission is not responsible for any use that may be made of the information it contains.

REFERENCES

[1] T. Kimmo, T. Joonas, K. Eetu y K. Virpi, «McSafer Proyect,» *Description of the MOTEL facility and instrumentation,* p. 37, 2021.

[2] Y. M. González, «Análisis Termohidráulico de la instalación experimental MOTEL mediante el código de sistemas TRACE,» Universidad Politécnica de Madrid, End-of-degree Project, 2024.

[3] J. C. Queral Salazar, E. Redondo Valero, J. Sánchez Torrijos, D. Canal, G. Jiménez Varas, S. Larriba del Apio, D. Cuervo Gómez, O. L. Cabellos de Francisco, L. F. Durán Vinuesa y L. G. Herranz, «Spanish research related to SMRs projects,» 2023.

[4] O. Ali, B. Kenneth J y C. J. Berry, «Forced Convection boiling inside helically-coiled tubes,» International Journal of Heat and Mass Transfer, pp. 1779-1793.

[5] Joonas, Telkkä; Antti, Räsänen; Eetu, Kotro; Heikki, Suikkanen;, "McSafer Proyect," 2021, p. 48.

LOCA sequence analysis in NuScale SMR with TRACE code

Yago Martinez-Gonzalez
Universidad Politécnica de Madrid, Madrid, Spain, yago.mgonzalez@alumnos.upm.es
ORCID: 0009-0006-5485-9891

Jorge Sanchez-Torrijos
NFQ Advisory Services, Madrid, Spain, jorge.sanchez@nfq.es
ORCID: 0000-0001-6561-3551

Cesar Queral
Universidad Politécnica de Madrid, Madrid, Spain, cesar.queral@upm.es
ORCID: 0000-0001-9384-2474

Cite this paper as: *Martinez-Gonzalez, Y, Sanchez-Torrijos, J, Queral, C. LOCA sequence analysis in NuScale SMR with TRACE code. 7th Int. Conf. on Nucl. and Ren. Energy Res. 27-30 October 2024, Antalya, Turkey*

Abstract: NuScale is a light-water cooled SMR with an integral design of the reactor pressure vessel which relies on natural circulation to establish the mass-flow rate in the primary side. This work focuses on the simulation of a LOCA caused by the rupture of the CVCS discharge line within the containment vessel. To do so, a 3D thermal-hydraulic model of NuScale has been developed for the TRACE code including the modelling of the Emergency Core Cooling System (ECCS) together with the containment vessel and the reactor pool. The selected scenarios are based on the event trees presented in the PSA analysis included in the DCA report of NuScale. In this study, the base case is developed considering that the Decay Heat Removal System actuation is not credited, the ECCS performance is limited to the opening of two out of three reactor vent valves and none of the reactor recirculation valves (RRVs). Additional simulations were performed involving the opening of one and two RRVs and the time available is computed for each case and the final comparison of results is made. The study concludes that there is sufficient time to perform the opening of at least one RRV.

Keywords: *TRACE, SMR, NuScale.*

1. INTRODUCTION

Given the growing interest in the Small Modular Reactors (SMR) along with the assessment of the performance of the Advanced Technology Fuels (ATFs) concepts, the simulation of a LOCA sequence in NuScale leading to core damage and based on the rupture of the discharge line of the Chemical Volume Control System and in the failure to open of the Reactor Recirculation Valves is addressed. By doing so, it has been possible to preliminarily assess the coping time under the postulated scenario. To achieve this, the necessary models were implemented in the TRACE code by means of modifying the source code. At this point, it should be mentioned that this work has been developed as a result of a close collaboration between the NFQ Advisory Services S.L. company and the Universidad Politécnica de Madrid (UPM).

Additionally, the same LOCA case but considering the availability of the different configurations of Reactor Recirculation Valves (RRVs) have been also investigated and the results show that the opening of two out of three Reactor Vent Valves (RVVs) along with one out of two reactor recirculation valves in of the emergency cooling system is enough to deal with the postulated scenario.

2. NUSCALE POWER MODULE DESCRIPTION

NuScale is a Light Water Reactor (LWR)-SMR in which highly distinctive features have been introduced, such as natural circulation and fully passive safety systems. In this sense, it can be said that it is a completely disruptive design when compared to currently operating Pressurized Water Reactors (PWRs). Furthermore, NuScale remains the only design to have obtained the official approval from the U.S NRC and is intended to operate in plants where multiple units coexist simultaneously, with a maximum of twelve units in its most powerful version.

The smallest independent unit in a NuScale power plant capable of producing electrical power is called the NuScale Power Module (NPM). In that sense, it should be noted that each NPM has its own Chemical and Volume Control System (CVCS) responsible for controlling the boron concentration in the primary circuit (RCS), the RCS water inventory, the RCS pressure by means of the PZR spray and the amount of non-condensable gases accumulated in the PZR steam bubble during normal operation [4], steam supply system including the dedicated turbine and, the passive safety systems, except for the reactor pool, which is common to all the units forming the plant.

The functioning of each Nuclear Power Module (NPM) is based on a simple and elegant natural circulation system. The NPM uses an integral Pressurized Water Reactor (PWR) with a compact core made of 37 17x17 fuel assemblies, each two meters long, generating 160 MW of thermal power. Unlike traditional reactors, the Reactor Coolant System (RCS), including the Pressurizer, is integrated into the reactor pressure vessel (RPV), and the RCS pumps are eliminated.

Cooling is achieved through natural circulation, driven by density differences and height differences between the core and steam generators. Water heated in the core rises through the central riser (hot leg) and is then cooled in the surrounding helically coiled steam generators (HCSGs), where it becomes denser and flows downwards through the downcomer, completing the loop. The entire process is gravity-driven, without the need for mechanical pumps. Finally, the flow path within the RCS can be seen in Figure 1.

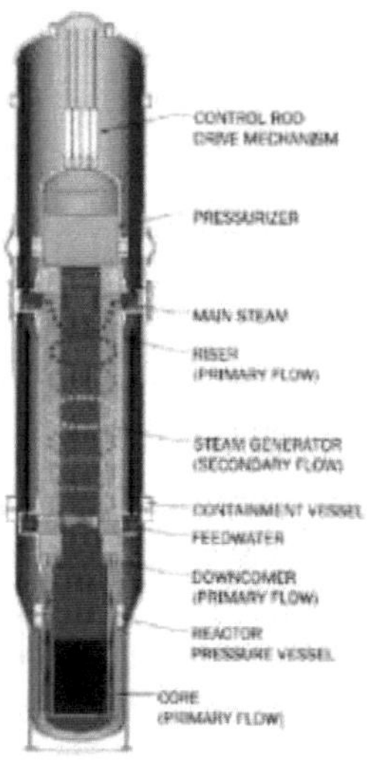

Figure 1. General arrangement of a NPM including the CNV vessel.

Additionally, the reader should keep in mind that the RPV is allocated within the interior volume of a steel Containment Vessel (CNV) partially submerged in the reactor pool which is designed to remove the heat from the core under accidental conditions acting as the ultimate heat sink [5], see Figure. 1. Furthermore, CNV works under vacuum condition during the normal operation of the NPM to minimize the heat losses and it is used as an additional safety system able to provide an enhanced protection against overpressure transients and a heat transfer path between the RPV and the reactor pool [5].

The passive safety systems in each NPM include the Decay Heat Removal System (DHRS) and the Emergency Core Cooling System (ECCS). The DHRS is designed to remove decay heat when the reactor is shut down and the normal core cooling system is unavailable, using the secondary side of the system. It consists of two independent trains, each with a heat exchanger connected to the steam and feedwater lines, mounted on the outer surface of the Containment Vessel (CNV) and submerged in the reactor pool. During normal operation, the DHRS is isolated by actuation valves.

When activated, the Main Steam and Feedwater Isolation Valves (MSIVs and FWIVs) close, and the DHRS valves open, allowing natural circulation to occur. Steam from the Heat Exchanger Steam Generators (HCSGs) drives the circulation due to the density gradient and height difference between the DHRS heat exchangers and the SG inlets, with condensate being re-injected into the feedwater lines of the HCSGs.

In contrast, the ECCS is designed to handle failures in the normal core cooling system, particularly Loss of Coolant Accidents (LOCA) or low-temperature overpressure transients. It comprises three Reactor Vent Valves (RVVs) and two Reactor Recirculation Valves (RRVs) located at the top of the Reactor Pressure Vessel (RPV) and above the Top of Active Fuel (TAF). Each ECCS valve is equipped with an Inadvertent Actuation Block (IAB) to prevent accidental opening during normal operations. The ECCS criteria require at least two of the three RVVs and one of the two RRVs to open. When these valves open, coolant is released into the CNV, where it condenses on the walls. As the CNV liquid level rises, coolant is redirected from the CNV to the RPV, refilling the core and riser areas through the RRVs.

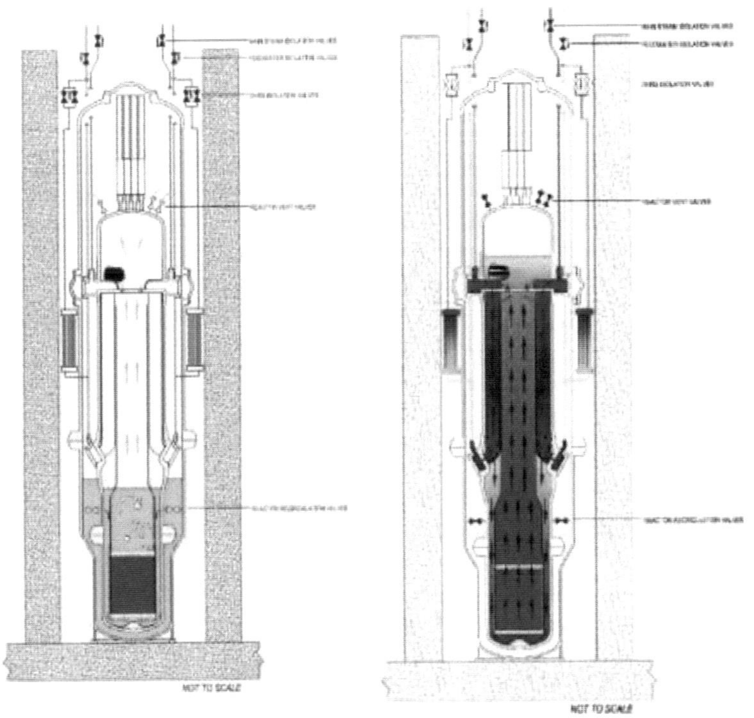

Figure 2. ECCS (left) and DHRS (right) performances.

3. DESCRIPTION OF THE LOCA SEQUENCES IN NUSCALE

In NuScale, the Large-Break LOCA is fully avoided by the integral design of the RPV and the most relevant are the ones involving ruptures of lines inside the CNV since they cannot be isolated through the actuation of the module protection system. In that sense, there are only pipes with small diameters within the RCS going through the CNV such as the CVCS makeup and letdown lines, the RPV high point vent line, or the pressurizer spray supply line. For those reasons, in addition to the break size, the LOCA accidents in NuScale can be classified in terms of the fluid condition, steam or liquid, in direct contact with the rupture.

The evolution and consequences of the LOCA accidents in NuScale are similar, only slight differences depending on the location of the break are found, more in particular, if it took place in a steam space such as in the case of the RPV high point vent or the pressurizer spray supply lines, the ECCS actuation is commonly demanded by 'low PZR pressure' signal. On the other hand, if a liquid space is in direct contact with the break, as in the case of the rupture of the CVCS makeup or discharge lines, the most likely ECCS signal to be triggered the first is 'High CNV level'.

A LOCA accident in NuScale starts with the rupture of one of the previously commented lines causing a discharge of the RCS inventory into the CNV. For that reason, the pressure along with the liquid level within the CNV internal volume is increased and the ECCS actuation is demanded. Once the IAB differential pressure threshold is reached, the ECCS valves (RVV and RRV) will open allowing the pressure within the RPV and CNV to be equalized. As the RRVs are now open, the inventory within the CNV can be re-injected in the RCS and a natural circulation loop is established between the RPV and the CNV driven by the decay heat produced within the core and the cooling capability of the CNV walls in contact with the reactor pool.

4. MODELLING OF THE NUSCALE POWER MODULE USING TRACE

A full-plant model of the NuScale SMR has been built for the system code TRACE using a combination of 1D and 3D hydraulic components and HEAT STRUCTURE components to model the heat transfer and the thermomechanical features of the fuel rods with Zr-4. In this model, the RCS is simulated by means of three 3D VESSEL components, two for the RCS regions and one for the core, with the following features, see Figure 3:

- The core region is modelled using a 3D Cartesian VESSEL with 20 axial levels and a square matrix of 7x7 allowing to achieve a 1:1 ratio between the thermal-hydraulic channels in the VESSEL and the number of fuel assemblies. In that sense, each fuel assembly has been implemented in the model using the corresponding HEAT STRUCTURE components in which the thermomechanical parameters related to the pre-transient fuel rod geometry, internal pressure and materials are implemented.
- The respective data regarding the fuel rod geometry of the different cases simulated can be seen in Table 1.
- Upper RPV region includes the regions associated with the upper riser, the annular region with the primary side of the HCSGs and the upper downcomer section. It is simulated using the 'Tube Bank Crossflow' 3D cylindrical VESSEL component and it is built with 45 axial levels, 2 radial rings (1 for the downcomer and 1 for the riser region) and 4 azimuthal sectors.
- Lower RPV region simulates the annular region with the lower part of the downcomer, the lower ple-num, the core and the lower and transition riser regions. This VESSEL is built with 18 axial levels, 2 radial rings (1 for the downcomer and 1 for the core) and 4 azimuthal sectors.
- The PZR is implemented in the TRACE model using a 1D 'Pressurizer' kind of PIPE component.

As it can be seen in Figure. 3, the secondary side is explicitly modelled up to the turbine stop valve. The interior of the HCSGs has been modelled using the 'curved pipe' PIPE components to model the special geometry of those tubes and, the CNV is simulated using a 3D cylindrical VESSEL with 24 axial levels, 3 radial rings and 1 azimuthal sector to allow the flow patterns after the rupture to be developed. Finally, the piping and heat exchangers of the DHRS have been explicitly modelled along with the reactor pool using 1D PIPEs components.

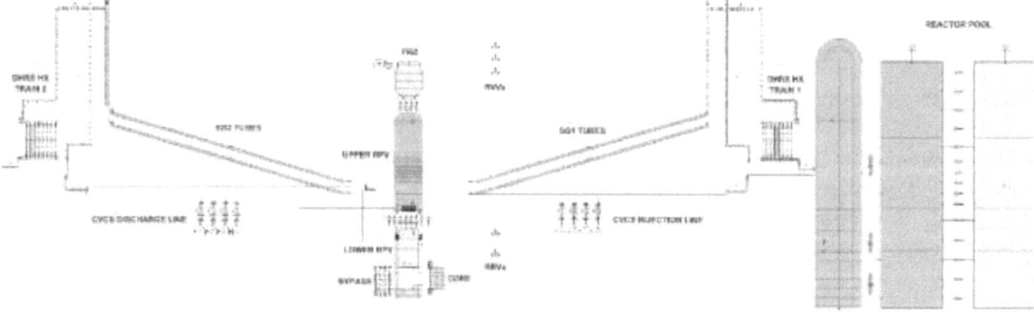

Figure 3. Nodalization scheme of the full-plant TRACE model.

Table I. Fuel rod data for the different cases.

Parameter (SI units)	UO2/Zr-4
Fuel pellet Outer Diameter (mm)	8.115
Cladding Inner Diameter (mm)	8.280
Cladding Outer Diameter (mm)	9.499
Cladding thickness (mm)	0.610
Fill Gas	He
Fill Gas Pressure (MPa)	1.482

4.1. STEADY-STATE CALCULATION

A null-transient calculation of 6000 seconds is simulated considering the following assumptions:

- The initial status of the reactor is Hot Full Power (100% of the nominal power) at the Beginning of Cycle (BOC) condition determining the radial and axial power profiles, see [2];
- Nominal values are assumed for the plant parameters except for the RCS mass flow rate which is restrained to its minimum value (535.24 kg/s);
- The CVCS letdown mass flow rate is assumed to be equal to the CVCS makeup mass flow rate which is restrained to its maximum value (3.15 kg/s);
- The CNV pressure under normal operation condition remains at 0.21 bar.

A summarizing table with the comparison of the results and the reference values is included below.

Table II. Comparison of steady-state calculation parameters with the reference values from the DCA.

Parameter (SI units)	DCA	TRACE (Error %)
RCS pressure (MPa)	12.755	12.732 (-0.18)
RCS mass-flow rate (kg/s)	535.24	536.98 (0.32)
Core mass-flow rate (kg/s)	496.17	501.52 (1.08)
Core inlet temperature	531.48	531.59 (0.02)
Steam temperature (K)	580.04	585.45 (0.93)
SG pressure (MPa)	3.447	3.443 (-0.11)

4.2. MODELLING OF A LOCA LEQUENCE IN THE CVCS DISCHARGE LINE OF A NUSCALE-LIKE SMR

In this work, the simulated scenario consists of a LOCA sequence caused by the rupture of the CVCS discharge line within the CNV, in which the opening of one out of the two RRVs and two out of the three RVVs is achieved. The LOCA transient begins 50 seconds after the steady-state calculation (with a duration of 6000 seconds) with the discharge of the RCS liquid inventory into the CNV due to the break of the CVCS discharge line. For this reason, the collapsed water level along with the pressure rapidly increases (Figure 4 and Figure 5). After another 50 seconds of simulation, the reactor trips as the "High CNV Level" signal is triggered, and ECCS actuation is demanded. When the IAB pressure threshold falls below 62 bar, the ECCS valves are allowed to open. As the opening of one RRV and two RVVs is considered, the pressure within the RPV and CNV is equalized.

In contrast, a similar LOCA sequence was also simulated assuming the failure of both RRVs to open, leading to a scenario where only two RVVs are available. In this case, without the ability to re-inject liquid water from the CNV into the RPV through the RRVs (Figure 6), the core becomes uncovered, and the cladding temperature increases until it reaches the acceptance criterion of 1477 K around 5450 seconds after the beginning of the LOCA transient, leading to core damage conditions (Figure 7).

A very similar performance regarding the evolution of the CNV level and cladding temperatures has been obtained in the cases where two RRVs are available (Figure 7). According to the results, the opening of two RVVs and one RRV is sufficient to cope with the postulated LOCA scenario. Finally, a new analysis was performed to determine the time available in the event of an RRV opening failure (Figure 7). The simulation results show that the main control room crew has a time available of 3750 seconds to recover the opening of one RRV and avoid core damage.

Table III. Sequence of events (LOCA with 2 RVV – 1RRV)

Event	Time (s)
Line Break in CVCS discharge line	6050
High CNV Pressure Signal (0.655 bar): SCRAM Signal	6054
Low PZR Pressure (137.9 bar): Isolation of Steam generators	6059
High CNV water level (6.4 m): ECCS signal.	6411
IAB Differential Pressure (ECCS). Begin of RVV and RRV valve opening	6505
RPV and CNV pressure reach same value	6530
Stabilization of the CNV level (5.8 m)	8000

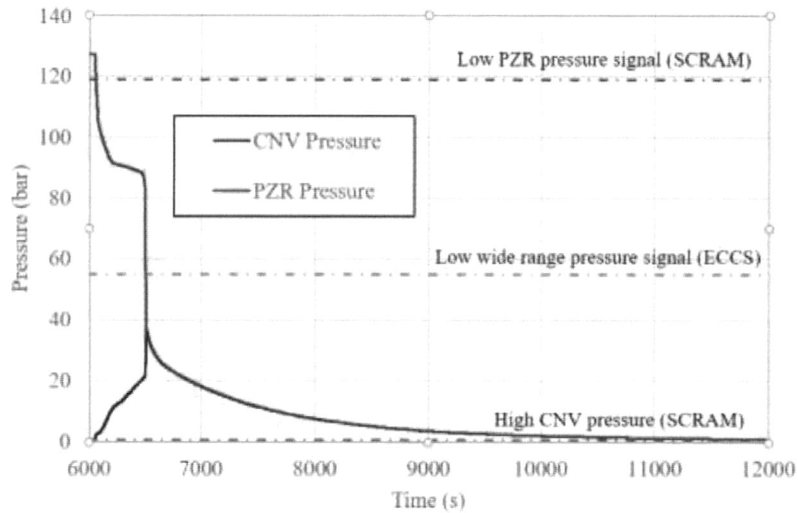

Figure 4. Evolution of the PZR/CNV pressure

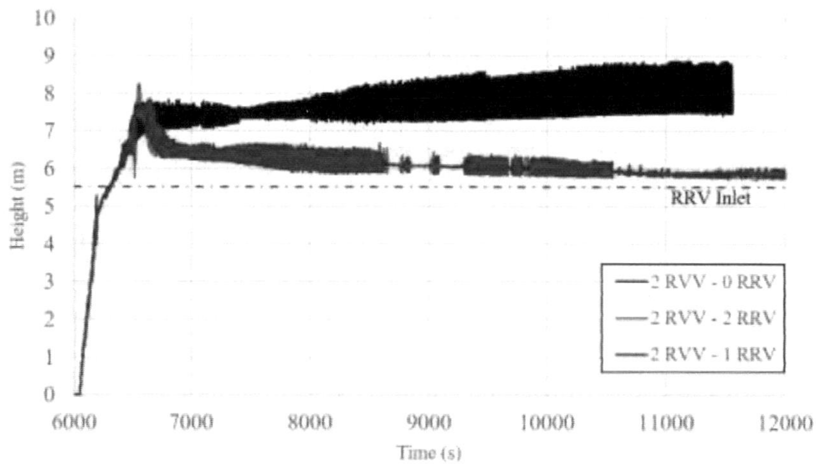

Figure 5. Collapsed water level in CNV

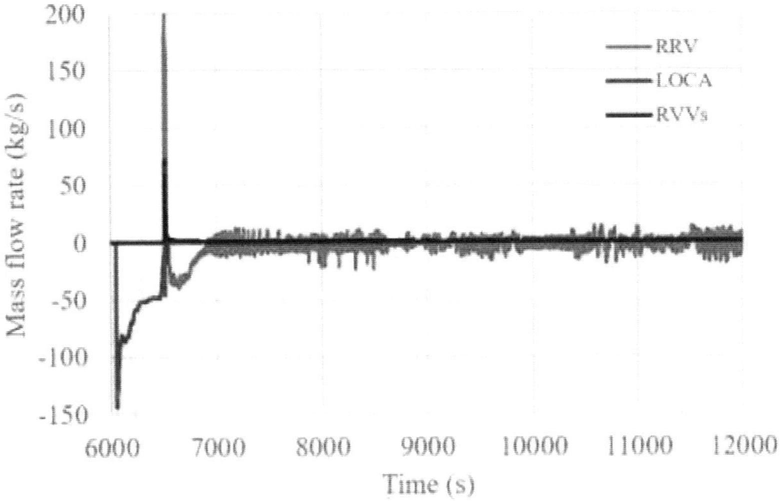

Figure 6. Mass Flow rate in RVVs, CVCS discharge Line (LOCA) and RRV (2 RRV – 1 RVV)

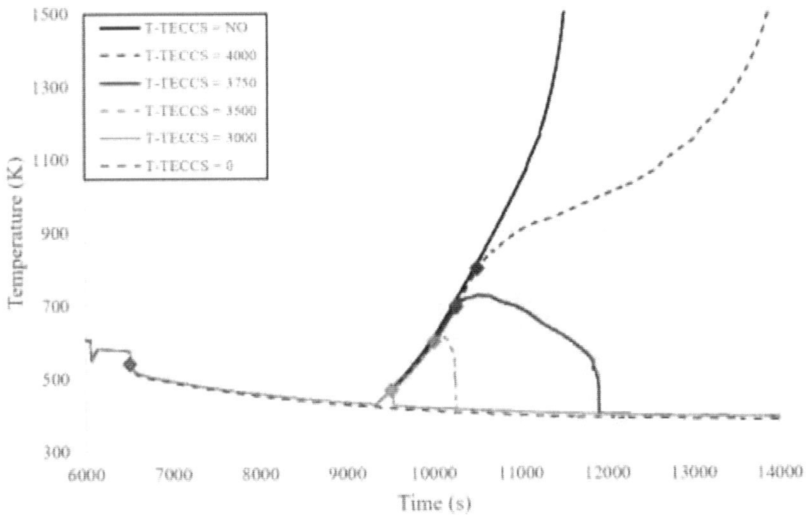

Figure 7. Peak Cladding Temperature (cases with/without RRV recovery)

5. CONCLUSIONS

The main conclusions drawn from this preliminary analysis are listed below:

- In this study, a LOCA accident in a NuScale- like SMR based on the rupture of the CVCS discharge line has been simulated with a full-plant TRACE model in which the fail to open of the RRVs leads directly to core damage after around 5000 seconds.
- Two additional cases have been simulated considering the opening of the different configurations of the RRVs and, it has been demonstrated that with only one RRV available, the NuScale-like SMR is able to deal with the postulated scenario conditions.
- The simulation results show that the main control room crew has a time available of 3750 seconds to recover the opening of one RRV and avoid core damage.

6. ACKNOWLEDGEMENTS

The authors would like to express their gratitude to the Spanish Nuclear Safety Council (CSN) for the funding to perform this work within the GODED chair.

7. REFERENCES

[1] NuScale Power, LLC, "NuScale Standard Plant: Design Certification Application: Chapter 1: Intro-duction and General Description of the Plant," Rev.5, (2020).

[2] NuScale Power, LLC, "NuScale Standard Plant: Design Certification Application: Chapter 4: Reactor," Rev.5, (2020).

[3] NuScale Power, LLC, "NuScale Standard Plant: Design Certification Application: Chapter 5: Reactor Coolant System and Connection systems," Rev.5, (2020).

[4] NuScale Power, LLC, "NuScale Standard Plant: Design Certification Application: Chapter 9: Auxiliary Systems" Rev.5, 2020).

[5] NuScale Power, LLC, "NuScale Standard Plant: Design Certification Application: Chapter 6: Engineering Safety Features" Rev.5, (2020).

[6] NuScale Power, LLC, "NuScale Standard Plant: Design Certification Application: Chapter 19: Probabilistic Risk Assessment and Severe Accident Evaluation" Rev.5, (2020).

[7] NuScale Power, LLC, "NuScale Standard Plant: Design Certification Application: Chapter 15: Transient and Accident Analyses" Rev.5, (2020).

NURER 2024
Nuclear and Renewable Energy Resources

Optimized Fuel Cycle and Burnup Analysis for Pebble-bed Reactors

Sefa Bektas
Istanbul Technical University, Istanbul, Turkiye, sefa.bektas@itu.edu.tr,
ORCID: 0000-0002-0427-3220

Volkan Seker
University of Michigan Nuclear Engineering and Radiological Science, Ann Arbor, MI, USA, vseker@umich.edu,
ORCID: 0000-0002-8945-0205

Uner Colak
Istanbul Technical University, Istanbul, Turkiye, unercolak@itu.edu.tr,
ORCID: 0000-0001-9293-6065

Thomas Downar
University of Michigan Nuclear Engineering and Radiological Science, Ann Arbor, MI, USA, downar@umich.edu,
ORCID: 0000-0002-5810-6727

Cite this paper as: *Bektaş Sefa, Seker Volkan, Çolak Üner, Downar Thomas. Optimized Fuel Cycle and Burnup Analysis for Pebble-bed Reactors. 7th Int. Conf. on Nucl. and Ren. Energy Res. 27-30 October 2024, Antaya, Turkey.*

Abstract: In a PBR core, hundreds of thousands of densely packed fuel pebbles flow slowly downward. This complicates fuel movement and extends computational time for fuel cycle analysis. To overcome this problem, a quasi-static pebble flow is combined with batch-wise refueling. Achieving an equilibrium core state requires a burn-up sensitivity analysis to assess how methods for coupling neutronics with depletion impact the accuracy of burn-up calculations. Numerical burn-up calculations face a fundamental challenge of nonlinearity: the burn-up matrix, which determines fuel depletion, varies over time due to its dependence on the neutron flux, which itself is influenced by the evolving nuclide density distributions. While the explicit Euler method is commonly used for coupling neutronics with fuel depletion, its low accuracy can be problematic in PBR applications. In contrast, the predictor-corrector method enhances accuracy but requires twice as many transport calculations, increasing computational demands. To address these challenges, this study performed a time-step optimization using the SERPENT Monte Carlo code on the HTR200 design under the once-through-then-out (OTTO) scheme. The study highlighted significant runtime reductions, from approximately 10 hours to about 4 hours, while analyzing the effective multiplication factor (*k-eff*) and key isotopes, such as Xe-135, U-235, and Pu-239.

Keywords: *HTR-PBR, Equilibrium Cycle, Burnup Analysis, SERPENT, Predictor-Corrector*

1. INTRODUCTION

Pebble-Bed Reactor, orginated from Forschungszentrum Jülich (FZJ) national research institute in Germany [1], has undergone extensive research at different institutes facilities and companies with iterative version of it so far [2]–[6]. Pebble-bed reactors (PBRs) use pebbles as both fuel carriers and moderators, each containing hundreds of thousands of TRIstructural-ISOtropic (TRISO) fuel particles. This design enables continuous circulation of fuel pebbles and online refueling, providing high operational flexibility with low excess reactivity and a notable inherent safety advantage through the use of TRISO particulate fuel. However, unlike light water reactors, the continuous flow of pebbles demands the development of new methods for studying depletion and burn-up [7], enabling more accurate predictions of reactor behavior, quicker achievement of equilibrium core configuration, and more efficient fuel cycle management.

On the other hand, due to the slow movement in PBRs, the pebble flow is classified as quasi-static dense granular flow, where particles in a random dense packing move together with their nearest neighbors over short distances [8], [9]. This flow has been studied using two frameworks: Eulerian and Lagrangian. The Eulerian approach assumes force-free movement, while the Lagrangian framework employs the Discrete Element Method (DEM) to precisely analyze contact forces [10]. However, DEM is computationally intensive, particularly when coupled with deterministic or stochastic codes for neutronics and depletion calculations, further complicating the process. The concept of region-wise pebble movement, which represents a dynamic core in a quasi-static manner, is a useful approach to adapt and alleviate the computational burden within the Eulerian framework. In this approach, the pebble-bed reactor core is divided into distinct regions based on material properties and temperature profiles. For certain periods, these regions remain stationary before advancing to the next region.

This approach links the refueling of pebble-bed reactors (PBRs) to the batch-wise refueling of light water reactors (LWRs), but the duration of refueling periods differs significantly, with PBRs operating on a timescale of days compared to the months typically required for LWRs. But, similar to light water reactors (LWRs) [11], burn-up calculations in pebble-bed reactors (PBRs) face challenges concerning accuracy and stability. Firstly, this arises from the need to solve the neutron transport or diffusion equation along with the Bateman equation, resulting in a coupled, non-linear system. Secondly, the pebbles spend significant time within the reactor, encountering different spatial and spectral zones and undergoing higher burn-up as they move through the core, even though their depletion time steps are shorter compared to those in LWRs. Accuracy is particularly significant and has been the subject of extensive analysis in this paper because previous studies employed step-wise movement for evaluating burn-up calculations but did not address this issue [12], [13]. Furthermore, when a multi-pass fuel cycle is implemented, errors can accumulate, as each pass contributes additional discrepancies to the burn-up calculations.

In light of the aforementioned reasons, this work will provide fresh insights into coupling depletion with the Serpent Monte Carlo transport code for OTTO (Once-Through-Then-Out) fuel cycle scheme. It will be compared two time integration methods in terms of accuracy and run-time—one low-order and the other high-order—embedded in the Serpent Monte Carlo code. Additionally, it will be sought to balance the trade-off between accuracy and computational run time for burn-up analysis.

2. METHOD

It is beneficial to begin by outlining the selection of the reactor core before investigating coupling methods for steady-state neutronics and Bateman solutions. This foundational understanding will provide context for the subsequent analysis and facilitate a more effective examination of the coupling techniques employed.
An optimization was performed, varying the burnup coupling methods alongside with different sub-burnup time steps for the HTR-200 pebble-bed-design reactor [14], operating under the once-through-then-out (OTTO) scheme. This reactor is designed to provide a thermal power output of up to 200 MWth, featuring an outlet temperature of 700°C and a nominal inlet temperature of 250°C. It employs TRISO (tri-structural isotropic) particle fuel encased in a graphite matrix, with chemically stable, mid-pressure helium (He) serving as the coolant. Table 2.1 presents the key nuclear design data for the HTR-200 under consideration.

Table 2.1. *The main parameters of the HTR-200.*

Thermal Power MW_{th}	200
Height of the reactor core (m)	5.5
Diameter of the reactor core (m)	3.8
Height-to-diameter ratio of the core	1.45
Number of fuel elements (pebble)	3.37E5
Fuel type	UO_2
Fuel matrix	Pebble

Pebble diameter (cm)	6
Mean residence time of the pebbles (day)	932
Helium pressure (MPa)	5
Gas inlet temperature (°C)	250
Gas exit temperature (°C)	700
Pressure drop (MPa)	0.05
Flow (kg/sec)	85.6
Refueling Scheme	OTTO (Once Through Then Out)
Neutron Spectrum	Thermal

The HTR-200 fuel is composed of tri-structural isotropic (TRISO) fuel particles, which are embedded within a carbon matrix pebble. One of the key safety-related functions of the fuel is to act as the primary barrier, ensuring functional containment. The TRISO particles have a porous carbon buffer layer between the kernel and the IPyC layer, which creates space for fission gases and helps control pressure buildup. The reactor pressure vessel acts as the secondary barrier in the HTR-200's functional containment system for the fuel within the reactor core. Figure 2.1 depicts both the pebble and the particle and Table 2.2 gives fuel particle properties. The geometric configuration of the reactor design and the loading condition of the spherical fuel elements can be seen in Figure 2.2.

Table 2.2: *Fuel particle properties.*

Coating	Thickness x 10^{-2} cm	Density g/cm^3
Buffer	0.9	1.0
IPyC	0.4	1.85
SiC	0.35	3.2
OPyC	0.35	1.85
Radius of fuel kernel cm	0.025	

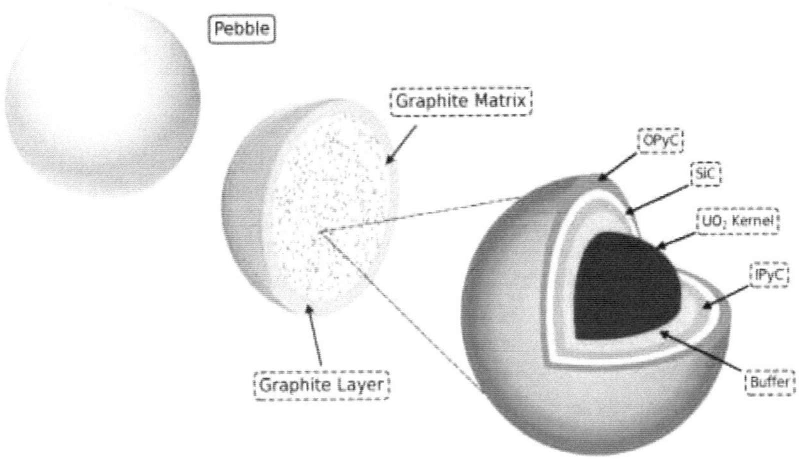

Figure 2.1. *Fuel pebble and triso particle.*

In the neutronic calculations, the entire reactor is divided into 170 zones with different compositions. The core is partitioned into 150 small regions of equal volume. The effective upper, lower, and side reflectors are 1.5 m, 1.0 m, and 1.0 m thick, respectively. In the fresh reactor, five types of spherical fuel elements with different enrichment levels are used to simulate the fuel and heat distribution of the burned-in core. Fresh fuel elements, with three different enrichment levels, are continuously loaded from the top and moved through the core. The fresh fuel condition, the core's fuel cycle, and the temperature profile used for the calculations are illustrated in Figure 2.2. In the upper reflector zones 1, 2, and 3, B_4C is uniformly distributed to simplify the representation of control elements that regulate reactor power. Figure 2.2 also shows the control rod positions at the start of the power change.

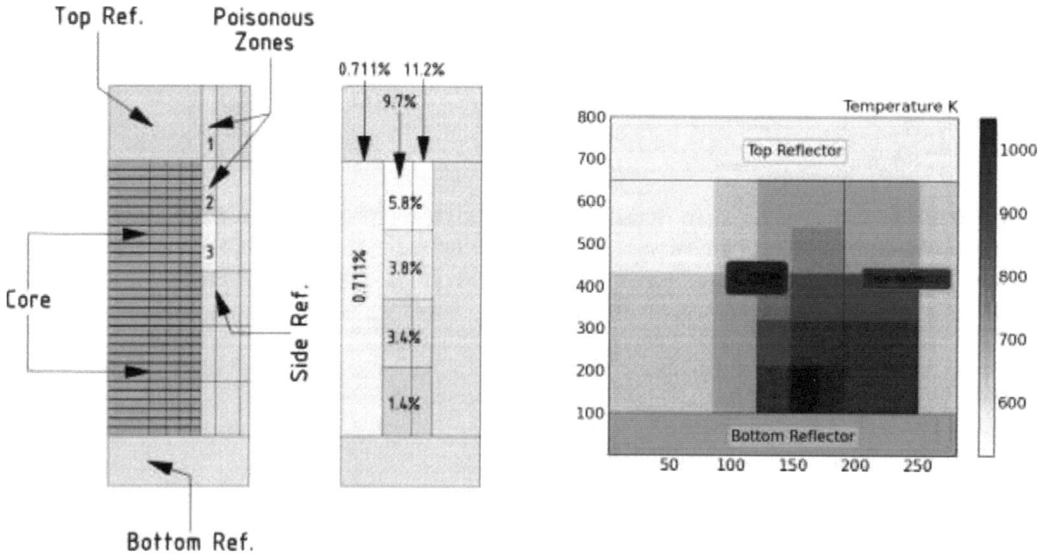

Figure 2.2. *Neutronic calculation model of HTR-200.*

Integrating pebble movement with online refueling involves multiple axial fuel channels, each containing distinct layers that contain pebbles pebbles and shift throughout the core in discrete steps. This approach combines quasi-static pebble flow with batch-wise refueling. These calculations are essential because the steady-state neutronics and Bateman solutions are coupled with a linking algorithm [15]. This algorithm uses one or more neutronics solutions to predict changes in cross-sections, flux, or reaction rates over a single step and then selects constant reaction rates to represent these predictions in the Bateman solutions. Because, the burn-up matrix, which determines fuel depletion, varies over time due to its dependence on the neutron flux, which itself is influenced by the evolving nuclide density distributions. This nonlinearity complicates the burn-up determination which can be addressed by the explicit Euler method which is commonly used to couple neutronics with fuel depletion. This simplest and most straightforward method for performing burnup calculations involves a systematic approach. Initially, neutronics is solved using the beginning-of-period (BOP) material compositions. Changes in the material compositions are then calculated for the interval, assuming the reaction rates remain constant at their initial values. The resulting end-of-period (EOP) compositions then serve as the BOP compositions for the subsequent interval, so-called predictor or low order method [16].

Using the beginning-of-period (BOP) reaction rates for the entire interval is a rather poor approximation, which requires relatively short periods for improved accuracy. As a result, several predictor-corrector methods are typically employed instead. These methods involve obtaining two neutronics solutions for each interval instead of just one, while the extra computational time is more than offset by the improved accuracy, allowing for the use of longer intervals. These advanced time-stepping algorithms, known as the predictor-corrector or high-order methods, provides enhanced accuracy in time compared to the explicit Euler method [15]–[18]. The predictor-corrector method for burnup calculations proceeds as follows: First, the beginning-of-period (BOP) cross-sections and flux are calculated by solving the neutronics. Next, materials are depleted using the BOP reaction rates for the desired interval length. Following this, the end-of-period (EOP) material compositions obtained from the previous period are used to estimate the EOP cross-sections and flux. Subsequently, the BOP materials are re-depleted using weighted of the BOP and EOP cross-sections and flux to adjust the reaction rates. Finally, the new EOP compositions become the initial compositions for the next period.

Higher-order estimates allow the use of linear and quadratic fits to supersede the constant and linear approximations employed in the predictor and corrector steps, respectively. The average derived from a polynomial fit simplifies to a weighted sum of the values at the fitted points, specifically incorporating the beginning-of-period (BOP) values from the previous period, as well as the BOP and end-of-period (EOP) values from the current period. The possible settings for the time integration method in burn-up calculation given in Table 2.3

Table 2.3. *Possible setting for mode in Serpent Monte Carlo code* [19].

Acronym	Predictor method	Corrector method
CE	Constant extrapolation	N.A.
CE/LI	Constant extrapolation	Linear interpolation
LE	Linear extrapolation	N.A.
LE/LI	Linear extrapolation	Linear interpolation
LE/QI	Linear extrapolation	Quadratic interpolation
N.A.	Constant or linear extrapolation	Linear or quadratic interpolation
CE/CE	Constant extrapolation	Constant backwards extrapolation

Although these methods have their advantages and disadvantages, the developers opted for LE/QI as a higher-order approach to compare with other advanced algorithms implemented in different codes, such as CE/CM (Constant Midpoint), which involves solving for the middle of the step (MOS) in MCNPX and MCNP6, and CE/BE (Backwards Extrapolation) by MCODE.Considering these trade-offs [15], a time step sensitivity analysis for the fuel cycle time of the Pebble Bed Reactor (PBR) using the SERPENT Monte Carlo code was performed in the work presented here for the coupled neutronic and depletion calculation using the SERPENT Monte Carlo code with the ENDF/VII.0 data library. Given that the average transit time of the pebbles through the core is approximately 932 days, a running period of 31 days was chosen before recycling, discharging layers from the bottom, and refueling at the top. In terms of coupling neutronics and depletion, the Linear Extrapolation/Quadratic Interpolation (LE/QI) predictor-corrector method was used as the higher-order approach, while the explicit Euler predictor method was utilized as the lower-order method for comparative analysis. The impact of each method, with varying sub-burn-up-time steps during the 31-day shuffling period, was examined for the effective multiplication factor, *k-eff*, and different isotopes such as Xe-135, U-235, and Pu-239. This investigation involved tracking the layers in 5 different flow channels throughout 1 pass period, and the PC (LE/QI) method with 6 sub-burn-up depletion steps, was utilized as the reference case and other cases we investigated were given in Table 2.4. However, the first 31-day depletion step was divided into 8 sub-steps for all cases to capture the same xenon effect at the beginning

Table 2.4. *Use of predictor–corrector method for moving cells.*

Case	Method	Time-Step	Designation	Days
1	Predictor-Corrector	6	PCC4/Ref	(5,5,5,5,5,6)
2	Predictor-Corrector	3	PCC4/3	(5,13,13)
3	Predictor-Corrector	2	PCC4/2	(5,26)
4	Predictor-Corrector	1	PCC4/1	(31)
5	Predictor	2	PCC0/2	(5,26)

3. RESULTS

In this study, our focus was on nuclide densities rather than comparing time-dependent reaction rates. This is because reaction rates are integral parameters influenced by fluxes, cross sections, and nuclide densities, all of which change at different rates during burn-up. In the predictor method, reaction rates are based on initial values, whereas in predictor-corrector methods, they are determined at both the beginning and the end of intervals. We, therefore, focused on nuclide densities, as they can be calculated at fixed time points, enabling straightforward comparisons. Additionally, we considered the eigenvalue, *k-eff,* which reflects the overall behavior of the reactor as a global parameter.

During the simulation, even though the models exhibit nearly identical asymptotic behavior for the eigenvalue, discrepancies between the reference case and the proposed cases tell a different story, as shown in Figure 3.1. According to Figure 1, the difference between the reference case and the PCC0/2 gradually increased, declining by more than 600 pcm, while the difference between the reference case and the PCC4/1 remained relatively lower, peaking towards the end at around 200 pcm. Looking at more fine steps, such as PCC4/2 and PCC4/3, these models fluctuated around the reference value with a difference of less than 100 pcm.

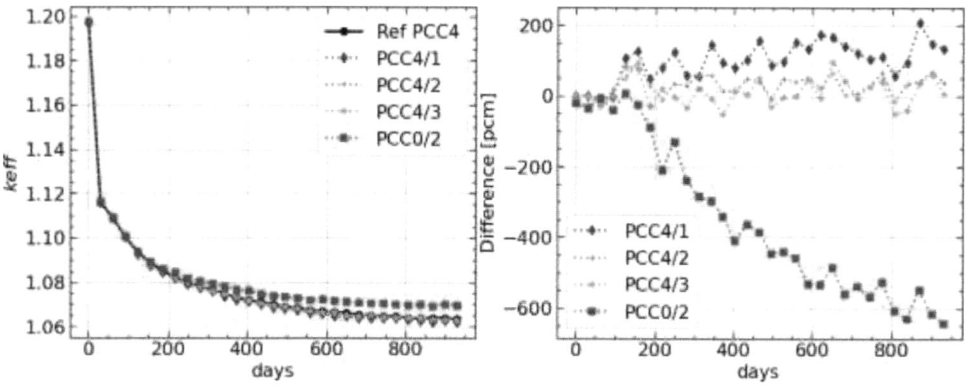

Figure 3.1. *Comparison of k-eff eigenvalues for different coupling models.*

On the other hand, the qualitative behavior of the difference remains consistent across both long and short intervals in the predictor-corrector method, when considering the total concentration of short-lived and long-lived nuclides in the core. As anticipated, the magnitudes of these differences increase or decrease in relation to the length of the intervals. The only significant exception is the explicit Euler predictor method (PCC0/2), which was shown to be unstable for various nuclides. The improvement for long-lived nuclides is relatively small and not significantly impacted by finer steps when comparing the total concentration for the predictor-corrector method, making the effects of sub-steps on them less decisive.

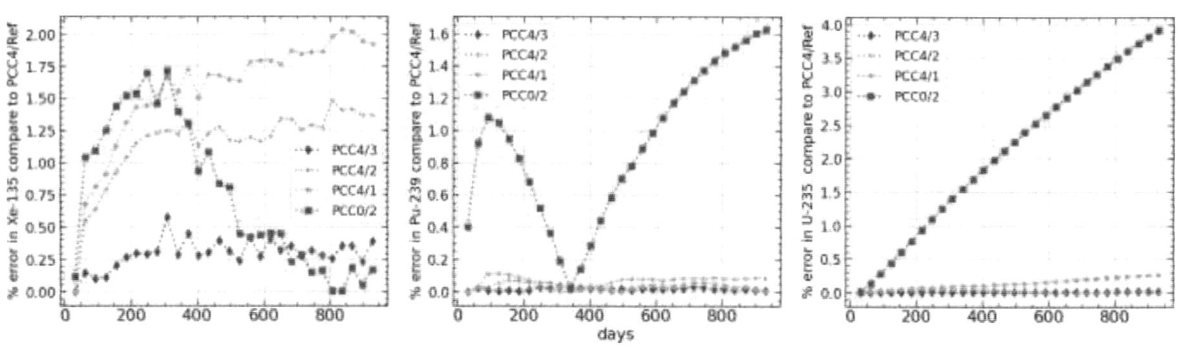

Figure 3.2. *Percentage error of various total isotope concentrations and models compared to the reference case, PCC4-Ref.*

When tracing the percentage error for a cell moving from top to bottom in each of the five radial channels, the difference for the predictor-corrector methods in long-lived nuclides remained below 1%, while the predictor method showed an error of less than 5%. However, when comparing the error in total concentration, the percentage error in local cell concentration was slightly higher.

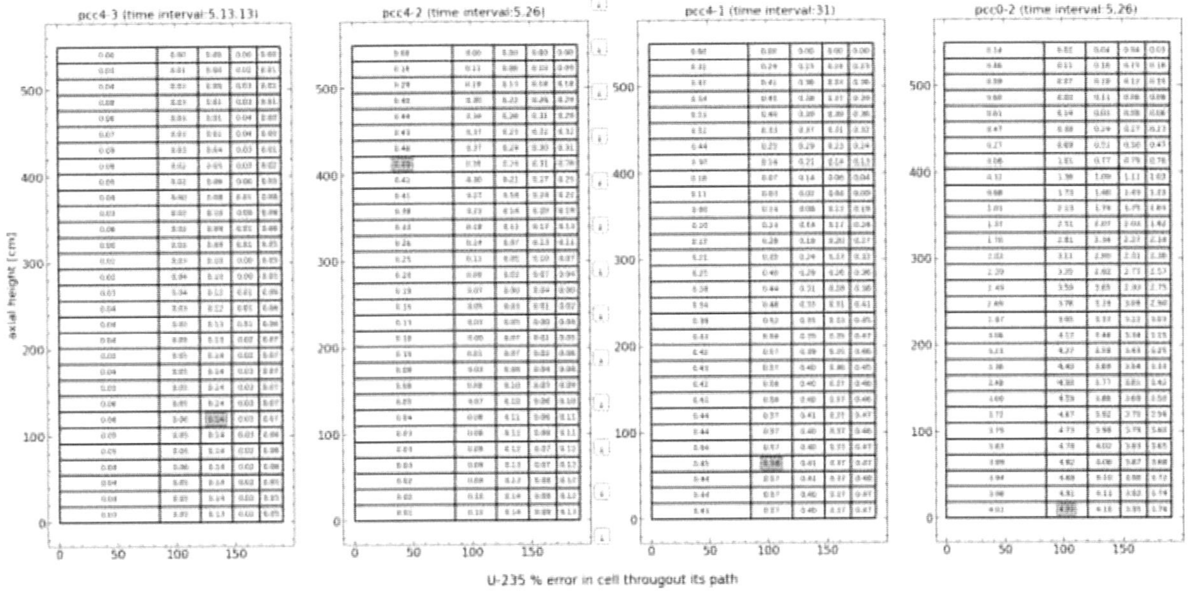

Figure 3.3. *Percentage error of U-235 concentrations and models compared to the reference case, PCC4-Ref, for cells moving along their path in each channel.*

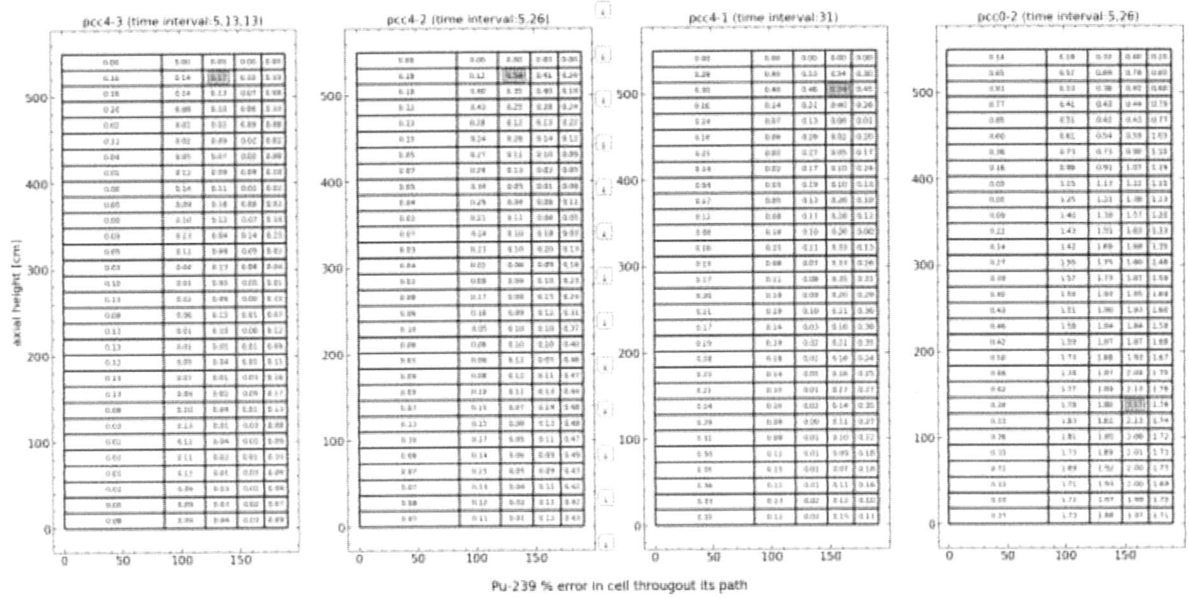

Figure 3.4. *Percentage error of Pu-239 concentrations and models compared to the reference case, PCC4-Ref, for cells moving along their path in each channel.*

In contrast, for the short-lived nuclide Xe-135, the scenario is notably different. In this case, the peak error for each case exceeds 5%, with the exception of the PCC4/3, which reported an error of only 4.45%.

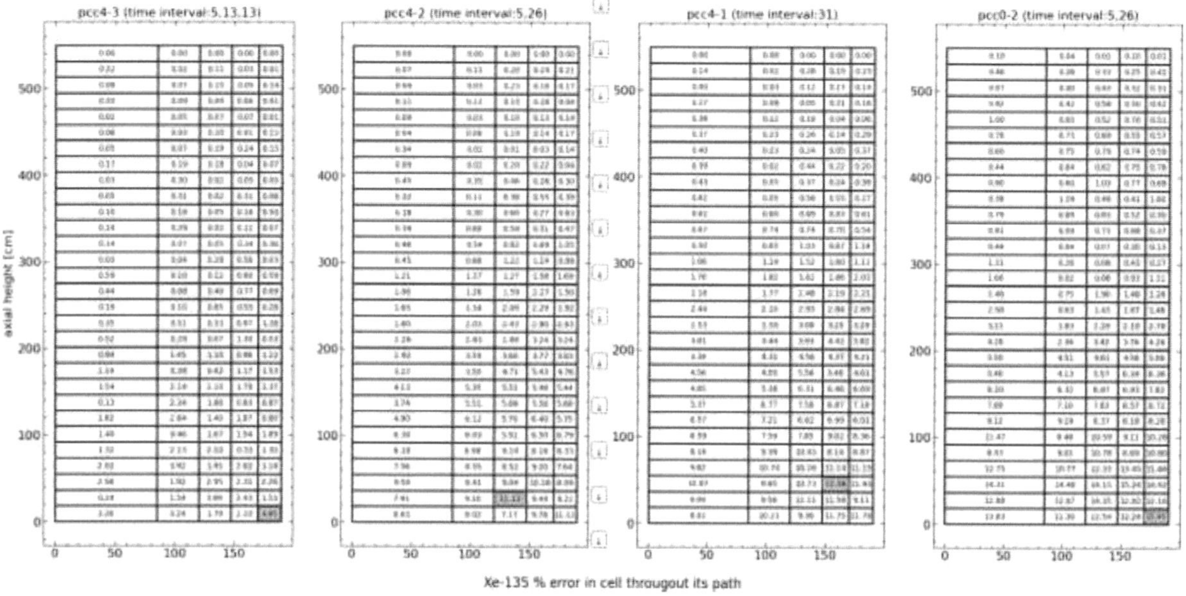

Figure 3.5. *Percentage error of Xe-135 concentrations and models compared to the reference case, PCC4-Ref, for cells moving along their path in each channel.*

Overall, we observed that the *k-eff* in the PC method with 3 burnup steps differs by less than 100 pcm, while the infinite norm of the relative difference for isotopic concentration of all nuclides is less than 5% in each layer. Additionally, the runtime for the 31-day refueling period significantly decreased from approximately 10 hours to about 4 hours with 36-core processors.

Table 3.1. *The maximum difference compared to the reference case across various parameters for moving cells.*

Parameter	Maximum difference (Ref- PCC4/3)	Maximum difference (Ref- PCC4/2)	Maximum difference (Ref- PCC4/1)	Maximum difference (Ref- PCC0/2)
k-eff	97 pcm	94 pcm	208 pcm	642 pcm
Xe-135	4.45%	11.12%	12.36%	15.45%
U-235	0.14%	0.49%	0.58%	4.93%
Pu-239	0.17%	0.56%	0.59%	2.17%

4. CONCLUSIONS

Returning to the question posed at the beginning of this study, it is now possible to conclude that employing the predictor-corrector method instead of relying solely on the predictor has significantly enhanced the results for all examined parameters. This study also aimed to improve the understanding of how finer depletion steps dramatically enhance outcomes for short-lived nuclides like Xe-135. For fuels that move throughout the core in a quasi-batch manner, the effective half-life of a nuclide needs to be comparable to the duration of the depletion interval.

Additionally, non-short-lived heavy metals, including Pu-239 and U-235, similarly reflect inaccuracies in the fissile inventory and are consequently affected.

However, finer steps increase the number of depletion calculations required, which can significantly slow down the process and double the runtime, leading to higher computational costs. Since subdivisions of the depletion interval and the higher orders of the predictor-corrector method tackle different sources of error, they should be optimized together to achieve the best results and faster calculations. When operating pebble-bed reactors using quasi-batch movement, these optimizations need to be undertaken before reaching equilibrium. Nevertheless, further studies are required to gain a better understanding of the mechanisms underlying numerical stability and Xe-135 oscillations in pebble-bed reactors.

ACKNOWLEDGMENT

This work was supported by Scientific Research Projects Department of Istanbul Technical University. Project Number: 45892.

REFERENCES

[1] G. H. Lohnert, "Technical design features and essential safety-related properties of the HTR-module," *Nucl. Eng. Des.*, vol. 121, no. 2, pp. 259–275, Jul. 1990, doi: 10.1016/0029-5493(90)90111-A.

[2] H. Gottaut and K. Krüger, "Results of experiments at the AVR reactor," *Nucl. Eng. Des.*, vol. 121, no. 2, pp. 143–153, Jul. 1990, doi: 10.1016/0029-5493(90)90099-J.

[3] A. Koster, H. D. Matzner, and D. R. Nicholsi, "PBMR design for the future," *Nucl. Eng. Des.*, vol. 222, no. 2–3, pp. 231–245, Jun. 2003, doi: 10.1016/S0029-5493(03)00029-3.

[4] E. J. Mulder and W. A. Boyes, "Neutronics characteristics of a 165 MWth Xe-100 reactor," *Nucl. Eng. Des.*, vol. 357, p. 110415, Feb. 2020, doi: 10.1016/J.NUCENGDES.2019.110415.

[5] N. A. W. Walton, R. Crowder, N. Satvat, N. R. Brown, and V. Sobes, "Demonstration of a random sampling approach to uncertainty propagation for generic pebble-bed fluoride-salt-cooled high temperature reactor (gFHR)," *Nucl. Eng. Des.*, vol. 395, p. 111843, Aug. 2022, doi: 10.1016/J.NUCENGDES.2022.111843.

[6] Z. Zhang *et al.*, "Loss-of-cooling tests to verify inherent safety feature in the world's first HTR-PM nuclear power plant," *Joule*, vol. 8, no. 7, pp. 2146–2159, Jul. 2024, doi: 10.1016/J.JOULE.2024.06.014.

[7] E. M. Duchnowski and N. R. Brown, "A review of multiphysics tools and methods to evaluate high temperature pebble bed reactors," *Prog. Nucl. Energy*, vol. 171, p. 105164, Jun. 2024, doi: 10.1016/J.PNUCENE.2024.105164.

[8] M. Z. Bazant, "The Spot Model for random-packing dynamics," *Mech. Mater.*, vol. 38, no. 8–10, pp. 717–731, Aug. 2006, doi: 10.1016/J.MECHMAT.2005.06.016.

[9] C. H. Rycroft, G. S. Grest, J. W. Landry, and M. Z. Bazant, "Analysis of granular flow in a pebble-bed nuclear reactor," 2006, doi: 10.1103/PhysRevE.74.021306.

[10] S. Jiang, J. Tu, X. Yang, and N. Gui, *Multiphase Flow and Heat Transfer in Pebble Bed Reactor Core*. 2021.

[11] P. Cosgrove, E. Shwageraus, and G. T. Parks, "A simple implicit coupling scheme for Monte Carlo neutronics and isotopic depletion," *Ann. Nucl. Energy*, vol. 141, p. 107374, Jun. 2020, doi: 10.1016/J.ANUCENE.2020.107374.

[12] C. Lu and Z. Wu, "A New Method to Efficiently Estimate the Equilibrium State of Pebble Bed Reactors," *Nucl. Technol.*, vol. 208, no. 10, pp. 1577–1590, Oct. 2022, doi: 10.1080/00295450.2022.2049966.

[13] Y. Robert, T. Siaraferas, and M. Fratoni, "Proof of concept for hyper-fidelity depletion of full-scale pebble bed reactors," *Ann. Nucl. Energy*, vol. 183, p. 109648, Apr. 2023, doi: 10.1016/J.ANUCENE.2022.109648.

[14] Z. Gao, "Auslegungsuntersuchung für den HTR zur Erschließung von Schwerstöllagerstätten," 1977.

[15] A. Isotalo and V. Sahlberg, "Comparison of Neutronics-Depletion Coupling Schemes for Burnup Calculations," *Nucl. Sci. Eng.*, vol. 179, no. 4, pp. 434–459, Apr. 2015, doi: 10.13182/NSE14-35.

[16] D. Kotlyar and E. Shwageraus, "On the use of predictor–corrector method for coupled Monte Carlo burnup codes," *Ann. Nucl. Energy*, vol. 58, pp. 228–237, Aug. 2013, doi: 10.1016/J.ANUCENE.2013.03.034.

[17] A. E. Isotalo and P. A. Aarnio, "Higher order methods for burnup calculations with Bateman solutions," *Ann. Nucl. Energy*, vol. 38, no. 9, pp. 1987–1995, Sep. 2011, doi: 10.1016/J.ANUCENE.2011.04.022.

[18] P. Cosgrove and E. Shwageraus, "Stability analysis of higher-order neutronics-depletion coupling schemes and Bateman operators," *J. Comput. Phys.*, vol. 448, p. 110702, Jan. 2022, doi: 10.1016/J.JCP.2021.110702.

[19] A. E. Isotalo, J. Leppänen, and J. Dufek, "Preventing xenon oscillations in Monte Carlo burnup calculations by enforcing equilibrium xenon distribution," *Ann. Nucl. Energy*, vol. 60, pp. 78–85, Oct. 2013, doi: 10.1016/J.ANUCENE.2013.04.031.

NURER 2024
Nuclear and Renewable Energy Resources

Hydrogen technology development during a revolutionary era in Europe

Virginie Cartier

UR CONFLUENCE : Sciences et Humanités [EA 1598], UCLy, Lyon, France, vcartier@univ-catholyon.fr ORCID: 0000-0002-0154-2157

Elisabeth A. Cazier

GEPEA [UMR 6144], Nantes Université, Oniris, Nantes, France, elisabeth.cazier@univ-nantes.fr ORCID: 0000-0002-1136-9415

Cite this paper as: *Cartier, V. Cazier, E.A. Hydrogen technology development in a French revolutionary era. 7th Int.Conf. on Nucl. and Ren. Energy Res. 27-30 October 2024, Antalya, Turkey*

Abstract: After a reminder of the European Commission's policy on renewable energies, this article details an overview of the current landscape of hydrogen in Europe as a game changer. The challenges encountered in hydrogen development regarding regulatory parameters and technical issues are deepened. Hydrogen production processes are detailed with a focus on the economic aspects. Current hydrogen end-users as an energy vector or a chemical are described with a break-even price correlation with production. To promote the use of renewable hydrogen, decreasing its production cost is unavoidable. Additionally, this article highlights the importance of improving the energy efficiency of hydrogen use in comparison to conventional fuel, particularly in the mobility sector. Considering the location of hydrogen end-users and producers, it is essential to have flexibility in the electrical network, as well as in hydrogen transmission and storage infrastructures to ensure the security of hydrogen supply. As an illustration, this article discusses European projects that involve the conversion of natural gas networks.

Keywords: *Hydrogen; Energy transition; European market; French market; Low-Carbon Infrastructures*

1 INTRODUCTION

Climate change is a real challenge for our generation and for those to come. The decarbonization of energy use is unavoidable, and the development of renewable energies is key to ensure this. The objective of limiting global warming to 1.5°C by 2030 can only be achieved if we - citizens, industrialists and institutions - succeed in decarbonizing our uses (Mukherji et al., 2023). One way to achieve this is by developing quickly renewable energies, such as solar power, wind power, anaerobic digestion, and, as we will explore in this article, the use of hydrogen (H_2) as an energy source. H_2 is a widely used chemical reagent, and its decarbonated origin is also a valuable asset in the process of reducing greenhouse gases. H_2 is part of the various decarbonization plan with various origins (blue, green, grey, black, white) and various uses (transports, chemical production, refining…). However, renewable H_2 production tends to be still too expensive for its everyday uses.

This article will explore the H_2 sector in Europe, detailing the various technologies used for H_2 production with the associated costs, its end-users, and the infrastructures projects in Europe to supply electricity and H_2, with a specific focus on the French situation.

2 CONTEXT

H_2 and its derivatives like ammonia and methanol play a crucial role in the European Green Deal and the RePowerEU plan (European Commission, Secretariat-General, 2022). These plans aim to facilitate the decarbonization of the energy system and reduce dependence on fossil fuels. The development of renewable H [1] production in recent years has been bolstered by supportive European policies.

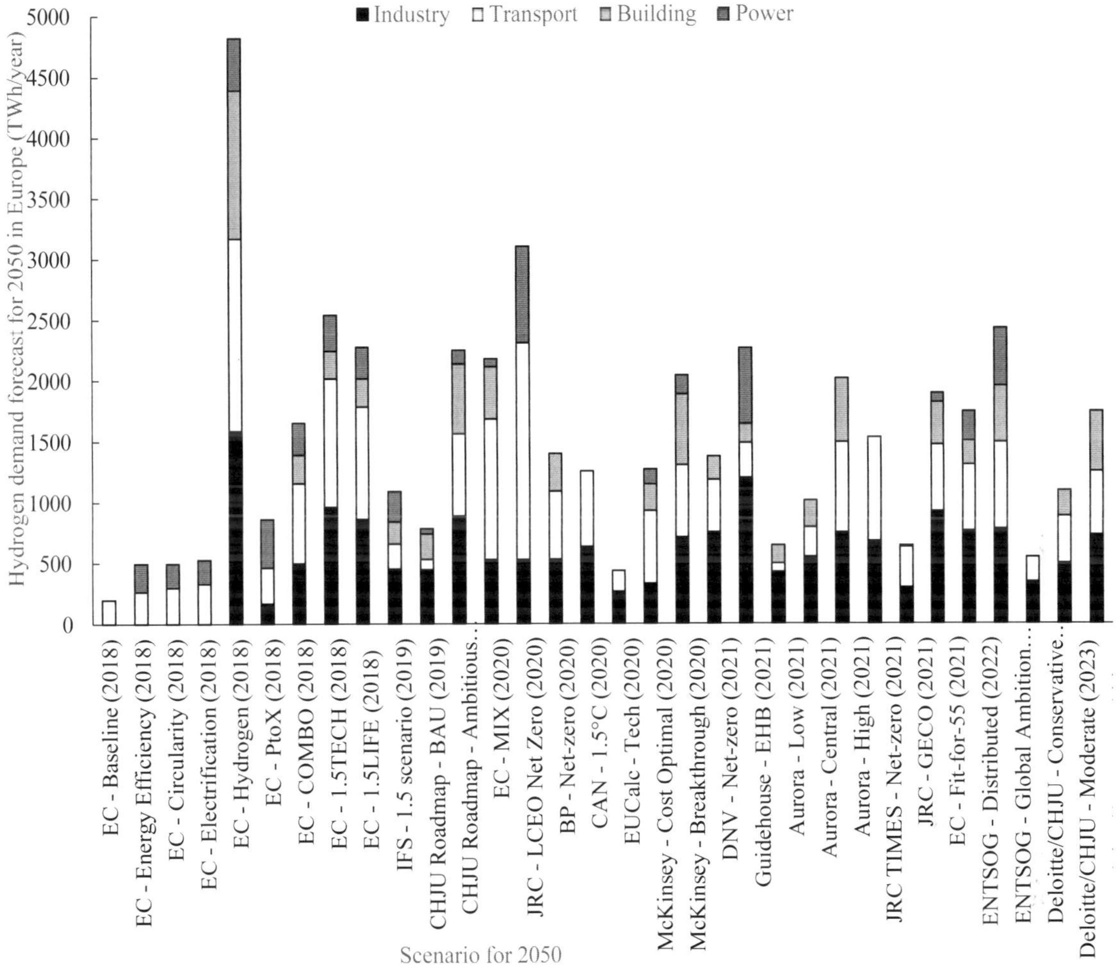

Figure 1: H_2 demands forecast for 2050 based on various scenarii in Europe. EHB: European Hydrogen Backbone, CHP: Clean Hydrogen Partnership, Aurora: Aurora Energy Research, CAN: Climate Action Network; EC: European Commission, (European Hydrogen Observatory, 2023a)

Since 2018, a high number of renewable energy road maps for Europe of various origins were shared including forecasts for H_2 demands in 2050 (Figure 1). They evolve from 198 to 4,818 TWh/year, with a median at 1,386 TWh/year. Most scenarii predict that H_2 will be used in four sectors: industry, mobility, building and power. The primary uses are expected in industry (ranging from 0 to 1,584 TWh/year) and mobility (ranging from 85 to 1,782 TWh/year). Its use in buildings (for heating and cooking) is uncertain (ranging from 0 to 1,221 TWh/year, with a median of 203 TWh/year), as is its use in the power sector (ranging from 0 to 792 TWh/year, with a median of 93 TWh/year). The European Commission's REPowerEU plan (European Commission, Secretariat-General, 2022) aims to produce between 500 and 550 TWh of renewable electricity by 2030 and 10 million tonnes of non-biological renewable fuels. This would correspond to 14% of the EU's total electricity consumption. More broadly, the Commission envisions increasing the renewable energy target to 45% by 2030. Under this plan, 10 million tonnes of renewable H_2 is allocated for both domestic production and importation by 2030 (European Commission, Secretariat-General, 2022).

Due to its high production cost, decarbonated H_2 should primarily be reserved for sectors that are hard to decarbonize, such as industry, sea shipping, and air transports. The European H_2 strategy outlined in the EU Green Deal, is an ambitious plan consisting of three phases (European Hydrogen Observatory, 2023b). In the first phase, by 2024, the

[1] According to the European Commission regulations, renewable H_2-based fuels referred to as renewable fuels of non-biological origin (RFNBO), are generated from renewable energy sources excluding biomass.

installation of 6 GW of renewable H_2 electrolysers in the EU is targeted (i.e. 1 million tonnes of renewable H_2 production). In the second phase, by 2030, the aim is to install 40 GW of renewable H_2 electrolysers, corresponding to the production of 10 million tonnes of renewable H_2. The final phase, extending until 2050, involves the large-scale deployment of renewable H_2 production to reach all hard-to-decarbonize sectors where other alternatives might not be feasible or would incur higher costs. This phase envisions using one-quarter of the EU's renewable electricity production for H_2 production. Clean H_2 production is projected to make up over 23% of the energy mix by 2050 (European Hydrogen Observatory, 2023b).

In 2020, Europe set up the European Clean Hydrogen Alliance to stimulate clean H_2 production and use, to help developing new technologies and to analyze the existing barriers. Its members are industrials (i.e. electrolysers manufacturers, clean H_2 producers, energy companies…), public authorities, civil society organizations, financial institutions, research organizations (European Clean Hydrogen Alliance, 2021). The critical points identified include: *(1)* the availability and affordability of renewable and low-carbon H_2, *(2)* the stimulation of H_2 use in the industry, mobility and building sectors to estimate the costs and demand, *(3)* the availability and readiness of transport infrastructure, as well as sufficient storage capacities (European Clean Hydrogen Alliance, 2021). Furthermore, it is crucial to synchronize H_2 value chains; it requires a supportive regulatory framework encompassing trading, supply, and certifications. Additionally, aligning subsidy schemes is essential to support this synchronization effectively (European Clean Hydrogen Alliance, 2021). These demands were incorporated in the Fit-for-Package in 2021, which included legislative proposals to improve the European H_2 regulatory framework. Targets were set from 2030 for renewable H_2 in the industrial and transport sectors. Further advancements were planned for June 2023, including the introduction of two delegated acts within the Renewable Energy Directive. These acts would delineate the definition of renewable H_2 and establish methodologies for calculating the life-cycle emissions of renewable H_2 and recycled carbon fuels (European Commission, 2023). Investments for the infrastructure with IPCEI Hy2Use in 2022, industry and mobility sectors with IPCEI Hy2Tech in 2022, research and innovation with the Clean Hydrogen Partnership (2021-2027), were implemented to address the critical points mentioned earlier.

3 H_2 PRODUCTION PROCESSES

The production of decarbonized or at least low-carbon H_2 can be achieved in five main ways: *(1)* extracting H_2 directly from underground deposits, the so-called elementary or geologic or white H_2, *(2)* carbon-based processes using methane pyrolysis, called grey H_2, *(3)* which may be combined with capture of the emitted CO_2, called blue H_2, *(4)* water electrolysis using low-carbon electricity from nuclear, wind, solar or hydroelectric power, called green H_2, and *(5)* alternative pyrolysis process using the conversion of biomass-type wastes, which are also considered as green H_2. There are other types of H_2 such as black H_2 when using bituminous coal for H_2 production or brown H_2 when using lignite, however they are not considered as renewable H_2, since they are not decarbonized process.

3.1 White H_2

Although work is being carried out on the subject, geologic H_2 is not widely available for industrial uses (Halas et al., 2021). Despite H_2 being the most common atom in the universe, it is primarily found on Earth in the form of water (H_2O) or hydrocarbons. Consequently, the industrial use of geological H_2 is currently impractical. Nevertheless, in certain geological contexts, H_2 can exist in a gaseous state produced during the serpentinization process with the oxidation of ferrous cations (Fe^{2+}) into ferric cations (Fe^{3+}) (Zgonnik, 2020). Other origins could be mentioned such as radiolysis (process that produce oxygen and H_2), decomposition of hydroxyls within the structure of minerals or biological activity. But, Zgonnik demonstrates that, deep-seated H_2, coming from the Earth's core and mantle, could represent an inexhaustible source of energy (Zgonnik, 2020). Regardless of the transformation phenomenon considered, numerous H_2 emission sites have been discovered and explored in Oman, the Philippines and Turkey, Mali, Russia, South Carolina and Kansas in the United States, Brazil and Australia. For example, the Canadian company Hydroma Inc. discovered a geologic H_2 field in Mali (Bourabougou) between 2008 and today (Diallo et al., 2022; Prinzhofer et al., 2018). As Prinzhofer concludes in 2018, producing geologic H_2 is certainly a technical challenge but, the Malian case is obviously a first succeeding pathway to an industrial exploitation. Prinzhofer also mentions that: "The cost associated with the exploitation of 1 kg of natural H_2 may be estimated to be two to ten times smaller than the cost of manufactured H_2, rendering this new source of energy quite attractive for the future of our energy consumption" (Prinzhofer et al., 2018).

In Europe, several French regions has been explored and identified as potentially suitable H_2 production sites. These include Lorraine, the Western and Eastern Pyrenees, the Centre region, and the Jura region. A large source of white H_2 was for example found in Folschviller and estimated by la Française De l'Energie (FDE) at 46 milliards of tonnes

(Hydrogen +, 2024). In 2022, as a result of several months of consultation involving close collaboration with the relevant ministry services, and incorporating input from numerous stakeholders, including environmental protection associations, metropolitan and overseas communities, industrial federations, the mining code was amended. One notable addition was the inclusion of native H_2, among other substances (Ministry of ecological transition, 2022). In November 2023, a first bylaw (Ministry of the economy, finance, industrial and digital sovereignty, 2023) has granted TBH2 Aquitaine an exclusive permit for the exploration of native H_2, helium, and other substances in the Pyrenees-Atlantic department. Five other requests are pending. Other countries in Europe such as recent discoveries of natural H_2 field in Albania (Truche et al., 2024) can be a game changer even if there are still uncertainties. Moving from the exploration phase to the pilot phase and then to industrial-scale production will likely take time. Pesson, from IFP Energies Nouvelles, predicts that large-scale production of natural H_2 for the global market probably cannot be expected before 2035-2040 (Yannick Pesson, 2024). Blay-Roger, underlines the need for additional research to move from laboratory studies to a large-scale exploitation stage. This includes studying reservoir potential, extraction and treatment processes throughout their life cycles, adopting a techno-economic approach, and addressing regulatory aspects (Blay-Roger et al., 2024).

3.2 *Grey and blue H_2*

The most common method of supplying large H_2 consumers is through on-site captive H_2 production. This is primarily achieved through steam methane reforming (SMR), resulting in the production of grey H_2. This method is used by refineries, as well as in ammonia, methanol, and H_2 peroxide production plants. The high production (95% of all H_2 production in Europe in 2022) (European Hydrogen Observatory, 2022) is due to the lower cost of its production (6,23 euros/kg) compared to the others methods (Table 1). However, this H_2 is produced from oil, coal and other fossils fuels and therefore cannot be considered as renewable H_2 but as low-carbon H_2 production (Tashie-Lewis and Nnabuife, 2021). SMR involves heating methane to produce syngas, which is composed of H_2 and carbon monoxide (Equation 1). Carbon monoxide, due to its high toxicity, may be converted into carbon dioxide using water–gas shift (WGS) reactor for example, followed by a separation of H_2 using various methods such as pressure swing absorption or membranes (Simpson and Lutz, 2007). Based on stoichiometric, 4,9 ton of water transformed using 6.2 MWh into steam and 2,2 ton of methane (gas) are needed for 1,1 ton of H_2 produced using methane steam reforming, which will also produce 6 ton of carbon dioxide (gas). However, 9 to 10 tonnes of carbon dioxide equivalent are produced using SMR for 1 tonne of H_2 produced (Ghavam et al., 2021).

H_2 **production method**	**Colours**	**Costs (euros/kg)**
SMR	"Grey"	6.23
SMR +carbon capture system	"Blue"	6.38
Electrolysis + grid electricity	"Green", "pink" if nuclear enectricity	9.85
Electrolysis + renewable energy	"Green"	6.86

Table 1: H_2 production cost in Europe in 2022 (European Hydrogen Observatory, 2022). SMR: steam methane reforming

$$CH_{4(g)} + H_2O_{(g)} \rightarrow 3\,H_{2(g)} + CO_{(g)}$$
$$CH_{4(g)} + 2\,H_2O_{(g)} \rightarrow 4\,H_{2(g)} + CO_{2(g)}$$

Equation 1: SMR with carbon monoxide or carbon dioxide production.

Other processes for producing grey H_2 use coal, such as gasification or cracking, or oils, such as partial oxidation or cracking (Tashie-Lewis and Nnabuife, 2021). Cracking refers to the breakdown of carbon-carbon bonds in complex hydrocarbons (such as kerogens and long-chain hydrocarbons) to obtain smaller molecules like light hydrocarbons, alkanes, and alkenes. This process typically requires catalysts, high temperatures, and solvents. As of 2021, approximately 30% of H_2 is produced using coal, while 18% is derived from oils (Tashie-Lewis and Nnabuife, 2021). Around 60% of the carbon dioxide produced in the process is captured in a high-concentration stream. This CO_2 is primarily formed from natural gas oxidation in the reforming and shift reactors. The remaining CO_2 is emitted from the reformer furnace in a lower concentration (IEA, 2022).

The production of H_2 using SMR coupled with carbon capture and storage is known as blue H_2. It results in a decrease of the carbon footprint by approximately 18% compared to conventional grey H_2 production (Mynko et al., 2023). A study showed that 90% the CO_2 capture can be increased up to 99% without significant energy requirement and cost increase (IEA, 2022). As showed in Table 1, the production cost of blue H_2 (6.38 euros/kg) and grey H_2 (6.23 euros/kg) are close. This carbon capture process is only developed in France (1.63% of H_2 produced), Italy (0.14%) and Netherland (0.33%) in Europe (European Hydrogen Observatory, 2022).

3.3 Turquoise H_2

Another pathway is the pyrolysis of methane from natural gas into H_2 and solid carbon at high temperature (>300°C). This method is also known as turquoise H_2 production. Pyrolysis involves the partial thermal degradation process of material without oxygen (as in combustion) or water (as in hydrolysis or SMR) (Equation 2). The process results in the production of carbon solids known as char, liquid oils called tars, and gases such as H_2 and/or gaseous hydrocarbons (Yukesh Kannah et al., 2021). The quality of the pyrolysis will depend of its operational condition the temperature, the decomposition time, heating rate, working pressure, and the composition of the substrate. A fast heating rate (10-200°C/s) and shorter decomposition time will give produce 60-75% of oils, 15-20% of biochar and 10-20% of syngas, with up to 45% of H_2 with 127°C/s and 5-7 min of decomposition time (Yukesh Kannah et al., 2021).

$$CH_{4(g)} \rightarrow 2\,H_{2(g)} + C_{(s)}$$

Equation 2: Pyrolysis using methane as the substrate

According to Fulcheri, this process is currently of growing interest. The advantage of this process is that it is thermodynamically much less energy-intensive than water dissociation, requiring approximately seven times less energy per mass (or mole) of H_2 (38 kJ vs. 285 kJ per mole of H_2) and is therefore cheaper in energy than SMR. Energy used for pyrolysis will cost around 15-20% of the global price of H_2 production, with a global price estimated at 1.25-2.57 USD/kg of H_2 (Yukesh Kannah et al., 2021). Another advantage is that it enables the co-production of carbon black, a carbonaceous nanomaterial used extensively in industry (12 million tonnes per year) and offering high added value (Fulcheri, 2020). Pyrolysis may also be used with renewable biomass, which will decrease the global price of H_2 production using pyrolysis, from 1.25-2.20 USD/ kg of H_2 using methane and 1.77-2.05 USD/kg for biomass (Yukesh Kannah et al., 2021).

3.4 Green H_2

Green H_2 primary refers to H_2 production through water electrolysis which involves splitting water into oxygen and H_2 using electricity. Heat is also generated as a byproduct of this process (Equation 3):

$$H_2O + Electricity\ (237.2\ kJ/mol) \rightarrow H_2 + \frac{1}{2}O_2 + Heat\ (48.6\ kJ/mol)$$

Equation 3: Water electrolysis for H_2 production (Shi et al., 2020)

Using grid electricity to produce green H_2 is more expensive than using renewable energy (Table 1). The cost is approximately 9.85 euros/kg on average and varies depending on the country ranging from 1.61 euros/kg of H_2 in Sweden to 10.68 euros/kg in Italy with an average of 7.07 euros/kg for Europe (European Hydrogen Observatory, 2022). However, the cost of H_2 from SMR vs electrolysis with renewable energy is not so different in Europe with a cost at 6.23 euros/kg vs 6.86 euros/kg respectively. There are two main technologies used for H_2 production through water electrolysis. The primary method is alkaline water electrolysis, accounting for 51% of the total water electrolysis capacity. It has a capital expenditure (CAPEX) of 1,666 euros/kW and an operating expenditure (OPEX) of 43 euros/kW/year (European Hydrogen Observatory, 2022). Alkaline water electrolysis corresponds to the transport of hydroxide ions through the electrolyte from the cathode to the anode using electricity, resulting in the generation of H_2 on the cathode side (Simoes et al., 2021). However, new technologies are emerging, including an increase in the use of proton exchange membrane electrolysers, which accounted for 73% of total water electrolysers sold in 2022 (European Hydrogen Observatory, 2022). Proton exchange membrane (PEM) electrolysers uses solid special polymeric material as electrolyte and are useful for intermittent power supply (Simoes et al., 2021). However, their capital cost and operating cost are higher than for alkaline water electrolysis, with a CAPEX of 1,970 euros/kW and an OPEX of 64 euros/kW/year (European Hydrogen Observatory, 2022). Two technologies for H_2 production are still in the research phase due to high costs or low yields: *(1)* the solid oxide electrolysis, which uses a solid ceramic material as the electrolyte, and selectively conducts negative-charged oxygen ions at high temperatures using electricity (Simoes et al., 2021), and *(2)* the microbial electrolysis cells which use electricity to generate H_2 from organic matters (Simoes et al., 2021). Another research aspect is the catalyst which may also be expansive. As an example, a new high performance and low cost catalyst for alkaline electrolysers was created: NiCoP/CoP sponge-like structure grown on stainless steel mesh (Gebreslase et al., 2023).

Currently, Europe has 187 MW of electrolysers capacity in operation for green H_2 production (Figure 2). The main producers are Germany (64 MW), Sweden (9 MW), France (6 MW), and the Netherlands (4 MW). Additionally, there are numerous electrolysers under construction in Europe, with a projected capacity of 832 MW by 2025. This will increase the European electrolysers capacity to 1,368 MW by 2025 (European Hydrogen Observatory, 2022). The main contributing countries to this increase are Sweden (520 MW), France (252 MW), and the Netherlands (205 MW). Actually, the world's first high-power fuel cell plant was just built in France by HDF Energy. This plant has a production capacity of 100 MW of power fuel cell per year, which are going to be used for heavy maritime and rail mobility (Serena Boni, 2024).

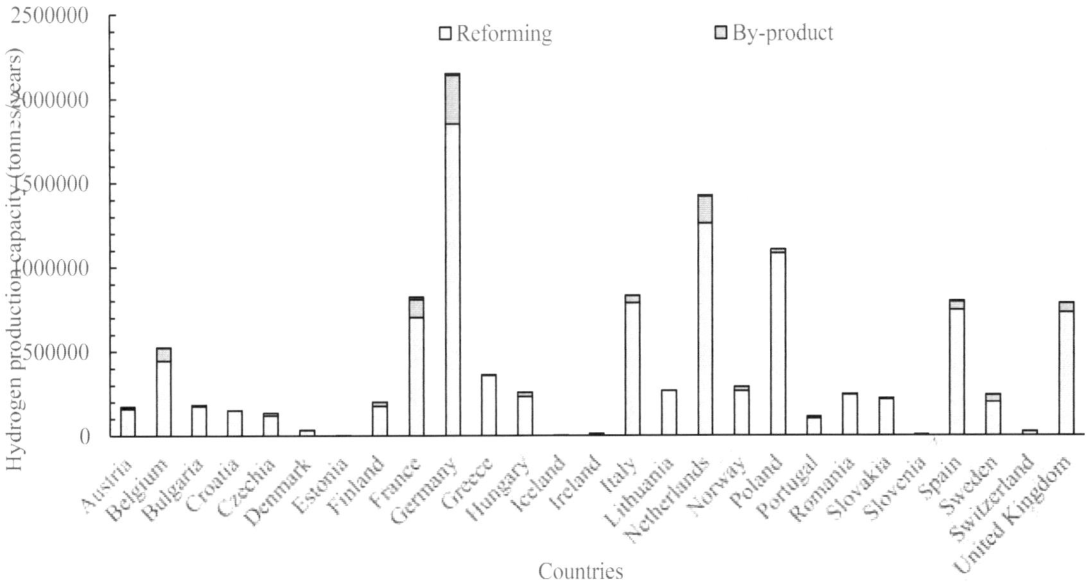

Figure 2: H_2 capacity production methods per countries in Europe in 2022 (European Hydrogen Observatory, 2022)

A challenge in green H_2 production through water electrolysis is the significant amount of water required. According to stoichiometry, 9 kg of water is needed per kg of H_2 (Shi et al., 2020), but depending on the type of electrolysers used, the requirement can range from 10 to 22.4 kg of water (Ghavam et al., 2021; Simoes et al., 2021). Water availability is declining due to climate change, and the use of water for electricity production could become an issue in the future. When considering different water sources, fresh water from public grids is the most suitable for water electrolysis, although seawater and recycled water could also be used to lessen transport and treatment costs, and to reduce competition with other uses (Simoes et al., 2021). Indeed, competition of other waters uses need to be considered when constructing H_2 production site. For example, with marine wind mill, it may be more interesting to directly use sea water for H_2 production. Research are actually on-going to demonstrate direct H_2 production without desalinization using sea water and electricity, for example a pilot project from Schaeffler in 2021 (IEA, 2022). In this process, the renewable electricity needs to be produced near the electrolysis plant to meet the H_2 user's needs and be used directly. One option is to store the renewable energy produced into batteries and use the energy for H_2 production when needed. However, as the overall energy efficiency of li-ion batteries is 83% corresponding to 17% loss, according to Pellow, this combination will decrease the yield of the global system. Another usual scenario is the use of H_2 as a storage solution for renewable energy to enable use throughout the year. A modeling study of the use of a regenerative H_2 fuel cell, which involves an alkaline water electrolysers coupled with a PEM fuel cell, revealed an even lower overall energy efficiency of 30% (Pellow et al., 2015). Therefore, H_2 needs to be used (gas infrastructure, direct use, compression…) without being converted into electricity to be more energy efficient (Pellow et al., 2015).

Finally, there are several techniques for producing H_2 from organic compounds or waste which may be considered also as green H_2: the reforming of biogas into syngas through thermochemical processes similar to pyrolysis (Megret et al., 2015), the gasification of waste, which involves decomposing a carbonaceous combustible solid with heat. H_2 may also be produced from microalgae (cyanobacteria) (Oey et al., 2016) or from biomass using dark fermentation (Guo et al., 2010) . These processes are not widely used due to their high costs and low yields.

3.5 H_2 production costs

When comparing all the H_2 production methods in Europe, the energy cost tends to be the main cost component in the H_2 production overall cost (Table 2). Indeed, for SMR, natural gas has the highest cost in the production, corresponding to 86 ± 2.7% of the global production without carbon capture, and 85 ± 3 % with carbon capture (due to the higher CAPEX). For water electrolysis, electricity consumption has the highest cost with 73± 22. 94% using grid and 63 ± 25% using renewable energy. Therefore, H_2 production cost will be extremely dependent of the evolution of energy costs in Europe, and especially natural gas and/or electricity, depending on the production methods. The high standards deviation for the electricity cost is due to high variation of its price in Europe, due to the various origin of electricity production. Therefore, to decrease the high cost of renewable H_2 production, decreasing electricity costs in Europe is necessary. Another possibility is to use H_2 for difficult to decarbonize industries; this approach will be presented in the next paragraph.

Table 2: Distribution of the H_2 production cost in Europe in 2022 (European Hydrogen Observatory, 2022), CAPEX: Capital expenditures, OPEX: Operational expenditure, SMR: steam methane reforming

% of total cost in Euro/kg	SMR			SMR with carbon capture			Water electrolysis with grid electricity			Water electrolysis with renewable energy		
Colours	Grey			Blue			Green					
CAPEX	3.33	±	0.64	5.88	±	1.20	13.31	±	5.10	33.00	±	7.06
Other OPEX	1.53	±	0.29	2.12	±	0.43	5.12	±	1.96	9.43	±	1.69
Natural gas cost	85.91	±	2.69	85.47	±	2.96						
CO_2 cost	9.24	±	1.77	6.53	±	1.33						
Electricity cost							73.01	±	22.94	63.06	±	24.85
Grid fees and taxes							8.55	±	30.42	-5.49	±	26.35

4 END-USE OF H_2

4.1 Generalities

There are various applications of H_2 (Leonhardt, 2022) as illustrated in Figure 3. H_2 can be used as a main reagent in refineries (52%) to desulfurize hydrocarbons and in the production of nitrogen fertilizers through ammonia (32%). It also finds application in the synthesis of methanol (3.52%) and the production of other chemicals (7.94%) (Figure 3). The use of renewable and low-carbon H_2, called clean H_2 has been developed in the steelmaking industry to reduce carbon emissions (Birat, 2022), or in the cement production industry. While its use in metallurgy is currently low (0.04% in 2022), the break-even price for using clean H_2 in primary steelmaking is lower than for other applications at 2.97 euros/kg, making it an attractive option for decarbonizing the industry (Table 3). H_2 can also be used as an energy carrier, offering an interesting alternative to fossil fuel such as methane, coal, alternative fuels… In 2022, researchers showed that the primary industrial energy consumers - including chemicals, metallurgy (mainly steel production), and the building materials industry (mainly cement) - collectively contribute to 75% of the sector's greenhouse gas emissions (CITEPA, 2021; The Shift Tank, 2022).

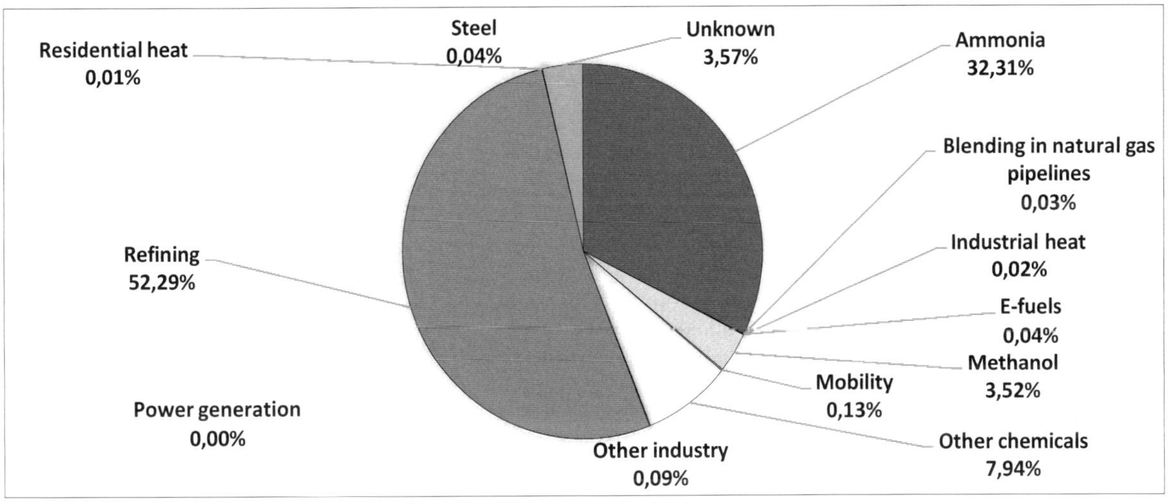

Figure 3: H_2 end-uses in Europe in 2022 (European Hydrogen Observatory, 2022).

4.2 H_2 uses in refineries

In industry, H_2 is considered as a key building block in chemical processes. H_2 is often used to upgrading hydrocarbon by removing sulfur, halide, oxygen, metals and nitrogen impurities (Moradpoor et al., 2023; Rambhujun et al., 2020). This use of H_2 increased due to the rising demand for diesel fuel and the more strict regulation on sulfur content (Moradpoor et al., 2023). Another possibility is the cracking of heavier to lightweight hydrocarbons, by reducing the size of the long carbon chain, to produce other chemicals (Moradpoor et al., 2023; Rambhujun et al., 2020). H_2 is also used in the hydrogenation process for fats and oils, with nickel catalyst, to reduce the degree of unsaturation, tastes and odors or produce trans fats (Rambhujun et al., 2020). The oil refining industry corresponded to a demand of 40 million tons of oil equivalent of pure H_2 in 2020 (Moradpoor et al., 2023).

H_2 is also often used in industry for the production of ammonia, using Haber-Bosch process. This process uses H_2 and nitrogen with a catalyst at high temperature (400-500°C) and pressure (150-300 bar) (Ghavam et al., 2021; Rambhujun et al., 2020). However, this method produce a high quantity of greenhouse gases emission, corresponding to 2.16 kg CO_2 equivalent /kg NH_3; and a high energy consumption with 30 GJ/tonnes of NH_3 (Ghavam et al., 2021). Actually, ammonia is produced in large plants (1,000-1,500 tonnes/days) with 90% from fossil fuels with the Haber-Bosh method, and 72% of the H_2 used for ammonia production come from SMR, with 4% from electricity (Ghavam et al., 2021). Ammonia, directly as fertilizer in agriculture or as ammonium nitrate, is produced from the catalytic oxidation of ammonia as nitric acid, or as urea with carbon dioxide (Rambhujun et al., 2020). Ammonia production corresponded in 2018 to a demand of 31 million tons of oil equivalent of pure H_2 (Rambhujun et al., 2020). Ammonia may also be used in mobility (4.3 H2 uses in mobilities).

Methanol correspond to a building block in the chemical industry to produce more complicated chemicals compounds (ex acetic acid, methyl tertiary butyl ether, dimethyl ether, methylamine) (Dalena et al., 2018).. Methanol may also be used in mobility such as in maritime application (4.3 H2 uses in mobilities). Methanol is produced using carbon dioxide and H_2 with various process such as partial oxidation, steam reforming, auto-thermal reforming, or methanol-water solution electrolysis (Dalena et al., 2018). Therefore, the production of methanol does not need to use pure H_2 and may use syngas for example. Methanol production correspond to a world demand of 12 million tons of oil equivalent of mixed H_2 in 2018 (Rambhujun et al., 2020).

4.3 H_2 uses in mobilities

Mobility corresponds to 0.13% of H_2 uses in Europe (Figure 3), which should increase in the coming years. In Europe, shipowners are required to start replacing fossil fuels by 2023 to comply with the VLSFO[2] regulations. They will need to replace grey H_2, grey ammonia, grey methanol and marine diesel by 2030, and LPG, LNG, CNG[3] by 2035 (Council of the European Union, 2023). For marine transportation, two cost-efficient options for H_2 are available. The first option involves using compressed H_2 with PEM fuel cells for small ships. This requires frequent refueling to limit the amount of fuel stored onboard. The second option is using liquefied H_2 with PEM fuel cells for medium size ships,

[2] Very Low Sulfur Fuel Oil (VLSFO).
[3] LPG – Liquid Petrol Gas, LNG – Liquid Natural Gas, CNG – Compressed Natural Gas

which needs optimization between cost and energy density (Hydrogen Europe, 2021). For deep sea maritime, H_2 is not the best solution. Ammonia, used with solid oxide fuel cells, is more promising due to its higher energy density for storage and may be produced from H_2 (4.2 H_2 uses in refineries). However, this option is limited by the higher cost of the fuel (Hydrogen Europe, 2021). As a global vision for clean H_2 uses, maritime application (small and medium ships) is interesting since it is one of the uses with the lowest break-even price (1.16 to 2.21 euros/kg of H_2 in 2022, Table 3).

Table 3: Break-even price for clean H_2 uses in Europe (European Hydrogen Observatory, 2022).Oil refining corresponds to the use of H_2 as feedstock, produced in Steam Methane Reforming with natural gas as input, Primary steel making corresponds to the production in blast furnace with basic oxygen furnace and coking coal as the reducing agent, Maritime application correspond to the use of H_2 in small and medium size ships. Heavy duty truck corresponds to H_2 fuel cell trucks.

End-use	Min (euro/kg)	Max (euro/kg)	Years
Oil refining	3,94	8.14	2022
Primary steel making	2.97	2.97	2022
Maritime applications	1.16	2.21	2022
Heavy-duty trucks	2.4	5.8	2023

Another parameter to consider when using alternative fuels is the variation in volume and weight for the same energy use, which is important for transportation (Table 4). Indeed, H_2 has a higher volume than marine diesel (6.6 times greater for liquid, 21,9 times higher for compressed) (Larrieu, 2024). However the weight differs with liquid blue H_2 weighting 0.36 times less and compressed blue H_2 10.3 times heavier (Larrieu, 2024). Therefore, this difference will have its importance on the energy consumption of the boat, with 1.1 to 1.4 time more energy intensive than using marine diesel when comparing well to wake use. As an example, Feadship has been built a superyacht called Feadship 821. It uses liquid H_2 (92 m^3 at -253°C, equivalent to 4 tonnes of H_2) and 16 fuel cells of 185 kW. This allows the yacht to travel at 18.5 km/h (10 miles/h) for a full week without needing to refuel (Michaël Torregrossa, 2024). Another problem is the high cost compared to the actual fuel used for marine transport (Table 4). Blue or green H_2 costs 3.1 to 7.1 times more than conventional marine Heavy Fuel Oil when used in shiping (Table 4). Other alternative such as ammonia (3.7 to 7.6), methanol (2.2 to 5.2) and electricity (53) have higher costs (Table 4). This higher cost is mainly due to the expensive fuel. Therefore, advanced mature technology such as improving energy efficiency in production and the use of motors is a necessary step to enhance low-carbon H_2 use in mobility.

Only 0.04% of H_2 is used to produce e-fuels, which corresponds to the production of hydrocarbons using H_2 and carbon dioxide (Ueckerdt et al., 2021). One example is the production of ammonia or methanol from H_2 which would be used for maritime applications for deep-sea shipping (Table 4). In 2022, IHI developed the first gas turbine powered by 100% liquid ammonia (IHI Groupe, 2022). They also introduced a co-generation ammonia/natural gas turbine powered by an ammonia ratio of 20% (Shintaro Ito et al., 2020). These examples illustrate the increase of ammonia and methanol production projects up to 33 projects in 2025 (IEA, 2022). However, using ammonia as fuel produces NOx which are considered as air pollutants causing eutrophication, acid rain, and respiratory tract irritations. Therefore, reducing the production of NOx in ammonia turbines is compulsory (Shintaro Ito et al., 2020). Another problem is the higher volume of ammonia (5.03 times) and weight (2.3 times) required for the same uses compared to marine diesel (Table 4), which will increase the energy consumption of the boat (1.7 to 1.8 time more energy-intensive in well to wake comparison) (Law et al., 2021; Larrieu, 2024).

In 2022, LNGV, the German railway company, conducted experiments with H_2-fueled passenger trains. They ordered 14 fuel-cells trains to Alstom to gradually replace existing diesel fleets (IEA, 2022). But, after one year, LNGV switched to battery electrics due to security concerns, difficulties in sourcing green H_2, and high costs (Keith Fender, 2024). In addition, H_2 trains were found to be 80% more expensive than battery hybrid trains (Leigh Collins, 2022). The high cost, ranging from 800 to 1,200 € per tonne of CO_2, limits their potential use (Ueckerdt et al., 2021).

Table 4: Comparison (x times) between alternative marine fuels and conventional marine Heavy Fuel Oil of relative volume, weight, life cycle energy intensity from well to wake and relative well to wake cost comparison (overall cost of various alternative pathways relative to the cost of HF0) , based on (Law et al., 2021; Larrieu, 2024). Well-to-wake refers to the entire process of fuel production, delivery and use onboard ships, and all emissions produced therein.

Compared to conventional marine Heavy Fuel Oil	Volume comparison	Weight comparison	Motors used (energy efficiency)	Life cycle energy intensity (well to wake) comparison	Relative well-to-wake cost comparison
Liquid blue H_2	6.6	0.36	Internal Combustion Engine (45%)	1.4	3.1
Compressed blue H_2	21.9	10.3	Fuel cell (60%)	1.1	7
Compressed green H_2 (solar)	21.9	10.3	Electric motor (92.5%) and Fuel cell (60%)	1.4	7.1
Blue ammonia	5.03	2.3	Internal Combustion Engine (45%)	1.8	3.7
Blue ammonia	5.03	2.3	Fuel cell (60%)	1.4	5.2
Green ammonia (solar)	5.03	2.3	Electric motor (92.5%) and Fuel cell (60%)	1.7	7.6
Grey methanol	2.3	2.15	Internal Combustion Engine (45%)	1.8	2.2
Blue methanol with carbon capture	2.3	2.15	Internal Combustion Engine (45%)	2.2	2.4
Green methanol (solar)	2.3	2.15	Internal Combustion Engine (45%)	2.1	5.2
Electricity (solar and directly used)	12.1	35.9	Electric motor (92.5%)	0.5	53

H_2 fuel-cell electric cars saw a 38% increase in Europe in 2022, reaching a total of 5,570 cars. Of these, 86% were passenger cars (European Hydrogen Observatory, 2022). Moreover, H_2 demand in road transport increased by 60% between 2020 and 2021, reaching 30 ktonnes in 2021, which is 0.03% of the total H_2 demand (IEA, 2022). In the transport sector, Germany has the highest number of H_2 fuel cell vehicles, mostly passenger cars and buses, while France has a significant number of H_2 vans (Table 5). In 2023, there was 178 publicly available H_2 refueling stations in Europe, which is insufficient for a widespread deployment of H_2 use in passenger cars, lorries, or buses. The distribution of these stations is uneven, with most located in Germany (96 stations), followed by France (21 stations) and the Netherlands (14 stations) (European Hydrogen Observatory, 2023c). For comparison, the fuel efficiency per 100 km is 11.43L for a gasoline car, 21.6 kWh for an electric car and 0.76 kg of H_2 for an H_2 fuel-cell car (Kolbe, 2019).

Table 5: Types of H_2 fuel cell vehicles per countries in Europe in 2022 (European Hydrogen Observatory, 2022) .

Countries	Passenger cars	Buses	Vans	Trucks	Total
Austria	268				268
Belgium	401	40	9		450
Czech Republic	22				22
Denmark	1,059		4		1,063
Estonia	3				3
Finland	11				11
France	1,794	2	1763		3,559
Germany	5,458	267	48	27	5,800
Iceland	117				117
Italy	218	102			320
Latvia	108	30			138
Lithuania	1				1
Luxembourg	14	15			29
Netherlands	1,789	128	63	73	2,053
Norway	1,186	45		16	1,247
Poland	199				199
Portugal	6	4			10
Slovakia	5				5
Spain	54	6			60

Countries	Passenger cars	Buses	Vans	Trucks	Total
Sweden	319	4			323
Switzerland	901	114	55	147	1,217
United Kingdom	1,119	189	37	1	1,346
Total	**14,944**	**946**	**1,979**	**264**	**18,241**

Sustainable aviation plays a key role in reducing climate impacts. Aviation's future fuel might involve fuel cell or H_2 gas turbine propulsion systems to replace kerosene. While H_2 turbine technology has been used in the past, there are still needs to develop and improve lightweight, high-performance equipment (tanks for example), electronics, airports' infrastructures, to change certification, and legislation, and to encourage heavy investments (Tiwari et al., 2024).

4.4 Correlation end-uses and production process for H_2

Moreover, there seems to be a correlation between the end-use and the production processes in Europe (Table 6). For instance, ammonia and methanol are predominantly produced by hydrocarbon reforming (99.8%), with only 0.2% of H_2 production from water electrolysis. On the other hand, e-fuels, industrial heat, mobility, power generation, residential heat, and steel are exclusively produced using H_2 from water electrolysis. Other chemicals and refining rely on H_2 as a by-product, with percentages from hydrocarbon reforming being 68.1% and 93.2%, respectively, and from water electrolysis being 0.4% and 0.1%, respectively.

Table 6: Production capacity in tonnes/years according to the production methods and the end-use in Europe in 2022 (European Hydrogen Observatory, 2022). SMR: Steam Methane Reforming

End-use/production method	By-product	SMR	SMR with carbon capture	Water electrolysis
H_2 Colors		**Grey**	**Blue**	**Green**
Ammonia		1,704,650		3,306
Blending in natural gas pipelines				1,446
E-fuels				2,240
Industrial heat				1,107
Methanol		185,724		331
Mobility				7,094
Other chemicals	132,227	285,922		1,653
Other industry			4,746	
Power generation				83
Refining	171,776	257,6242	13,394	2,595
Residential heat				496
Steel				2,268
Unknown	85,447	96,957	1,182	5,273

H_2 production and end-use can be reorganized, as seen in the example of France. The country has established seven major French H_2 basins, which include the South-West (Spanish-French border), Greater West (Nantes Saint-Nazaire), Seine Vallee (from Le Havre to Paris and its airports), North (Dunkirk), Moselle-Rhine (Germany-Belgium-French border), Rhône-Alps (the chemical valley near Lyon), and the Mediterranean (Fos Marseille). These basins cluster together the main industrial hubs such as chemicals, steelworks, and cement plants, as well as transportation and mobility hubs like cities, ports, airports, and logistics centers (France Hydrogène, 2021). The existence of these H_2 consumers' concentration zones offers the advantage of consolidating large-scale production facilities around these key industries (Figure A 4 in the Appendixes). These centralized production sources can also competitively supply a significant proportion of the diffuse consumption in these basins. This optimization also requires the deployment of resources such as pipelines and geological storage capacity to ensure security of supply for end customers, and to exchange within Europe (Figure A-3). The existence of these H_2 consumer hubs offers the advantage of generating economies of scale by pooling massive production units around these major industries. These centralized production sources can also competitively supply a significant proportion of the diffuse consumption within these basins. Outside these basins, projects are being developed, especially for mobility uses, corresponding to 15% of H_2 consumption (France Hydrogène, 2021). In these specific cases, a local production of H_2 may be a good option.

In the following paragraph, the authors develop the deployment of resources such as pipelines and geological storage capacity in order to optimize the global H_2 system and its supply chains in Europe.

5 ELECTRICAL NETWORK AND NATURAL GAS TRANSMISSION NETWORK FOR H_2 AVAILABILITY

RTE, operator of the French electricity transmission network, in its Energy Pathways to 2050 report underlines the necessary coupling between the gas and the electricity sectors to enable the development of the French power system. The expected increase of electricity consumption by 2050 will need a flexible availability of H_2. The operation of the electrolysers can be modulated using the H_2 transport and storage infrastructures in order to meet the conditions of supply-demand balance (RTE, 2021).

As previously mentioned, an interesting option for H_2 is to use existing natural gas infrastructures. A first scenario is to add 5-10% of H_2 to natural gas. In this case, there is no modification need of the equipment for the operator (pressure reducing valves, compressors) nor for the end-user (burner) (Kovač et al., 2021). If the addition of H_2 in the natural gas mix is up to 20%, the injection station will require minor modifications at a cost of 0.3 to 0.4 USD/kg of H_2 (IEA, 2022).

As published by the IEA, the complete repurposing of the natural gas pipeline to switch to pure H_2 will save 50 to 80% of investment compared to a new pipeline (IEA, 2022). The following tables are presenting North European projects (Table 7), Southern projects (Table 8) and a focus on the French projects (The MosaHYc, RHYN projects, the Franco-Belgium corridor WHHYN, the connection between Marseille and Obergailbach on the German border HY-FEN, and the BarMar subsea pipeline between Barcelona and Marseille have been recognized by the Commission as "Projects in the Common Interest" (**Erreur ! Référence non valide pour un signet.**).

Table 9), retrieved from the websites of the infrastructure's operators and from the literature indicating the new pipelines, repurposed and converted pipelines, converted underground storages (salted cavities, depleted and, in one project, an aquifer), the involved operators and the main characteristics (type of infrastructure, distance, due date). These projects illustrate the approach of the "33 European gas infrastructure operators" contributing to the European Hydrogen Backbone (EHB) initiative (European Hydrogen Backbone, 2024).

Table 7: Projects of low-emission H_2 infrastructures in Northern Europe (IEA, 2024). Legend: T&S: Transport and Storage, TN: New-build Transportation Network, TNc: Converted Transmission Network, UAS: Underground Aquifer Storage, USCS: Underground Salted Cavity Storage.

Project	Operators	Situation	Infrastructure	Distance/due date	References
Austria (predictions: 730 km by 2030)					
H2 collector East	Austria Gas	Northern Burgerland to Lower Asutria and Vienna	TN, TNc	200 km new 140 km converted By 2050	(Austria Gas Grid Management AG, 2024)
H2 backbone WAG Penta-West	Gas Connect Austria GmbH	Austria between Baumgarten (Slovakia) and Überackern (Germany)	TN, TNc	200 km new 140 km converted	
H2 Readiness TAG	Trans Austria Gasleitung	Austria between Baumgarten (Slovakia) and Klagenfürt (Italy)	TNc	380 km by 2030	
Belgium (600 km of existing operational H_2 pipelines)					
Backbone H2 connecting ports	Fluxys	Zeebrugge, Ghent, Antwerpen	TNc		(Fluxys, 2024)
BE-HyStore	Fluxys, Geostock	Loenhout storage near Antwerpen	UAS	1.4 billion m^3	
Denmark					
Danish-German H2 network	Energinet, Gasunie	Lille Torup (USCS) in the north of Denmark, to Heidenau, in Hamburg (Germany)	TN, USCS	550 km by 2028 2 cavities	(IEA, 2024)
Danish Backbone West project	TSO	Lille Torup (USCS) in the north of Denmark, to Ellund (Germany)	TN, TNc	360 km	(IEA, 2024; Mads Damsø Nielsen, 2023)
Germany (370 km of existing operational H_2 pipeline, predictions: 1,800 km by 2027/28, 9,700 km by 2032)					
Making H_2 Happen	GASCADE, Ontras, terranets	North South backbone from Lubmin - Schwedt - Berlin - Leipzig - Leuna - Erfurt - Ludwigshafen - Karlsruhe - Stuttgart	TNc	1,100 km by 2025 then 2028	(Flow, 2024)

Project	Operators	Situation	Infrastructure	Distance/due date	References
HyCAV Mobil	EWE, DLR Institute	Rüdersdorf storage (Brandenburg)	USCS	1 cavern	(Ron-David Heinen, 2023)
HyPerLink 1-5	Gasunie	Connection with the Netherlands, Germany and Denmark including Nüttermoor, Jemgum, Huntorf, Etzel and Harsefeld storage facilities	TN, TNc, USCS	610 km	(Gasunie Energy Development GmbH, 2024)
H_2ercules	RWE, OGE	North South backbone to connect with Norway, the Netherlands, Belgium, France and the Czech Republic	TN, TNc	2,000 km by 2030	(RWE, 2024)
The Netherlands (370 km of existing operational H_2 pipeline)					
Yara-Dow pipeline	Gasunie	Zeeland province (between Dow chemicals and Yara)	TNc	12 km in service in 2018	(Gasunie Energy Development GmbH, 2023, 2021)
H2 network Netherlands	Gasunie	Connections between major industrial clusters (Eemshaven, North Sea Canal area, Rotterdam, Zeeland and Limburg), Germany and Belgium	TNc, USCS	1,200 km by 2030	(Gasunie Energy Development GmbH, 2023)
HySTOCK	Gasunie, EnergyStock	Zuidwending storage	USCS	1 cavern (0,2 TWh) in 2028 3 caverns after 2030	(HyStock, 2024)
United Kingdom					
H2 T&S Networks Pathway	In 2024 - analysis on the location of T&S needs; By 2025 - allocation of the first round of the T&S projects; By 2030 - support up to two H_2 storage projects and the associated regional pipeline infrastructure, whether in operation or under construction				(OGL, 2023)

A thorough review of the projects including transmission networks and storage facilities indicated that two third of the 18,000 km of H_2 pipelines in Northwest Europe could be repurposed by 2030 (IEA, 2024). Norway, as a natural gas producer, is not concerned by these projects.

According to the IEA database (IEA, 2024), the underground H_2 storage capacity in Europe could exceed 3 TWh by 2030. Currently, at least half of this capacity is currently under construction or undergoing feasibility studies. The technology required for H_2 storage should allow for quick and efficient injection and withdrawal cycles, making salt caverns an ideal choice for renewable electricity production. While aquifer reservoirs are a possibility, there is currently only one project in this area.

By 2040, it is expected to have a nearly 53,000-km-long pipeline network for pure H_2 in Europe – based predominantly on the conversion of existing gas infrastructure, covering 28 countries. It will run from Portugal (solar production) or the Nordic region (offshore wind production) to the main areas of consumption in Northwest Europe. Part of the EHB, the H2Med corridor which will transport 10% of expected H_2 consumption in Europe by 2030, between Portugal, Spain and France. It includes the following projects: CelZa (248 km), between Portugal and Spain and BarMar (450 km), from Barcelona to Marseille (Table 8).

Table 8: Projects of low-emission H_2 infrastructures in Southern Europe (IEA, 2024). Legend: TN: New-build Transportation Network, TNc: Converted Transmission Network, USCS: Underground Salted Cavity Storage.

Project	Operators	Situation	Infrastructure	Distance/due date	References
Portugal					
H2med-CelZa	REN-Gasodutos	between Portugal and Spain Part of H2med	TN, TNc	162 km by 2030	(H2med, 2024)
Spain					
H2med-CelZa	Enagas, GRTgaz, Terėga, REN and OGE	between Spain and France through the BarMar Part of H2med	TN, TNc	248 km by 2030	(H2med, 2024)

Project	Operators	Situation	Infrastructure	Distance/due date	References
Spanish H_2 Backbone	Enagás	Vía de la Plata Axis and the Cantabrian Coast, Ebro Valley and Levante Axes	TN, TNc	1,250 km by 2030 1,500 km by 2030	(H2med, 2024)
2 underground H_2 storage facilities	Enagás	in Cantabria and the Basque region	USCS		(H2med, 2024)

The MosaHYc, RHYN projects, the Franco-Belgium corridor WHHYN, the connection between Marseille and Obergailbach on the German border HY-FEN, and the BarMar subsea pipeline between Barcelona and Marseille have been recognized by the Commission as "Projects in the Common Interest" (**Erreur ! Référence non valide pour un signet.**).

Table 9: Projects of low-emission H2 infrastructures in France (IEA, 2024; Muzart and Martin, 2022; Moret, 2022). Legend: TN: New-build Transportation Network, TNc: Converted Transmission Network, USCS: Underground Salted Cavity Storage.

Project	Operators	Situation	Infrastructure	Distance/due date	References
		France			
AtlantHYc	GRTgaz	between Saint Nazaire and Paris			(IEA, 2024)
Bar Mar	REN, Enagás, GRTgaz and Teréga	Between Barcelona and Marseille Part of H2med	TN	455 km by 2030	(Fluxys, 2024)
DHUNE	GRTgaz	between Dunkirk industrial zone and Belgium	TN	50 km	(GRGTgaz, 2022a)
FrHyGe	Storengy & Co	Manosque then SaltHy (Harsefeld, Allemagne)	USCS	1 cavitie by 2027, 2 cavities by 2029	(Storengy, 2024)
HY-FEN	GRTgaz	between Fos-Marseille to Grand Est	TN, USCS	1,200 km by 2030	(GRGTgaz, 2020)
HyGreen Provence Project	Engie, DLVA	local renewable energy production system in Manosque	USCS	1 cavity by 2026 0,2 GWh	(IEA, 2024)
HYnframed	GRTgaz	between Fos sur Mer to Manosque	TN, USCS	150 km	(GRGTgaz, 2022b)
HyPSTer	Storengy	Etrez	USCS	1 cavity	(Moret, 2022)
HySoW	Teréga Teréga, SALINS	from the Atlantic coastline (Bordeaux and Bayonne) to the Mediterranean Sea (Port La Nouvelle) Storage	TN, TNc USCS	600 km by 2030 0,5 to 1 TWh	(Teréga, 2023)
mosaHYc	GRTgaz, Ceos Deutschland, Encevo	between Moselle (France) and Sarre (Germany)	TNc	70 km	(Muzart and Martin, 2022)
RHYn	GRTgaz	between Dessenheim and Chalampé-Ottermarsheim industrial zone	TN, TNc	100 km incl. 60 km TNc by 2028	(GRGTgaz, 2022c)
WHHYN	GRTgaz, Fluxis	between Valenciennes (France) and Mons (Belgium) will eventually link up with the Paris region	TN	40 km by 2030 465 km by mid-2030s	(GRGTgaz, 2022d)

Among all the projects listed in The MosaHYc, RHYN projects, the Franco-Belgium corridor WHHYN, the connection between Marseille and Obergailbach on the German border HY-FEN, and the BarMar subsea pipeline between Barcelona and Marseille have been recognized by the Commission as "Projects in the Common Interest" (**Erreur ! Référence non valide pour un signet.**).

Table 9, only the mosaHYc trans-border transport project (Muzart and Martin, 2022) and HyPSTER storage project (Moret, 2022) have been studied in the scientific literature.

In Moselle, the mosaHYc a cross-border H_2 network project involving the conversion of existing gas pipelines, linking the industrial regions of France, Germany and Luxembourg: the "Grande Région Hydrogène" (GRH) (Grande Region Hydrogen, 2024). This highly technical project aims to demonstrate on an industrial scale the feasibility of converting a natural gas pipeline for H_2 transport and to verify: *1)*; the evolution of the integrity of the steels used, *2)* the compliance of equipment (valves, regulators, etc.). *3)* the necessary adaptation of equipment dedicated to quality and measurement. This project is also essential for defining and implementing industrial safety measures, which will subsequently be used to update regulations (Muzart and Martin, 2022).

The HyPSTER project, launched in 2020, aims to store H_2 in a cavern at the Etrez storage site, which currently has 20 caverns in operation with natural gas. In 2023, H_2 has been injected. The goals are to validate several operational and technical aspects to ensure the large-scale industrialization of the process: *1)* the adaptation of equipment (piping and completion) to the specificities of H_2 storage subjected to frequent injection – withdrawal cycles, *2)* the tightness of the salt cavern to avoid any risk of H_2 leakage to the surface, *3)* the interaction of the stored H_2 with its underground environment (dissolution, chemical or biochemical reactions in the cavity) (Moret, 2022).

Repurposing a liquid natural gas (LNG) import terminal could be challenging because of the H_2 characteristics compared with LNG's (lower density and boiling temperature of liquefied H_2). Most of the equipment, especially the storage tanks, which accounts for around half of the terminal investment costs, but also boil-off system, pumps and piping, should be modified (IEA, 2022). For ammonia, it is estimated that converting a LNG terminal would cost 11 to 20%; in this case, ammonia storage tanks should be in stainless steel, in order to avoid corrosion (IEA, 2022) These infrastructures conversions are technically difficult and costly. However, in the context of a future low-carbon energy system, it will prevent the obsolescence of costly LNG infrastructures (Riemer et al., 2022). Another option is to produce blue H_2 from LNG using carbon capture via SMR. To achieve this, a new H_2 production plant needs to be integrated into the LNG terminal. This would require an investment cost of 1,500 USD/kW of H_2 and 15-25 USD/tonnes of carbon dioxide for transport and storage (IEA, 2022).

According to the Observatory of Clean Hydrogen, there were 178 H_2 refueling stations, as on March 2023 (European Hydrogen Observatory, 2023c). The majority of these stations are located in Germany (96 stations), France (21 stations) and the Netherlands (14 stations). It is also mentioned that 90% of these stations have car dispensers at 700 bars, and 39% have car and/or bus dispensers at 350 bars (Figure A-2).

6 CONCLUSION

This article provides a comprehensive overview of the H_2 market in Europe, covering production processes and their specificities, end uses as well as infrastructure development for transporting and storing H_2. From an economic point of view, clean H_2 is highly dependent of the price of energy (electricity and natural gas). End use of H_2 currently focuses mainly on refining and ammonia industries but futures additional users will depend on optimizing the production location, the cost, the volume and weight available and the yield of H_2 users process. Storing, and transporting H_2 in the European natural gas transportation and storage network is an interesting option and it could be a good competitor to the conversion of gas infrastructures to biomethane. Use of H_2 for difficult to decarbonize industries seems to be also currently the more interesting option due to the high cost of H_2 production compared to others energy options. Decreasing the H_2 production cost is necessary to improve its further use in mobility for example. To improve clean H_2 users, manufacturers, researchers, and local authorities need to cooperate. A good example is the French situation where the involvement of numerous players in H_2 and its role in the low-carbon national strategy, including the French Ministry of the Economy, the Energy Regulatory Commission (CRE), the Agency for Ecological Transition (ADEME), and France Hydrogène, the French H_2 association are working together. The policy needs also to be implemented at the local level, especially in areas where H_2 uses already exists or can be developed. Currently, H_2 and the corresponding technologies and infrastructures are well accepted, but, it is compulsory to communicate on this energetic vector in order to maintain its acceptance (Turmel, 2021).

The development of blue, green, or purple H_2 production and low-carbon H_2 use in Europe is at a turning point, enabled by a strong regulatory support, a strong industrial commitment of manufacturers, H_2 producers, and end-users as well as promising technological progresses, leading to a cleaner and more sustainable future. White H_2 could be a game changer as it could be cheaper to extract from the underground, very abundant an earth with a reduced environmental impact.

In conclusion, low-carbon and renewable H_2 should be included in the energy mix to contribute to Europe's carbon neutrality objective. Additionally, in France, according to National Low-Carbon Strategy in France (Ministère de la transition écologique et solidaire, 2020), local production of H_2 could contribute to energy sovereignty.

7 BIBLIOGRAPHY

Austria Gas Grid Management AG, 2024. H2 Roadmap: Hydrogen, the energy carrier of the future. URL https://www.aggm.at/en/energy-transition/h2-roadmap/ (accessed 5.18.24).

Birat, J.-P., 2022. La décarbonation de la filière sidérurgique : les enjeux du défi de l'« acier vert »: Annales des Mines - Réalités industrielles Novembre 2022, 77–80. https://doi.org/10.3917/rindu1.224.0077

Blay-Roger, R., Bach, W., Bobadilla, L.F., Reina, T.R., Odriozola, J.A., Amils, R., Blay, V., 2024. Natural hydrogen in the energy transition: Fundamentals, promise, and enigmas. Renewable and Sustainable Energy Reviews 189, 113888. https://doi.org/10.1016/j.rser.2023.113888

CITEPA, 2021. Données d'émissions et rapports d'inventaire. URL https://www.citepa.org/fr/telechargements/ (accessed 4.12.24).

Clean Hydrogen Partnership, n.d. Hydrogen Valleys. URL https://h2v.eu/hydrogen-valleys (accessed 6.7.24).

Council of the European Union, 2023. FuelEU Maritime, 2021/0210.

Dalena, F., Senatore, A., Marino, A., Gordano, A., Basile, M., Basile, A., 2018. Methanol Production and Applications: An Overview, in: Methanol. Elsevier, pp. 3–28. https://doi.org/10.1016/B978-0-444-63903-5.00001-7

Diallo, A., Cissé, C.S.T., Lemay, J., Brière, D.J., 2022. La découverte de l'hydrogène naturel par Hydroma, un « Game Changer » pour la transition énergétique: Annales des Mines - Réalités industrielles Novembre 2022, 154–160. https://doi.org/10.3917/rindu1.224.0154

European Clean Hydrogen Alliance, 2021. The European Clean Hydrogen Alliance reports of the alliance roundtables on barrieres and mitigations measures.

European Commission, 2023. Hydrogen. URL https://energy.ec.europa.eu/topics/energy-systems-integration/hydrogen_en (accessed 4.22.24).

European Commission, Secretariat-General, 2022. Communication from the commission to the European parliament, the European council, the council, the European economic and social committee and the committee of the regions. Document 52022DC0230. URL : https://eur-lex.europa.eu/legal-content/EN/ALL/?uri=COM:2022:230:FIN (accessed 4.12.24).

European Hydrogen Backbone, 2024. EHB Implementation Roadmap: Public support as catalyst for hydrogen infrastructure.

European Hydrogen Observatory, 2023a. Scenarios for future hydrogen demand. URL https://observatory.clean-hydrogen.europa.eu/tools-reports/scenarios-future-hydrogen-demand (accessed 4.22.24).

European Hydrogen Observatory, 2023b. EU Hydrogen Strategy under the EU Green Deal. European Hydrogen Observatory. URL https://observatory.clean-hydrogen.europa.eu/eu-policy/eu-hydrogen-strategy-under-eu-green-deal (accessed 4.12.24).

European Hydrogen Observatory, 2023c. Hydrogen Refuelling Stations.

European Hydrogen Observatory, 2022. Hydrogen-related datasets.

Flow, 2024. Making Hydrogen happen. URL https://www.flow-hydrogen.com/en/home-en/#:%7E:text=With%20Flow%20%2D%20making%20hydrogen%20happen,quantities%20of%20hydrogen%20to%20 industry. (accessed 5.21.24).

Fluxys, 2024. Hydrogen: Preparing to build the network. URL https://www.fluxys.com/en/projects/hydrogen-preparing-to-build-the-network (accessed 5.18.24).

France Hydrogène, 2021. Trajectoire pour une grande ambition Hydrogène - Contribution de la filière au déploiement à 2030.

Fulcheri, L., 2020. Production d'hydrogène décarboné : la troisième voie: Annales des Mines - Responsabilité et environnement N° 99, 93–100. https://doi.org/10.3917/re1.099.0093

Gasunie Energy Development GmbH, 2024. About hyperlink. URL https://www.hyperlink-gasunie.de/en/about-hyperlink

Gasunie Energy Development GmbH, 2023. Hydrogen network Netherlands. URL https://www.gasunie.nl/en/projects/hydrogen-network-netherlands (accessed 5.21.24).

Gasunie Energy Development GmbH, 2021. Hydrogen pipeline in Zeeland has been proving added value for three years. URL https://www.gasunie.nl/en/news/hydrogen-pipeline-in-zeeland-has-been-proving-added-value-for-three-years

Gebreslase, G.A., Martínez-Huerta, M.V., Sebastián, D., Lázaro, M.J., 2023. NiCoP/CoP sponge-like structure grown on stainless steel mesh as a high-performance electrocatalyst for hydrogen evolution reaction. Electrochimica Acta 438, 141538. https://doi.org/10.1016/j.electacta.2022.141538

Ghavam, S., Vahdati, M., Wilson, I.A.G., Styring, P., 2021. Sustainable Ammonia Production Processes. Front. Energy Res. 9, 580808. https://doi.org/10.3389/fenrg.2021.580808

Grande Region Hydrogen, 2024. Grande Region Hydrogen: an initiative that aim to develop a hydrogen ecosystem. URL https://grande-region-hydrogen.eu/en/initiative-and-vision/ (accessed 5.21.24).

GRGTgaz, 2022a. DHUNE. URL https://www.grtgaz.com/en/our-energy-transition/hydrogen-transport/dhune (accessed 6.21.24).

GRGTgaz, 2022b. HYnframed. URL https://www.grtgaz.com/en/our-energy-transition/hydrogen-transport/hynframed (accessed 6.21.24).

GRGTgaz, 2022c. Rhyn. URL https://www.grtgaz.com/en/our-energy-transition/hydrogen-transport/rhyn (accessed 6.21.24).

GRGTgaz, 2022d. Whhyn. URL https://www.grtgaz.com/en/our-energy-transition/hydrogen-transport/whhyn (accessed 6.21.24).

GRGTgaz, 2020. Hy-Fen. URL https://www.grtgaz.com/notre-transition-energetique/transport-hydrogene/hyfen (accessed 6.21.24).

Guo, X.M., Trably, E., Latrille, E., Carrère, H., Steyer, J.-P.P., Carrre, H., Steyer, J.-P.P., 2010. Hydrogen production from

agricultural waste by dark fermentation: A review. International Journal of Hydrogen Energy 35, 10660–10673. https://doi.org/10.1016/j.ijhydene.2010.03.008
H2med, 2024. Hydrogen Corridor: Building the European hydrogen Backbone. URL https://h2medproject.com/building-the-european-hydrogen-backbone/ (accessed 5.21.24).
Halas, P., Dupuy, A., Franceschi, M., Bordmann, V., Fleury, J.-M., Duclerc, D., 2021. Hydrogen gas in circular depressions in South Gironde, France: Flux, stock, or artefact? Applied Geochemistry 127, 104928. https://doi.org/10.1016/j.apgeochem.2021.104928
Hydrogen +, 2024. Gisement d'hydrogène en Lorraine. URL https://www.innovation24.news/2024/01/06/gisement-dhydrogene-en-lorraine/ (accessed 6.6.24).
Hydrogen Europe, 2021. Techno-economic assessment of low carbon hydrogen technologies for the decarbonization of shipping.
HyStock, 2024. HyStock. URL https://www.hystock.nl/en (accessed 5.21.24).
IEA, 2024. Northwest European Hydrogen Monitor 2024. URL https://iea.blob.core.windows.net/assets/b8ba8ad3-f135-4002-9e21-b8cbd213fb36/NorthwestEuropeanHydrogenMonitor2024.pdf
IEA, 2022. Global Hydrogen Review. URL https://iea.blob.core.windows.net/assets/c5bc75b1-9e4d-460d-9056-6e8e626a11c4/GlobalHydrogenReview2022.pdf
IHI Groupe, 2022. CO_2-free power generation achieved with the world's first gas turbine using 100% liquid ammonia – Reduction of over 99% greenhouse gases during combustion–.
Keith Fender, 2024. German hydrogen rail project funding suspended. International Railway Journal.
Kolbe, K., 2019. Mitigating urban heat island effect and carbon dioxide emissions through different mobility concepts: Comparison of conventional vehicles with electric vehicles, hydrogen vehicles and public transportation. Transport Policy 80, 1–11. https://doi.org/10.1016/j.tranpol.2019.05.007
Kovač, A., Paranos, M., Marciuš, D., 2021. Hydrogen in energy transition: A review. International Journal of Hydrogen Energy 46, 10016–10035. https://doi.org/10.1016/j.ijhydene.2020.11.256
Law, L.C., Foscoli, B., Mastorakos, E., Evans, S., 2021. A comparison of alternative fuels for shipping in terms of lifecycle energy and cost. Energies 14. https://doi.org/10.3390/en14248502
Leigh Collins, 2022. "Will no longer be considered" | Hydrogen trains up to 80% more expensive than electric options, German state finds. Hydrogen Insight.
Leonhardt, J.-M., 2022. Décarbonation de l'industrie : le joker Hydrogène: Annales des Mines - Réalités industrielles Novembre 2022, 113–118. https://doi.org/10.3917/rindu1.224.0113
Mads Damsø Nielsen, 2023. Danish Backbone West: Webinar. URL https://energinet.dk/media/khujfiny/energinet_presentation_hydrogen_maturation_project.pdf
Megret, Olivier, Hubert, Loreline, Calbry, M., Trably, Eric, Carrere, Hélène, Garcia-Bernet, Diana, Bernet, Nicolas, 2015. Production d'hydrogène à partir de déchets. Etat de l'art et potentiel d'émergence (No. ETUDE N° 13-0239/1A.).
Michaël Torregrossa, 2024. Unique au monde, ce superyacht à hydrogène est vendu à un prix démentiel. H2 mobile.
Ministère de la transition écologique et solidaire, 2020. Stratégie Nationale Bas-Carbon. La transition écologique et solidaire vers la neutralité carbone. URL Ministère de la transition écologique et solidaire (accessed 6.21.24).
Ministry of ecological transition, 2022. Ordonnance n° 2022-536 du 13 avril 2022 modifiant le modèle minier et les régimes légaux relevant du code minier.
Ministry of the economy, finance, industrial and digital sovereignty, 2023. Arrêté du 23 novembre 2023 accordant un permis exclusif de recherches de mines d'hydrogène natif, hélium et substances connexes dit « Sauve Terre H2 » (département des Pyrénées Atlantiques).
Moradpoor, I., Syri, S., Santasalo-Aarnio, A., 2023. Green hydrogen production for oil refining – Finnish case. Renewable and Sustainable Energy Reviews 175, 113159. https://doi.org/10.1016/j.rser.2023.113159
Moret, Y.-É., 2022. Le projet HyPSTER : émergence d'une nouvelle forme de stockage massif de l'hydrogène essentielle à l'atteinte de la neutralité carbone: Annales des Mines - Réalités industrielles Novembre 2022, 137–142. https://doi.org/10.3917/rindu1.224.0137
Mukherji, A., Thorne, P., Cheung, W.W.L., Connors, S.L., Garschagen, M., Geden, O., Hayward, B., Simpson, N.P., Totin, E., Blok, K., Eriksen, S., Fischer, E., Garner, G., Guivarch, C., Haasnoot, M., Hermans, T., Ley, D., Lewis, J., Nicholls, Z., Niamir, L., Szopa, S., Trewin, B., Howden, M., Méndez, C., Pereira, J., Pichs, R., Rose, S.K., Saheb, Y., Sánchez, R., Xiao, C., Yassaa, N., 2023. Synthesis report on the IPCC 6th assessment report (AR6).
Muzart, L., Martin, F., 2022. MosaHYc ou la conversion d'un réseau de transport du méthane au transport de l'hydrogène pur: Annales des Mines - Réalités industrielles Novembre 2022, 119–124. https://doi.org/10.3917/rindu1.224.0119
Mynko, O., Bonheure, M., Amghizar, I., Brown, D.J., Chen, L., Marin, G.B., Freitas De Alvarenga, R., Civancik Uslu, D., Dewulf, J., Van Geem, K.M., 2023. Electrification of steam cracking as a pathway to reduce the impact of the petrochemical industry on climate change. Journal of Cleaner Production 427, 139208. https://doi.org/10.1016/j.jclepro.2023.139208
Oey, M., Sawyer, A.L., Ross, I.L., Hankamer, B., 2016. Challenges and opportunities for hydrogen production from microalgae. Plant Biotechnology Journal 14, 1487–1499. https://doi.org/10.1111/pbi.12516
OGL, 2023. Hydrogen Transport and Storage Network.
Pellow, M.A., Emmott, C.J.M., Barnhart, C.J., Benson, S.M., 2015. Hydrogen or batteries for grid storage? A net energy analysis. Energy Environ. Sci. 8, 1938–1952. https://doi.org/10.1039/C4EE04041D
Pierre Yves Larrieu, 2024. Tableau de bord stratégique des carburants marins alternatifs pou la zone Europe. URL

https://media.licdn.com/dms/document/media/D4E1FAQEdwxj50987pg/feedshare-document-pdf-analyzed/0/1715530670173?e=1718236800&v=beta&t=CtxqedZn23kVW5fbvNxRgDdjZXpirY4S_xDMF6ZcmSA (accessed 5.31.24).

Prinzhofer, A., Tahara Cissé, C.S., Diallo, A.B., 2018. Discovery of a large accumulation of natural hydrogen in Bourakebougou (Mali). International Journal of Hydrogen Energy 43, 19315–19326. https://doi.org/10.1016/j.ijhydene.2018.08.193

Rambhujun, N., Salman, M.S., Wang, T., Pratthana, C., Sapkota, P., Costalin, M., Lai, Q., Aguey-Zinsou, K.-F., 2020. Renewable hydrogen for the chemical industry. MRS Energy & Sustainability 7, 33. https://doi.org/10.1557/mre.2020.33

Riemer, M., Schreiner, F., Wachsmuth, J., 2022. Conversion of LNG terminals for liquid hydrogen or ammonia. Fraunhofer ISI. https://doi.org/10.24406/PUBLICA-464

Ron-David Heinen, 2023. EWE stores hydrogen in gas storage for the first time. URL https://www.energate-messenger.com/news/237122/ewe-stores-hydrogen-in-gas-storage-for-the-first-time (accessed 5.21.24).

RTE, 2021. Energy Pathways to 2025. Key results. URL https://assets.rte-france.com/prod/public/2022-01/Energy%20pathways%202050_Key%20results.pdf

RWE, 2024. H2ercules. URL https://www.h2ercules.com/h2ercules (accessed 5.21.24).

Serena Boni, 2024. HDF Energy inaugurates the world's first high-power hydrogen fuel cell factory in Blanquefort, France. URL https://www.hdf-energy.com/en/investors/press-release/?ID=ACTUS-0-86051&CLIENT=ACTUS-0-478 (accessed 6.4.24).

Shi, X., Liao, X., Li, Y., 2020. Quantification of fresh water consumption and scarcity footprints of hydrogen from water electrolysis: A methodology framework. Renewable Energy 154, 786–796. https://doi.org/10.1016/j.renene.2020.03.026

Shintaro Ito, Masahiro Uchida, Toshiyuki Suda, Toshiro Fujimori, 2020. Development of Ammonia Gas Turbine Co-Generation Technology. IHI Engineering Review.

Simoes, S.G., Catarino, J., Picado, A., Lopes, T.F., Di Berardino, S., Amorim, F., Gírio, F., Rangel, C.M., Ponce De Leão, T., 2021. Water availability and water usage solutions for electrolysis in hydrogen production. Journal of Cleaner Production 315, 128124. https://doi.org/10.1016/j.jclepro.2021.128124

Simpson, A., Lutz, A., 2007. Exergy analysis of hydrogen production via steam methane reforming. International Journal of Hydrogen Energy 32, 4811–4820. https://doi.org/10.1016/j.ijhydene.2007.08.025

Storengy, 2024. Lancement du projet FrHyGe, démonstrateur de stockage souterrain d'hydrogène. URL https://www.storengy.com/fr/medias/actualites/lancement-du-projet-frhyge-demonstrateur-de-stockage-souterrain-dhydrogene (accessed 5.21.24).

Tashie-Lewis, B.C., Nnabuife, S.G., 2021. Hydrogen Production, Distribution, Storage and Power Conversion in a Hydrogen Economy - A Technology Review. Chemical Engineering Journal Advances 8, 100172. https://doi.org/10.1016/j.ceja.2021.100172

Teréga, 2023. European hydrogen backbone: Teréga and the SALINS group join forces on a storage project.

The Shift Tank, 2022. Décarboner l'industrie sans la saborder, dans le cadre du plan de transformation de l'économie française.

Tiwari, S., Pekris, M.J., Doherty, J.J., 2024. A review of liquid hydrogen aircraft and propulsion technologies. International Journal of Hydrogen Energy 57, 1174–1196. https://doi.org/10.1016/j.ijhydene.2023.12.263

Truche, L., Donzé, F.-V., Goskolli, E., Muceku, B., Loisy, C., Monnin, C., Dutoit, H., Cerepi, A., 2024. A deep reservoir for hydrogen drives intense degassing in the Bulqizë ophiolite. Science 383, 618–621. https://doi.org/10.1126/science.adk9099

Turmel, F.G.R., 2021. Doit-on s'attendre à un refus du déploiement de l'hydrogène ? Dynamique territoriale et transition énergétique. vertigo. https://doi.org/10.4000/vertigo.30350

Ueckerdt, F., Bauer, C., Dirnaichner, A., Everall, J., Sacchi, R., Luderer, G., 2021. Potential and risks of hydrogen-based e-fuels in climate change mitigation. Nat. Clim. Chang. 11, 384–393. https://doi.org/10.1038/s41558-021-01032-7

Yannick Pesson, 2024. Avis d'expert: l'hydrogène au naturel. URL https://www.ifpenergiesnouvelles.fr/article/avis-dexpert-lhydrogene-naturel

Yukesh Kannah, R., Kavitha, S., Preethi, Parthiba Karthikeyan, O., Kumar, G., Dai-Viet, N.Vo., Rajesh Banu, J., 2021. Techno-economic assessment of various hydrogen production methods – A review. Bioresource Technology 319, 124175. https://doi.org/10.1016/j.biortech.2020.124175

Zgonnik, V., 2020. The occurrence and geoscience of natural hydrogen: A comprehensive review. Earth-Science Reviews 203, 103140. https://doi.org/10.1016/j.earscirev.2020.103140

8 APPENDIXES

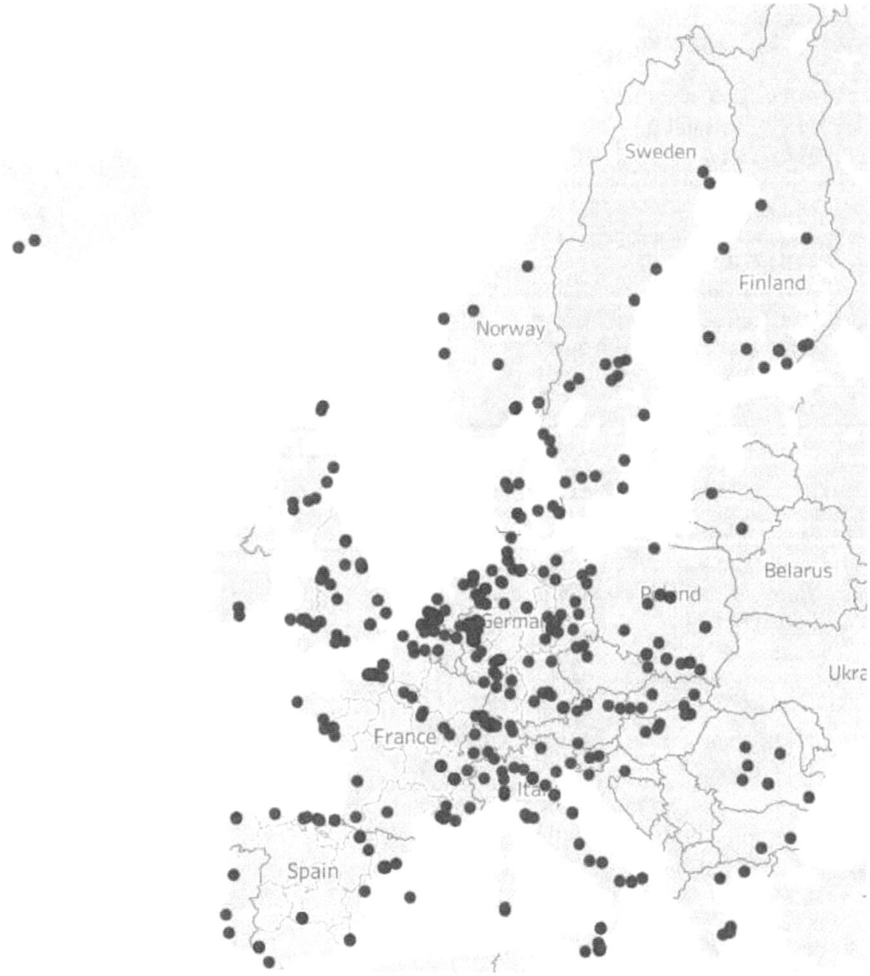

Figure A-1: Hydrogen production sites in Europe. (European Hydrogen Observatory, 2022)

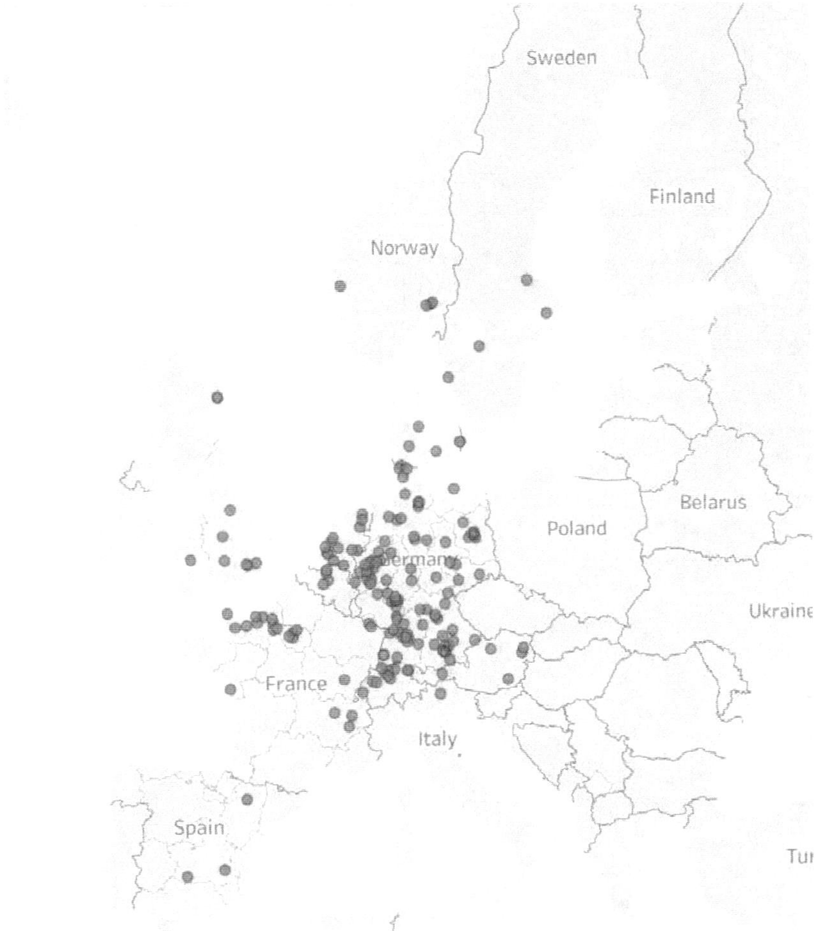

Figure A-2: Hydrogen refueling stations in Europe. (European Hydrogen Observatory, 2022)

Figure A-3: Hydrogen consumption sites in Europe. (European Hydrogen Observatory, 2022)

Figure A 4: Hydrogen valleys in Europe (Clean Hydrogen Partnership, n.d.)

NURER 2024
Nuclear and Renewable Energy Resources

Safety design technologies for sodium-cooled fast reactor coupled to thermal energy storage system with sodium-molten salt heat exchanger

Hidemasa YAMANO
Japan Atomic Energy Agency, Oarai, Japan, yamano.hidemasa@jaea.go.jp,
ORCID: https://orcid.org/0000-0002-2755-7680
Kazuya TAKANO
Japan Atomic Energy Agency, Oarai, Japan, takano.kazuya@jaea.go.jp,
ORCID: 0009-0007-6565-5605
Kenichi KURISAKA
Japan Atomic Energy Agency, Oarai, Japan, kurisaka.kennichi@jaea.go.jp,
ORCID: 0009-0000-7636-1441
Shin KIKUCHI
Japan Atomic Energy Agency, Oarai, Japan, kikuchi.shin@jaea.go.jp,
ORCID: 0000-0002-5936-5130
Toshiki KONDO
Japan Atomic Energy Agency, Oarai, Japan, kondo.toshiki@jaea.go.jp,
ORCID: 0000-0002-9585-453X
Masaaki HAYASHI
Hitachi-GE Nuclear Energy, Ltd, Hitachi, Japan, masaaki.hayashi.ak@hitachi.com,
ORCID: 0009-0008-3161-954X

Cite this paper as:	*Yamano, H., Takano, K., Kurisaka, K., Kikuchi, S., Kondo, T., Hayashi, M., Safety design technologies for sodium-cooled fast reactor coupled to thermal energy storage system with sodium-molten salt heat exchanger. 8th Int. Conf. on Nucl. and Ren. Energy Res. May 25-29, 2024, Ankara, Turkey.*

Abstract: A sodium-cooled fast reactor (SFR), which is one of next-generation innovative reactors, can make flexibility with variable renewable energy by coupling a thermal energy storage (TES) system with molten salt. New challenges for the SFR coupled with TES are to develop a safety design approach and a heat exchanger (HX) between sodium and molten salt. On that account, since 2022, a three-year project has been performed to develop 1) a safety design approach and risk assessment methodology of the SFR with TES, 2) a performance evaluation technology of an HX between sodium and molten salt, and proposal of heat transfer improvement measures, and 3) an evaluation technology of chemical reaction characteristics between sodium and molten salt, and proposal of safety improvement measures. This paper describes the project overview and recent progress. For 1), this study investigated the effect of sodium-molten salt heat transfer tube failure. For 2), the HX performance was evaluated by computational fluid dynamics (CFD) simulations with a partial model of selected heat exchanger. For 3), thermal analyses obtained preliminary data of sodium-nitrate molten salt reaction. No significant reaction was not observed.

Keywords: *Sodium-cooled fast reactor, thermal energy storage, safety design, molten salt, heat exchanger*

NURER 2024
Nuclear and Renewable Energy Resources

Analysis of SBO sequences in a Small Modular Reactor using the TRACE code

Jorge Arrabal de Haro
NFQ, Madrid, Madrid, jorge.arrabal@nfq.es,
ORCID: 0009-0007-9699-0588

David José Canal
Universidad Politécnica de Madrid, Madrid, Madrid, david.canal@upm.es
ORCID: 0009-0005-0626-2014

César Queral
Universidad Politécnica de Madrid, Madrid, Spain, cesar.queral@upm.es
ORCID: 0000-0001-9384-2474

Cite this paper as: *Arrabal de Haro, J, Canal, D., Queral, C. Analysis of SBO sequences in a Small Modular Reactor using the TRACE code. 7th Int. Conf. on Nucl. and Ren. Energy Res. 27-30 October 2024, Antalya, Turkey*

Abstract: This work is based on a Small Modular Reactor (SMR), specifically on the simulation of an SBO sequence in an integral Pressurized Water Reactor (iPWR) using the TRACE system code. The main objective is to analyze the behavior of the iPWR, verifying that the plant reaches a safe shutdown after 36 hours. To achieve this, different detailed models were applied. First, a model of the vessel and the isolation condenser with three-dimensional components was developed. Later, an improvement to the isolation condenser was implemented, aiming to reduce its coolant volume inside the containment. For this purpose, a model of thermally coupled pools inside and outside the containment was designed, yielding relevant results that allow for economic optimization and safety improvements in this type of reactor.

Keywords: TRACE, SMR, iPWR, isolation condenser, SBO

Influence of Reduced Graphene Oxide on Anode Functional Layer Performance in Solid Oxide Fuel Cells

Cigdem TIMURKUTLUK
Nigde Omer Halisdemir University, Prof. Dr. T. Nejat Veziroglu Clean Energy Research Center, Nigde, 51245, Turkey,
Nigde Omer Halisdemir University Department of Energy Science and Technologies, 51245, Nigde, Turkey,
cigdemtimurkutluk@ohu.edu.tr
ORCID: 0000-0002-8672-993X
Recep ZAN
Nanotechnology Application and Research Center, Niğde Ömer Halisdemir University, Niğde, Turkey
Department of Physics, Niğde Ömer Halisdemir University, Niğde, Turkey
recep.zan@ohu.edu.tr
ORCID: 0000-0001-6739-4348
Bora TIMURKUTLUK
Nigde Omer Halisdemir University, Prof. Dr. T. Nejat Veziroglu Clean Energy Research Center, Nigde, 51245, Turkey,
Nigde Omer Halisdemir University, Mechanical Engineering Department, Nigde, 51245,
boratimurkutluk@ohu.edu.tr
ORCID: 0000-0001-6916-7720
Furkan TORUNTAY
Nigde Omer Halisdemir University, Prof. Dr. T. Nejat Veziroglu Clean Energy Research Center, Nigde, 51245, Turkey,
Nigde Omer Halisdemir University, Mechanical Engineering Department, Nigde, 51245,
furkantoruntay@ohu.edu.tr
ORCID: 0000-0001-8764-9782
Sezer ONBILGIN
Nigde Omer Halisdemir University, Prof. Dr. T. Nejat Veziroglu Clean Energy Research Center, Nigde, 51245, Turkey,
Nigde Omer Halisdemir University, Mechanical Engineering Department, Nigde, 51245,
sezeronbilgin@ohu.edu.tr
ORCID: 0000-0002-5349-8936
Ali ALTUNTEPE
Nanotechnology Application and Research Center, Niğde Ömer Halisdemir University, Niğde, Turkey
alialtuntepe@sivas.edu.tr
ORCID: 0000-0002-6366-4125
Abdulkadir Muhittin DOĞAN
Nigde Omer Halisdemir University, Prof. Dr. T. Nejat Veziroglu Clean Energy Research Center, Nigde, 51245, Turkey,
kdrdgn16@hotmail.com
ORCID: 0000-0001-5390-8889

Cite this paper as: *Timurkutluk C., Zan R.,Timurkutluk B., Toruntay F., Onbilgin S., Altuntepe A., Effect of reduced Graphene Oxide (rGO) on Anode Functional Layer Performance in Solid Oxide Fuel Cells. 7th Int. Conf. on Nucl. and Ren. Energy Res. 27-30 October 2024, Antalya, Turkey*

Abstract: Reduced graphene oxide (rGO) powders are synthesized and added to the anode catalyst layer of SOFC (solid oxide fuel cell) to increase the cell performance and characterization tests are carried out. In this context, first, rGO production and characterization studies are performed. After that, the effects of rGO content are investigated by adding various amounts (0.1, 0.2 and 0.3 g) of rGO to the anode functional layer screen printing paste. It is seen that the maximum power densities of 0.542, 0.394 and 0.385 W/cm^2 are obtained from the cells prepared by adding 0.1, 0.2 and 0.3 g rGO to the AFL, respectively, under 700-800°C operating temperatures. The optimum rGO-doped cell performs ~17% higher peak performance than that of the reference cell.

Keywords: *Energy; Fuel cell; Solid oxide fuel cell; Anode; Reduced graphene oxide.*

1. INTRODUCTION

In the context of global decarbonization and the difficulty in meeting the rapidly growing energy demand, the demand for renewable energy and energy sources has been steadily increasing. Solid oxide fuel cells (SOFCs), one of the leading technologies in meeting this demand, stand out with the use of cheaper catalysts and improved catalytic activity due to high operating temperatures (600-1000 °C), compared to other fuel cells. The most widely used SOFC membrane electrode assembly (MEA) materials are Ni-YSZ anode, YSZ electrolyte and LSM-YSZ cathode. However, carbon-based nanomaterials have an important place today and are the subject of much research and are widely used in industrial fields. Reduced graphene oxide (rGO), one of them, has already started to be used in the development of solar cells [1] and transistors [2], and many recent scientific studies on the use of graphene, graphene oxide or doped graphene in fuel cells have been carried out due to having superior properties [3-5]. In the fuel cell context, rGO is used as a catalyst or electrode. However, studies on the development of rGO-doped solid oxide fuel cell electrodes are very limited. In this study, therefore, the addition of rGO is investigated to improve the anode functional layer (AFL) performance for SOFCs.

2. EXPERIMENTAL

rGO systhesis

The Modified Hummers approach was used for GO synthesis. Chemical ratio, reactivation times, and other synthesis parameters (temperature, spinning time etc.) were optimized according to our previous work [6]. Graphite, sulfuric acid (H_2SO_4), potassium permanganate ($KMnO_4$) and hydrogen peroxide (H_2O_2) were implemented in the GO synthesis step. For the rGO synthesis, hydrazine was utilized as a reduction agent. GO powders were stirred for 12 h after adding hydrazine and deionized water. rGO powders were then obtained following filtering and drying steps.

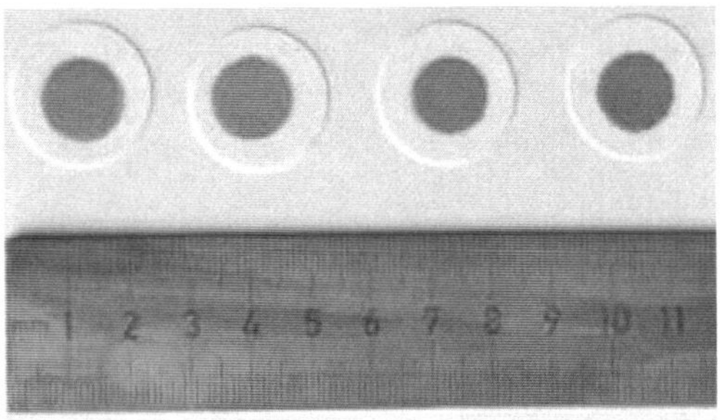

Figure 1. Photos of the cell fabricated (RC, rGO-C1, rGO-C2 and rGO-C3 from left to right)

Cell fabrication

The electrolyte supports were fabricated via tape casting method. First, YSZ powders were mixed with appropriate amounts of binder, plasticizer, solvent and dispersant to achieve a tape casting slurry. The YSZ solution was then ball milled for 24 hours with zirconia balls in polyethylene bottles. After that, the resulting YSZ slurry was cast with a doctor blade gap of 190 µm on a Mylar strip via a tape casting device. The tapes were then stacked and pressed to complete the lamination. The laminates were sintered after cutting with the desired geometry. Then, one side of YSZ support layer was coated with rGO (0.1, 0.2 and 0.3 g) added NiO-YSZ anode functional layer (Table 1.) and NiO-YSZ anode current collecting layer by screen printing. After sintering the anode layers, LSM-YSZ cathode functional and cathode current collection layers were coated and sintered similarly. Both electrodes have an active area of 1 cm^2. A cell without rGO addition as a reference cell (RC) was also fabricated and tested for comparison (Fig.1.). The electrochemical performance measurements were carried out under dry hydrogen and air at 700-800 °C.

Table 1. Properties of the cells fabricated.

Cell	Amount of rGO in AFL paste (wt. %)
RC	-
rGO-C1	1
rGO-C2	2
rGO-C3	3

3. RESULTS AND DISCUSSIONS

The performance and impedance results of the cells under 800 °C operating temperature are compared in the Fig. 2. The maximum power density of 0.542, 0.394 and 0.385 W/cm^2 was achieved from the cells prepared by adding 0.1, 0.2 and 0.3 g rGO into the AFL, respectively. It is seen that the peak cell performance shows a decreasing trend with increasing the amount of rGO addition. This may be attributed to the reduction in the active reaction zones and porosity of AFL despite increased electronic conductivity. Nevertheless, the maximum power density of the reference cell prepared without adding rGO is measured as 0.462 W/cm^2 under the same operating conditions. Similar performance improvements are also obtained from the cell with 0.1 g rGO added AFL at the operation temperatures of 700 and 750 °C.

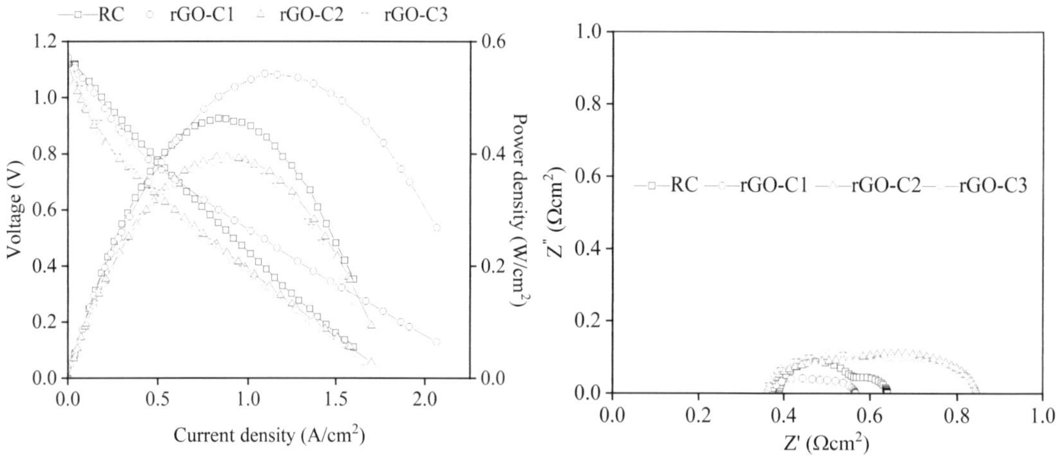

Figure 2. Effect of rGO amount on cell performance (left) and impedance (right) at 800 °C operating temperature

4. CONCLUSION

In this study, the effects of rGO addition to the anode functional layer on cell performance solid oxide fuel cell cells having 1cm^2 active area are investigated. The performance and impedance results reveal that the cell with reduced graphene added anode functional layer exhibits ~17% higher maximum power density than that of the reference cell at 800 °C operating temperature due to improved electronic conductivity.

REFERENCES

[1] Yin Z, Wu S, Zhou X, Huang X, Zhang Q, Boey F, et al. *Electrochemical Deposition of ZnO Nanorods on Transparent Reduced Graphene Oxide Electrodes for Hybrid Solar Cells*. Small. 2010;6:307-12. https://doi.org/10.1002/smll.200901968

[2] He Q, Wu S, Gao S, Cao X, Yin Z, Li H, et al. *Transparent, Flexible, All-Reduced Graphene Oxide Thin Film Transistors*. Acs Nano. 2011;5:5038-44. https://doi.org/10.1021/nn201118c

[3] Pareek A, Shanthi Sravan J, Venkata Mohan S. *Exploring chemically reduced graphene oxide electrode for power generation in microbial fuel cell. Materials Science for Energy Technologies*. 2019;2:600-6. https://doi.org/10.1016/j.mset.2019.06.006

[4] Antolini E. *Graphene as a new carbon support for low-temperature fuel cell catalysts. Applied Catalysis*. 2012;123-124:52-68. https://doi.org/10.1016/j.apcatb.2012.04.022

[5] A. Choudhury, H. Chandra, A. Arora *Application of solid oxide fuel cell technology for power generation—a review* 2013;20:430-442 https://doi.org/10.1016/j.rser.2012.11.031

[6] Timurkutluk C, Zan R, Timurkutluk B, Toruntay F, Onbilgin S, Hasret O, Altuntepe A. *Effect of reduced graphene oxide addition on cathode functional layer performance in solid oxide fuel cells, International Journal of Hydrogen Energy. 2023;* 23127-23135:48-60. https://doi.org/10.1016/j.ijhydene.2022.11.082

NURER 2024
Nuclear and Renewable Energy Resources

Role of Nuclear-Renewable Hybrid Energy Systems – A Case Study of Natrium™ Energy System for Power and Hydrogen Production

Hailei Wang
Utah State University, Logan UT, USA, hailei.wang@usu.edu
ORCID: 0000-0002-3037-1342

Cite this paper as: *Wang H, Role of Nuclear-Renewable Hybrid Energy Systems A Case Study of Natrium™ Energy System for Power and Hydrogen Production. 7th Int. Conf. on Nucl. and Ren. Energy Res. 27-30 October 2024, Antalya, Turkey*

Abstract: There is little doubt that our society is gaining the benefits and needs more clean and renewable energies such as wind and solar. As more and more solar and wind energy is added to our grids, however, we are also facing their consequences including greater fluctuation of electricity demand and higher price peaks. Moreover, greater penetration of renewables increasingly leads to negative electricity prices in some energy markets. Historically as the baseload energy source, nuclear energy is facing increased cost pressure and competition from the renewables. Yet, there are opportunities for nuclear and renewables to work together and achieve unprecedent synergy. In this talk, we will provide an overview of the Natrium™ energy system. It is designed to have a Nuclear Island that is powered by a sodium fast reactor and an Energy Island that integrates molten salt thermal energy storage and a reheat Rankine steam cycle. Given its energy storage and enhanced power dispatch capability, the Natrium™ is shown significantly better economics in fluctuating energy markets such as ERCOT (large amount of wind and rapid growth of solar) and CAISO (large amount of solar) compared to more stable energy markets such as MISO. The techno-economic analysis not only calculates the expected (mean) net present value and its distribution in a specific energy market, but also computes the optimal hourly dispatch profile. To help with the periods when electricity prices are low, the Natrium™ system is further coupled with a high-temperature steam electrolyzer to produce hydrogen. The results show that adding hydrogen production and storage can further improve the economics of the Natrium™ energy systems in those deregulated energy markets in the US.

Keywords: nuclear, renewable, hybrid energy system, Natrium, hydrogen

NURER 2024
Nuclear and Renewable Energy Resources

Synthetic Time Series Modeling and Sensitivity Study for Robust Techno-Economic Analysis of Advanced Nuclear Energy Systems

Hailei Wang
Utah State University, Logan UT, USA, hailei.wang@usu.edu
ORCID: 0000-0002-3037-1342

Cite this paper as:	*Wang H, Synthetic Time Series Modeling and Sensitivity Study for Robust Techno-Economic Analysis of Advanced Nuclear Energy Systems. 7th Int. Conf. on Nucl. and Ren. Energy Res. 27-30 October 2024, Antalya, Turkey*

Abstract: In performing techno-economic analysis (TEA) of renewable or hybrid energy systems, historical data of wind, solar, electricity demand and price is used. Increasingly, more and more rare events have occurred in recent years as an impact of climate change. Given the economics of advanced energy systems becomes progressively dominated by those rare events, accurate prediction of them becomes critical for a robust TEA. In this work, we have focused on various time series modeling of solar, wind, electricity load and price. Their performance is discussed in comparison with the historical data. In addition, various machine learning and deep learning neural network electricity price models are developed to enable dispatch optimization and TEA in price-maker scenarios. The results are compared with those of price-taker simulations. Furthermore, a sensitivity analysis is conducted using SOBOL indices, which includes various costs of building and operating nuclear reactors, as well as the time series of electricity price as it directly impacts plant revenue. The results show that the time series are important in energy markets such as CAISO and ERCOT despite large uncertainty exists for the costs of building advanced nuclear reactors.

Keywords: nuclear, time series modeling, sensitivity analysis, TEA, neural network, machine learning

NURER 2024
Nuclear and Renewable Energy Resources

Radiological Consequences of the SMRs - First Steps Towards Deployment

Adam Kecek
UJV Rez

Cite this paper as: *Kecek A., Radiological Consequences of the SMRs - First Steps Towards Deployment. 7th Int. Conf. on Nucl. and Ren. Energy Res. 27 30 October 2024, Antaya, Turkey*

Abstract: Nuclear safety may be in some situations understood as radiological consequences. In case of an accident, the nucelar power plant safety barriers should withstand the accidental load and reduce the possible causes as much as possible. For the existing units, the emergency planning zone diameter, depending on the country and unit, could reach dozens of kilometers. With the improvement of the newly planned concepts safety and SMR deployment, the emergency planning zone may be reduced to the area in the vicinity of the plant, often called to be "on the fence". This may open the possibility to build such units close to the urbanized areas, where proper prediction of fission product transport becomes very important and where the existing codes and models may not be sufficient. To investigate these issues, a national research project RACOPSAC (RAdiological COnsequences of Postulated SMR Accidents) was started. The aim of the project is to create apropriate experimental data for validation of the numerical tools used for fission product transport in the environment in conditions typical to SMR accident as well as development of a guideline how to asses the experimental data with the calculated values. An essential part of the project is a methodology, which should provide specific recommendations on radiological consequences calculations for SMRs.

Keywords: *International Energy Management, Nuclear Cooperation, Energy Security*

NURER 2024
Nuclear and Renewable Energy Resources

Probabilistic Risk Assessment of Molten Salt Reactors: Evaluating Safety Systems and Reliability

Ümmühan Barlak
Sinop University, Sinop, Türkiye, umbarlak@gmail.com
ORCID: 0000-0002-7365-8810

Huseyin Sahiner
Sinop University, Sinop, Türkiye, huseyinshnr@gmail.com
ORCID: 0000-0002-3191-1590

Cite this paper as:	Barlak, Ü, Sahiner, H. Risk Analysis In Molten Salt Reactors. 7th Int. Conf. on Nucl. and Ren. Energy Res. 27-30 October 2024, Antaya, Turkey

Abstract: Molten Salt Reactors (MSRs) are one of the six advanced reactor concepts selected by the Generation IV International Forum. Owing to their potential for sustainability, unique fuel properties, and safety functions, they are now the focus of a lot of research. Molten fuel and coolant properties allow some extra safety features beyond those for solid fuel. These safety systems should be considered in the risk analysis for licensing. Probabilistic Risk Assessment (PRA) is a tool to estimate systems reliability based on data and specialists' opinions. This method was employed to estimate the special safety systems. Reliability analysis of the safety functions starts with the definition of their components and connections (parallel or serial). By using an infinite number of backup systems, reliability of ≈100% could be achieved. However, this is commercially not practical. The study provided information about the basic passive and active safety functions. Safety systems were depicted schematically to create appropriate fault trees. Potential pathways of events with their probabilities of occurrence were illustrated using event trees. The results contribute to understanding MSRs improved safety and reliability better. This knowledge of MSRs will facilitate the licensing procedure for this cutting-edge nuclear reactor idea.

Keywords: *Molten Salt Reactor, Probabilistic Risk Analysis, Fault Tree, Event Tree*

Environmental Impacts of Energy Production and Alternatives

Huseyin Sahiner
Sinop University, Sinop, Türkiye, huseyinshnr@gmail.com
ORCID: 0000-0002-3191-1590

Cite this paper as: *Sahiner, H. Environmental Impacts of Energy Production and Alternatives. 7th Int. Conf. on Nucl. and Ren. Energy Res. 27-30 October 2024, Antaya, Turkey*

Abstract: Energy production has an impact on the environment, regardless of the energy kind. Understanding how energy production interacts with the ecosystem should be the first step in developing methods to reduce its environmental effect. Nuclear power is frequently promoted as a low-carbon energy source that can help the transition to a carbon-free energy system. The main environmental impact of nuclear reactors consists of heat introduced to the environment and radioactivity. Fossil fuels, including coal, oil, and natural gas, have long been the world's primary energy sources. The use of fossil fuels for energy production has a significant impact on greenhouse gas emissions, air pollution, and climate change. Renewable energy sources, such as solar, wind, hydropower, and geothermal, are frequently considered environmentally friendly alternatives. Despite being ecologically beneficial, renewables can still have environmental impacts, such as land use, water use, air quality, and effects on biodiversity. Solar and wind energy have lower energy densities, whereas solar needs sunlight, and wind turbines require consistent wind. Thus, they require more land and potentially harm the ecosystems. Wind farms endanger animals and generate noise, while solar panels are built from hazardous chemicals. On the other hand, nuclear reactors provide more long-term environmental advantages.

Keywords: *Ecology, Energy Production, Renewable Energy, Nuclear Energy*

KIT Investigations of the VVER-1000 plant behavior under design basis accidents

Victor Hugo Sanchez-Espinoza
Karlsruhe Institute of Technology, Karlsruhe, Germany, victor.sanchez@kit.edu
ORCID: 0000-0001-9397-7600

Nuri Beydogan
Karlsruhe Institute of Technology, Karlsruhe, Germany, nuri.beydogan@partner.kit.edu,
ORCID: 0009-0005-4068-1968

Gianfranco Huaccho-Zavala
Karlsruhe Institute of Technology, Karlsruhe, Germany, gianfranco.zavala@kit.edu,
ORCID: 0009-0005-6339-8508

Cite this paper as: *Sanchez-Espinoza, V. H., Beydogan, N., and Huaccho-Zavala, G. KIT Investigations of the VVER-1000 plant behavior under design basis accidents. 7th International Nuclear and Renewable Energy Resources Conference. 27-30 October 2024, Antalya, Türkiye*

Abstract: A large number of VVER nuclear power plants are under operation (440 and 1000) and under construction (1000 and 1200) in different countries. Investigations at KIT are focused on the validation and application of numerical tools for the analysis of design basis and severe accidents. The paper discusses the investigations ongoing on design basis accidents e.g., MSLB and boron dilution Benchmark Rostov-2 for VVER-1000 reactor designs using TRACE/PARCS and PARCS/SUBCHANFLOW. Key results will be presented and discussed. The paper will start with the motivation, a short description of the numerical tools used, and the models developed for the analysis of transients and accidents. An important aspect is the method to generate nodal few-energy group, condensed cross sections for 3D nodal diffusion solvers such as PARCS using different lattice physics codes. Modern core loadings consist of different fuel assembly types with different burnup and enrichment, which is challenging for the lattice physics tools as is also the modeling of the reflector. The investigations emphasize the need to use coupled codes including 1D/3D system thermal hydraulic and 3D core solvers to catch the physics of transients with strong non-symmetrical core behavior.

Keywords: *VVER, MSLB, LOCA, TRACE/PARCS, ROSTOV-2, Kozloduy-6*

1. INTRODUCTION

A large number of Water-Water-Energy-Reactors (VVER) are under operation (440 and 1000) and under construction (1000 and 1200) in different countries in Europe and worldwide e.g., VVER-1200 are being built in Turkey and Hungary. The core of new plants is loaded with modern fuel assembly types with increased heterogeneity within a fuel assembly. In addition, the VVER-1200 is equipped with additional passive safety systems compared to the predecessor VVER-1000. These technological developments and the advances in design and safety evaluation tools leads to the improvement and development of new tools for both the core (SP_N, transport, Monte Carlo) and the plants (3D coarse mesh thermal hydraulics, multi-scale thermal hydraulics) aiming to make possible a more accurate prediction of the safety margins for different types of transient scenarios such as rod ejection, and boron dilution.

The fuel assemblies of modern VVER-core loading are characterized by different enrichment of U-235, different number of burnable poisons (Gd-rods) located at different locations within the hexagonal fuel assemblies. In addition, the different fuel assembly (FA) types are distributed within the core so that the gradient of the neutron current between FAs of a core loading is considerable and hence it needs additional information such as Assembly Discontinuity Factors (ADF) to be provided to the classical diffusion solvers for an accurate prediction of the radial neutron flux distribution.

Moreover, the outer row of the FAs close to the reflectors are of different type and enrichment which makes necessary the estimation of Reflector Discontinuity Factors (RDF) for improve core simulations.

Transients such as Anticipated Transients without SCRAM (ATWS), Main Steam Line Break (MSLB), Boron Dilution (BD), etc. require the use of coupled neutronic/thermal hydraulic codes for a more realistic description of the multidimensional power distributions and the interactions between the core behavior and plant behavior. Hence, the validation of such numerical tools for safety evaluations for the prediction of key safety parameters e.g., the maximum pressure within the RPV (ATWS), margin to re-criticality and possible return to power (MSLB) is of paramount importance.

KIT-research is engaged in the validation of core and system thermal hydraulic tools for the analysis of VVERs under design basis and beyond design basis accidents [1]. The strategy of KIT is to combine the use of in-house codes with external reference codes e.g., the ones of the US Nuclear Regulatory Commission (US NRC) (TRACE, RELAP5, PARCS) or recently the German system thermal hydraulic tools ATHLET in the frame of the international Code Assessment and Maintenance Program (CAMP) and national cooperation with the GRS. KIT participated on different OECD/NEA benchmarks, where plant data of VVERs were provided for the validation of system thermal hydraulic codes coupled with 3D neutron kinetics such as the OECD/NEA V1-Coolant transient [2], [3], the V1-MSLB Benchmark [4], the Kalinin-3 [5] and the most recently Rostov-2 boron dilution benchmark [6]. In addition, KIT participated on the EU H2020 CAMIVVER project [7] devoted to validate and improve numerical tools for the analysis of the core physics of VVER-reactors, the 3D thermal hydraulics within the Reactor Pressure Vessel (RPV) by CFD-codes and the analysis of transients with system thermal hydraulic codes using point kinetics and 3D neutron diffusion solvers. In addition, high fidelity tools were used to predict the depletion behavior of the Temelín VVER-1000 core taking into account local feedback between the thermal hydraulic, neutronics and thermo-mechanics (Serpent2/SCF/TU/ICoCo [8]). The paper discusses the investigations of both the Rostov-2 Reactivity Compensation Benchmark and the academic VVVER-1000 MSLB Exercise defined for the Beginning of Cycle (BOC) in the CAMIVVER-project using the codes PARCS/SCF and TRACE/PARCS. Selected results will be presented and discussed. An important aspect is the method to generate nodal few-energy group, condensed cross sections for 3D nodal diffusion solvers such as PARCS using different lattice physics codes. Modern core loadings consist of different fuel assembly types with different burnup and enrichment, which is challenging for the lattice physics tools as is also the modeling of the reflector. The investigations emphasize the need to use coupled codes including 1D/3D system thermal hydraulic and 3D core solvers to accurately predict the underlying physics of transients with strong non-symmetrical core behavior. The paper starts with the motivation, a brief description of the Rostov-2 core and Kozloduy-6 plant. It is followed by a short description of the numerical tools, and the models developed for the analysis of two different transient scenarios. Finally, selected results are discussed and main conclusions are presented.

2. PECULIARITIES OF THE KOZLODUY-6 AND ROSTOV-2 VVER-1000 NUCLEAR POWER PLANTS

The Rostov-2 core consists of 163 fuel assemblies (FA) of different type e.g., 48 FA type U13 with U-235 enrichments of 1.3 %, 42 FA type U22 with U-235 enrichment of 2.2%, 37 FA U30Y9 with Gd-rods and an average U-235 enrichment of 2.98%, 24 FA type U39A9 with Gd-rods and an average U-235 enrichment of 3.9%, 12 FA type U39B6 with Gd-rods and an average U-235 enrichment of 3.9% [6]. In Figure 1, the radial distribution of the FAs and the control rod bank 10 is shown. In the core, the control rods are grouped in 10 banks. The core active height amounts 3.7 m and the length of the control rods is 3.55 m. The 12 spacer grids are distributed along the core height. The typical dimensions of the FAs can be found in [6].

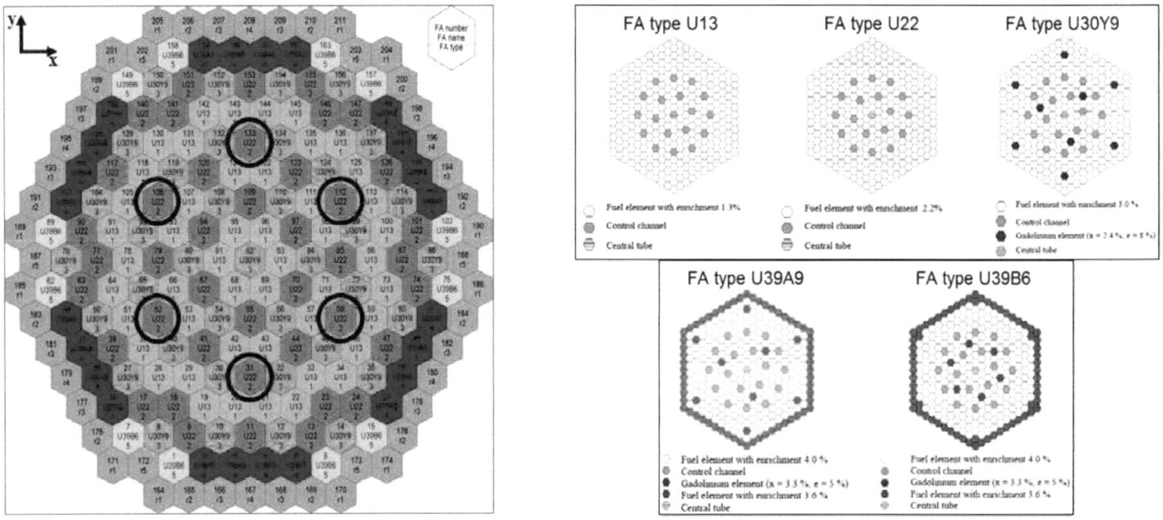

Figure 1: Rostov-2 core layout with the positions of the control rod bank nr. 10 (left) and the details of the different FA-designs of the core loading (right)

The Rostov-2 core mass flow rate is 18001.54 kg/s and the system pressure 15.7 MPa. The core inlet temperature amounts 279.8 °C. The nominal power is 3000 MW_{th}. The MSLB transient initial conditions and scenario is defined in the CAMIVVER-project for code-comparison purposes, [9]. It is based on the Final Specifications of the OECD/NEA V1000-MSLB Benchmark [4] but defined for BOC-conditions. The Kozloduy-6 is a VVER-1000 plant with a thermal power of 3000 MW_{th}, a system pressure of 15.7 MPa, and a core mass flow rate of 17600 kg/s. In Figure 2, the core loading with four different FA-types for BOC-conditions as well as the distribution of the control rod banks are presented [10].

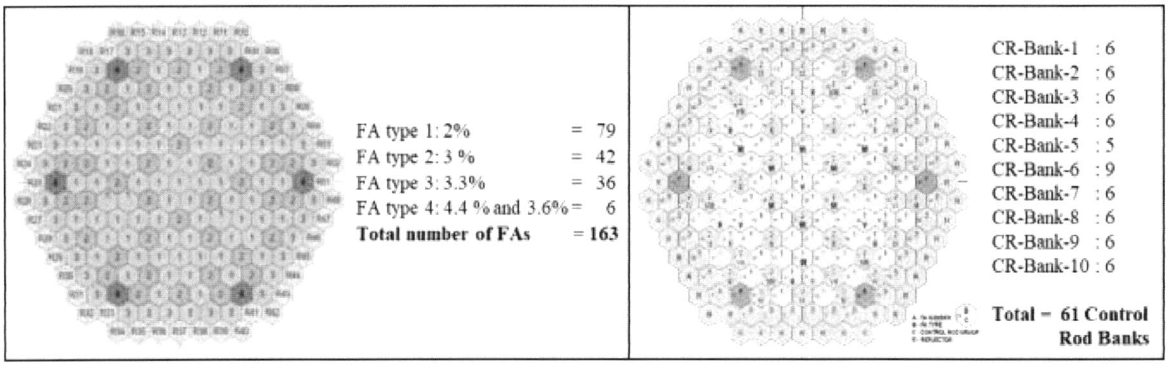

Figure 2: The Kozloduy-6 core layout with the positions of the FA-types (left) and radial distributions of the different control rod banks (right)

The details of the FA-designs are shown in Figure 3, where the different U-235 enrichment of the pins inside each FA-type is provided together with the location of the guide tubes and instrumentation tubes. It can be observed that the FA-type 4 contains fuel pin at the outer rows with a lower U-235 enrichment as the others.

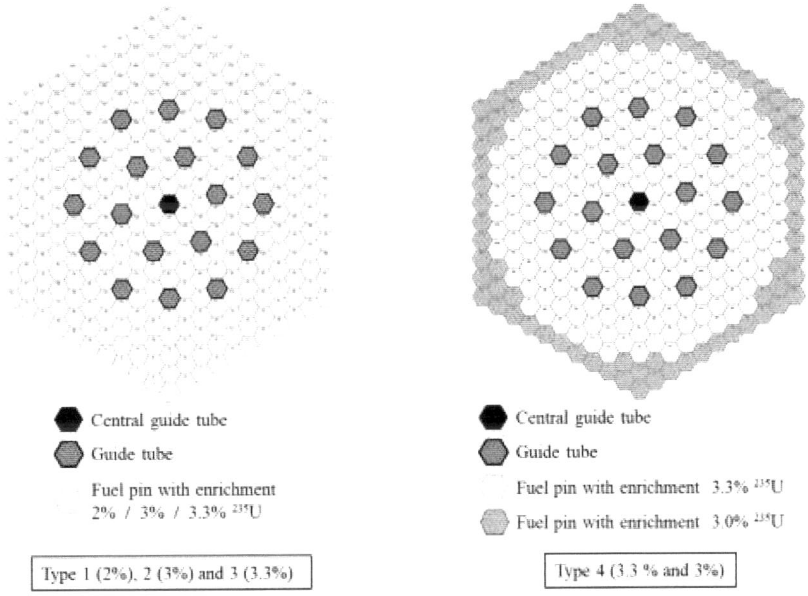

Figure 3: Layout of the two different FA-types of the core loading at the BOC

3. SELECTED TOOLS

KIT is using both in-house and external computer codes for the simulation of the core and plant behavior of Light Water reactors (LWR) including VVERs under design basis accident conditions [1]. The in-house codes under development are the subchannel code SUBCHANFLOW (SCF) [11], the porous-media 3D two-phase flow code TWOPORFLOW (TPF) and the core physics codes PARAFISH and KANECS. PARAFISH is a PN-based transport solver based on finite element for 3D Cartesian geometry and static core analysis [12]. KANECS is the new SPN-solver under development for diffusion and SPN simulation of reactor cores at nodal and pin-level [13], which is being developed for LWR including water-cooled SMRs. The subchannel code SUBCHANFLOW is a very flexible, adaptable, and fast running tool for the analysis of bundles and cores with square, hexagonal and plate-type fuel assemblies [14]. being developed and validated at KIT. It is coupled to both nodal neutron diffusion, SP3 and Monte Carlo codes. It is a single-phase flow code with additional models to predict the void fraction and the lateral flow. SCF is coupled to DYN3D [15], PARCS [16] [17], COBAYA4 [18] DYN3D-SP3 [15], MCNP5 [19] , Serpent2 [20], TRIPOLI [21], GUARDYAN [22] and to TRACE [23]. TWOPORFLOW is a new two-phase flow code that solves the mass, momentum and energy conservation in 3D Cartesian geometry with physical models for a vertical flow map including CHF, developed specifically for BWR but it can be applied to PWRs and VVER [24]. Currently, it is being validated and applied for the analysis of water-cooled SMRs. TPF is coupled with PARCS nodal solver [25]. Work is underway to couple it with KANECS and TRACE for Multiphysics and multiscale analysis of SMRs. In addition, for the plant behavior, KIT is using the codes PARCS [16] and TRACE [26] as well as the ATHLET [27] codes. The US NRC codes TRACE and PARCS are internally coupled for LWR with square and VVER with hexagonal fuel assemblies. For the analysis of PWR with square fuel assemblies, the user needs only to develop the respective input decks and provide few additional information; it works mainly automatically. But for coupled simulation of VVER-cores, the user needs to develop a large mapping file in addition to the input deck, which defined the data exchange of the neutronic with the thermal hydraulic domains so that the feedback parameters are exchanged in a consistent manner. Since the numbering of the fuel assemblies in the core done by PARCS is done according to the black-white logic, it is not trivial and often it is a source for errors. Consequently, KIT developed a ICoCo-based [28] for the coupling between TRACE and PARCS, which bypasses the internal coupling and it does not require additional information in the mapping file since the feedback parameters are stored in the mesh of the neutronic domain and of the thermal-hydraulic domain [1], where more details of this coupling approach are given. Figure 4 shows the coupling scheme between TRACE and PARCS using the modular and flexible ICoCo-approach for code coupling.

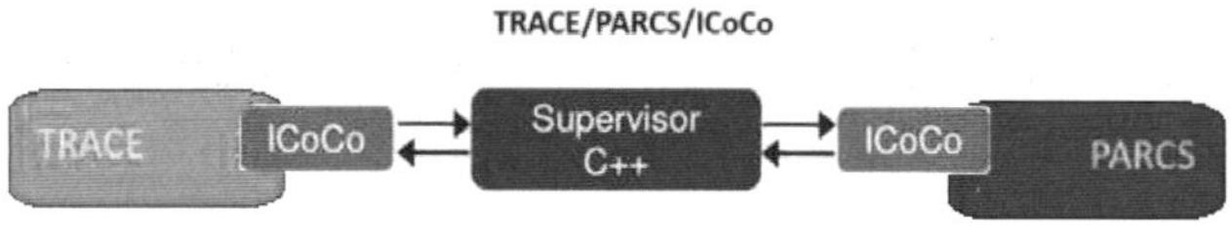

Figure 4: Scheme of the ICoCo-based coupling of TRACE with PARCS

4. ANALYSIS OF THE BORON DILUTION TRANSIENT

4.1 Short description of the boron dilution transient

A boron dilution transient experiment at the Rostov-2 VVER-1000 plant was defined in the OECD/NEA Benchmark [6] that consists on a reactivity compensation of boron dilution by stepwise insertion of control rod cluster into the core. It consists of different parts where part one is devoted to the analysis of the core depletion starting from 0 to 36.37 EFPDs (avg 1.45 MWd/kgU) using different numerical tools to predict the core conditions at the beginning of the boron dilution transient. The second part is devoted to the analysis of the boron dilution transient experiment. For both phases, experimental data is available for code validation purposes. The transient experiment was conducted when the plant was operated at 36.37 EFPDs with a thermal power of 69.37 % of the nominal power i.e., 2081 MWth. At the beginning of the transient the core is critical with all control rod banks extracted except the bank 10, which is 84.43 % extracted. The initial boron concentration of 4.6 g/kg gradually decreased to approximately 4.3 g/kg by the end of the transient. Throughout the transient, control rod group 10 was incrementally inserted into the core in steps of approximately 14 cm every 2-3 minutes in order to compensate the reactivity, starting from a position with 84.43% extraction and ending at a position with 20.64% extraction from the active core bottom.

The transient experiment began at 11:14:41 on 31 May 2010. The measured data for the core conditions is available from 11:00:00 to 12:59:59 on the date which the experiment was conducted. Therefore, analysis focused on from 11:00:00 to end of the first sixty minutes of the experiment based on benchmark.

4.2 Neutronic and thermal hydraulic core models

The KIT analysis of the boron dilution transient is focus on the core behavior considering thermal-hydraulic boundary conditions at the core inlet and outlet. For this purpose, a thermal-hydraulic core model defined at fuel assembly level for SCF and a 3D nodal neutronic model for PARCS are developed. Latter requires a set of condensed and homogenized nodal cross sections parameterized considering the state parameters and the burnup distribution at the beginning of the transient. These models are shortly described in the next subchapters.

4.2.1 SUBCHANFLOW model of the Rostov-2 core

The Rostov-2 core is represented in SCF-V371 by 163 parallel channels, where cross flow is considered between the adjacent channels, Figure 5 (left), in addition, fuel and coolant feedback mesh for SCF and PARCS is shown in Figure 5 (right). Axially 30 nodes with the same height are defined. In the center of each channel, a representative fuel rod is considered that represent the 312 pins of one FA. The pellet and cladding are radially discretized by 7 and two nodes. VVER-specific thermophysical properties of the fuel and cladding were used in SCF as defined in benchmark specifications including the Doppler temperature definition. According to it, a gap heat transfer coefficient of 3624 W/m^2K was used. The 12 spacer grids distributed along the core height are considered with the respective form loss coefficients, only. A lateral mixing coefficient of 0.0015 and a cross flow resistance coefficient of 5 are considered in the simulations.

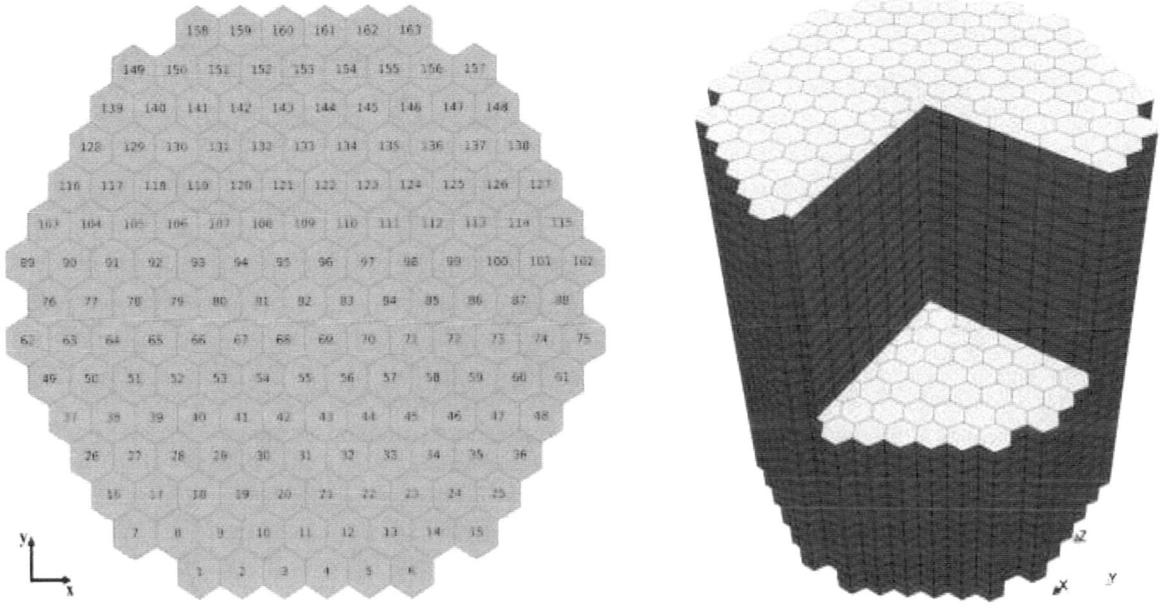

Figure 5: SUBCHANFLOW representation of the 163 thermal hydraulic channels of the core (left) and SUBCHANFLOW fuel and coolant feedback mesh for the core (right)

4.2.2 PARCS model of the Rostov-2 core

The core model consists of 211 radial nodes, 163 nodes represent the FAs and 48 nodes represent the radial reflector assemblies, Figure 6 (1 to 5 are FA-types and 10 to 14 are reflector elements). Moreover, there are 5 different bottom and 5 different top reflectors for the FA-types to develop more realistic core model. In total 32 axial nodes are considered in PARCS-V331, where the lowest and uppermost nodes represent the axial reflectors with a height of 0.60 m each and the remaining 30 nodes represent the active core with 0.12333 m node height.

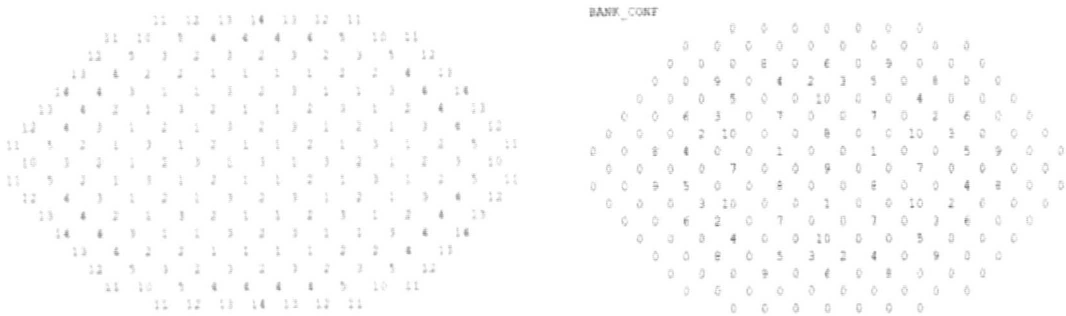

Figure 6: PARCS radial distribution of the five FA-types and 10 to 14 radial reflectors (left) and the distribution of the 10 control rod banks in the core

At the core outer radial and top/bottom axial reflector, zero flux boundary conditions are used. The higher order nodal coupling kernel, named TPEN, together with the nonlinear CMFD-solver is used to solve the neutron diffusion for a core with hexagonal FAs. ADF for the core and reflectors are considered in the XS-generation process.

Since all FA-types have the same material composition along the core height, there are five material sets for the core, five for the radial, and top and bottom reflector (in total 20 XS data sets). For each FA material set, branch calculations using CASMO5 were performed in 2D hexagonal fuel assembly model considering following burnup steps: 0.0, 0.1, 0.5, 1.0, 1.5, and 2.0 MWd/kgU. The considered branch states and reference state are given in Table 1. Moreover, 1D simplified slab models were developed to generate reflector assemblies while CASMO5 S5C card was employed to state case matrix for default branch calculations for reflectors.

Table 1. Branch states and reference state for FAs cross sections generation using CASMO5

	Reference State	Branch States				
Control Rod Position	OUT	OUT	IN-CR1	IN-CR2	IN-CR3	-
Fuel Temperature (K)	850	400	553	650	873	1273
Coolant temperature (K)	550	400	500	553	573	603
Boron concentration (ppm)	1300	0.1	733	1000	2100	-

In total 2406 branch calculations are performed with CASMO5 using the ENDF/B-VII.r1 nuclear data library for each fuel assembly type characterized by the same axial material composition to generate all needed homogenized and condensed nodal cross sections for the PARCS core simulations. In this approach, the nodal macroscopic cross sections for each material set were collapsed in two energy group (thermal < 0.625 eV, fast > 0.625 eV) and homogenized over the whole fuel assembly. The generation of core and reflector ADF was included in this process. Finally, the nodal condensed XS-sets were converted to PMAXS-format using the GenPMAXS-tool obtaining five PMAXS-files for the core and fifteen for the top, bottom and axial reflectors. The generated PMAXS-files are directly processed by PARCS during the static or transient simulation with the actual thermal-hydraulic parameters and control rod positions before to update the cross-section values before the nodal diffusion equations are solved. Then the 3D power profile is transferred to the thermal-hydraulic solver (SCF) using the mesh-to-mesh superposition approach of the ICoCo-method which is based on the MEDCoupling library.

4.3 Discussion of selected results from the analysis of the boron dilution transient experiment

Using the models of SCF and PARCS described before, the boron dilution transient experiment was analyzed with the coupled code PARCS/SCF/ICoCo. In Figure 7, 3D Power distribution (left) in the core as well as the axially averaged 2D fuel assembly power (right) at the beginning of the experiment is show. The radial power distribution is determined by the different enrichment of the fuel assemblies located within the core.

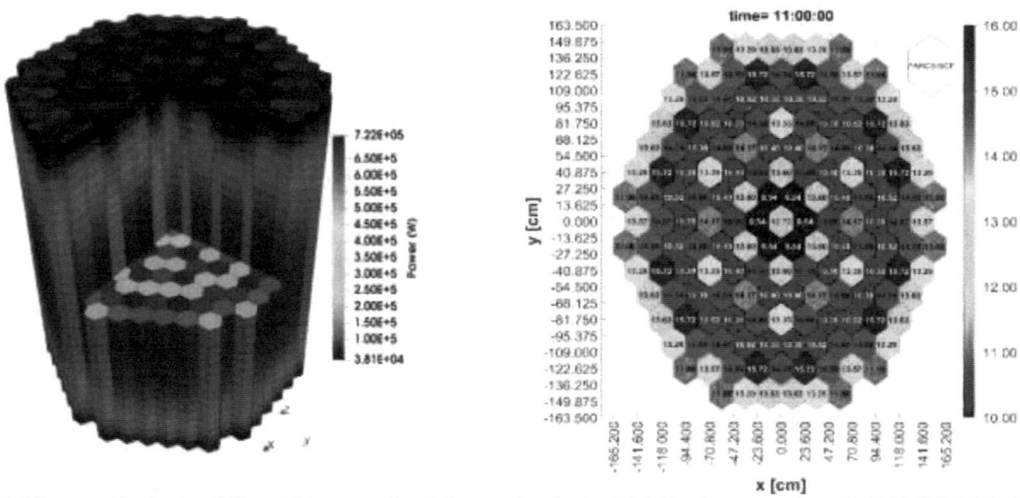

Figure 7: 3D power distribution (left) and FA averaged radial power distribution (right) in the core predicted by PARCS/SCF for the time just before the transient

Figure 8 shows the comparison of the coolant temperature predicted by PARCS/SCF/ICoCo with the measured coolant temperature at the core outlet of 91 fuel assemblies at 10 seconds after the beginning of the transient. These results show that the couple code is able to predict the local core outlet temperature in good agreement with the measured data.

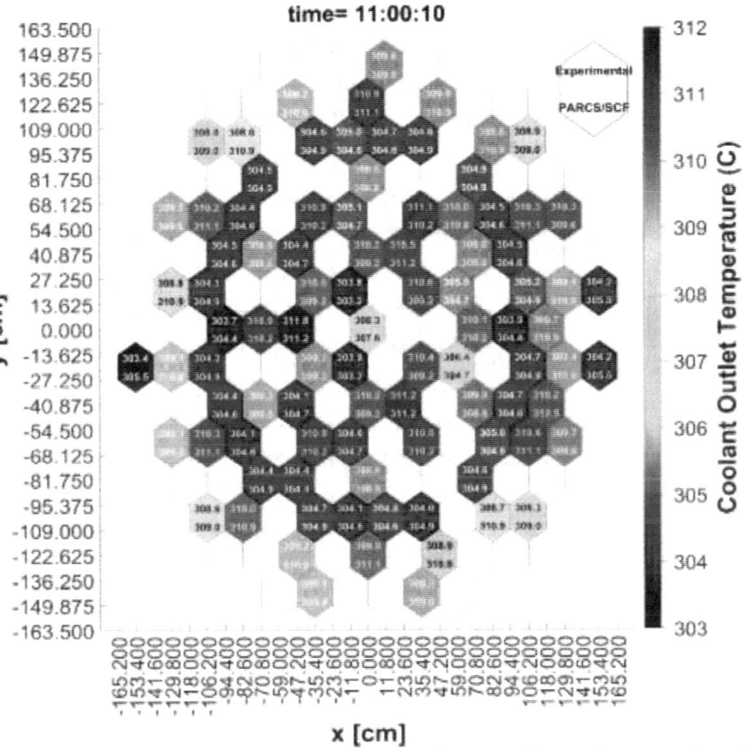

Figure 8: 2D coolant temperature distributions predicted by PARCS/SCF and measured during the tests at the outlet of selected FA-outlet

5. ANALYSIS OF THE MAIN STEAM LINE BREAK (MSLB) ACCIDENT

5.1 Short description of the MSLB-scenario at BOC

Based on the VVER-1000 coolant transient benchmark phase 2 devoted to the analysis of a MSLB [4], a slightly modified MSLB-scenario was defined in the H2020 CAMIVVER project [7] for BOC-conditions at Hot Full Power (HFP)-conditions to facilitate the cross section generation and avoid other source of discrepancies between the approach based on Serpent2 and the one based on deterministic lattice physics codes such as APOLLO3®, CASMO5, etc. to be used by the core simulators PARCS and APOLLO3® for the code-to-code comparison. It is assumed that the MSLB- transient occurs at HFP and BOC-conditions at the Kozloduy-6 plant initiated by a large break (size 580 mm diameter), located between the steam generator and the steam isolation valve, in the steam line of the loop-4, [9]. For simplification, SCRAM takes place just after the break initiation (normally SCRAM is caused either by low secondary side pressure, low primary circuit pressure or high thermal power) with the most reactive peripheral control rod -located close to the core sector with the highest overcooling- remain stuck out. Afterwards, many actions take place such as the closure of the turbine stop valves (MSIV), opening of the Turbine bypass to condenser (BRU-Ks) and switching to MSH-pressure control mode after closing MSIV, opening of all four BRU-K valves when the MSH-pressure is higher than 6.67 MPa, closing of the BRU-K valves when pressure reaches 5.79 MPa and re-opening if MSH-pressure achieve values higher than 6.67 MPa. No Makeup and Let down systems are operational during the transient phase. The PRZ-heaters are switched on after depressurization for some period until the primary pressure starts of increase due to the dry-out of SG-secondary side of loop-4. Finally, the MCP of the affected loop-4 is coast down within 55 s and the feed water valves fails remaining open. More details are given in [9].

5.2 Thermohydraulic and neutronic models for the MSLB-analysis

The KIT analysis of the MSLB-transient was performed using the coupled code TRACE/PARCS both internal and ICoCo-based coupling. For this purpose, the respective models were developed and will be shortly described hereafter.

5.2.1 PARCS model of the Kozloduy-6 core

The BOC-core of the Kozloduy-6 plant is represented in PARCS by 211 hexagonal nodes, where 163 of them are FAs and 48 are reflector assemblies. In Figure 9 and Figure 10, the distribution of the four FA-types (1 to 4) including the radial reflector assemblies (Nr. 5 to 52) and the distribution of the ten control rod banks for the BOC-core loading is exhibited. Axially, the core is subdivided in 30 axial levels and one bottom and upper level for the respective reflectors [29]. Additionally, the core is surrounded by 48 reflectors with the same hexagonal FA shape. The distinction between each reflector is considered to evaluate the influence of surface discontinuity factors. Zero flux boundary conditions were used at the radial and axial reflector borders. For the reactivity control of the core, ten control rod banks are distributed in the core, as illustrated in Figure 10. Control rod banks are only associated with FAs types 1 and 2. Axially the active core is 355 cm long and divided into 30 axial slices, and extra widths of 23.6 cm each are considered for the bottom (nr.53) and top (Nr.54) axial reflectors. The critical control rod position is defined as follows: CR-Banks 1 to 9: 100 % and CR-Bank-10: 80 % withdrawn, respectively. The critical boron concentration amounts 1630 ppm.

Figure 9: PARCS Kozloduy-6 core model developed in PARCS

Figure 10: PARCS Kozloduy-6 Control Rod Banks distribution in the core

For the solution of the neutron diffusion equation taking into account the thermal-hydraulic feedbacks and the control rod position, PARCS needed condensed and homogenized cross sections sets for all FA-types and reflector elements in 2-energy groups. In [30], the cross-section generation approach followed at KIT for the BOC-core loading of the Kozloduy-6 for the analysis of the MSLB-transient using TRACE/PARCS is described. The Serpent2 code was used for this purpose. Hence, detailed Serpent models for each FA-type and for the reflectors were developed. In order to cover the expected range of variation of core thermal hydraulic parameters in the cross-section generation process 9 values for the coolant density, 3 for the fuel temperature, 3 for the coolant temperature, and 3 for the boron concentration were considered. Considering these parameter range, a large number of branch calculations were performed for each FA-type which are characterized by the same axial material composition (material set) [29]. In

Figure 11, the different Serpent2 models developed for the generation of nodal cross sections for PARCS are shown. These models consider the position of each FA-type inside the core to properly generate the cross sections groups. Hence, for FA-type 1, a single assembly with reflective boundary conditions is considered while for FA-type 2, six FA-types are included in the model. Doing so, the homogenization and condensation of cross-sections are done only in the center where FA-type 2 is located. Similar approaches are considered for FA-type 3 and 4 as indicated in Figure 11. For the prediction of the 2-energy group homogenized cross-sections, a Serpent2 simulation using the leakage-corrected spectrum and the Fundamental mode (FM) with Cumulative Migration Method (CMM) diffusion coefficients was considered [31]. In addition, an intermediate 70-energy group structure was selected for the calculation of the leakage-corrected critical spectrum. The radial reflector cross sections are extracted considering a full 2D-core model developed for Serpent2. In this approach, each of the 48 radial reflectors is homogenized independently, and discontinuity factors are considered only in the surfaces in contact with the fuel assemblies. For the bottom/top axial reflectors, a 3D FA-type 1 infinite reflected in the X and Y directions, and black boundary conditions in the Z direction, was assumed. Then, the condensed and homogenized nodal cross sections are converted to the PMAXS-format as required in PARCS using the GenPMAXS v6.2-tool.

FA-type 1 FA-type 2 FA-type 3 FA-type 4

Figure 11: Serpent models for the generation of nodal cross sections for the different FA-types to use in PARCS 3D core analysis

With this neutron physical data, the critical boron concentration for the given control rod position was estimated using PARCS with internal thermal-hydraulic module. A boron concentration of 1630 ppm was predicted with the control rod position, where CRB 1-9 were 100 % and the CRB-10 was 80% extracted. This 3D core model developed for PARCS is used in the coupled TRACE/PARCS simulation of the MSLB-transient.

5.2.2 TRACE integral model of the Kozloduy-6 plants

A detailed model of the Kozloduy-6 plant was updated for the CAMIVVER-project based on RELAP5-models used for the analysis of previous VVER-1000 coolant transient benchmarks [32]. The overall TRACE model includes the 3D model of the Reactor Pressure Vessel (RPV), 1D models of the primary and secondary circuits, where all four loops are represented, and all safety systems expected to be activated during the MSLB-transient according to the plant specifications or assumptions made. The Main Coolant Pumps (MCP) characteristics provided in the benchmark specifications are considered as well as the Pressurizer and Steam Generator (SG). The secondary side includes a detailed modelling of the steam lines with the common header and the respective connections, valves to control the pressure, etc. while the feedwater lines are considered in a simplified manner to provide the mass flow rate and temperature of the coolant. The secondary pressure is fixed at the turbine, which is modelled in a simplified manner with a BREAK-component. Special attention is paid to model the different valves at the steam lines e.g., one steam dump valve connected with the containment (BRU-A), two safety valves (SV), one main isolation valve (BZOK), and a check valve (CHV). The steam dump valves (BRU-K) connect the Steam Header with the condenser. In addition, the Steam Header is connected with the atmosphere by the steam dump valves (BRU-SN). The horizontal SG-tubes are modelled by three representative groups at different elevations, they are connected with respective volumes of the SG-secondary side. At the upper part of the SG, a separator is considered and represented by SEPARATR-component of TRACE. There, the steam is separated from the entrained droplets and the liquid fraction flows back to the downcomer part of the SG. The 3D cylindrical RPV-model is subdivided in 26 axial nodes, 6 azimuthal sectors and 4 radial sectors. The core is represented by three rings, the next ones represent the bypass, core barrel and downcomer. It allows to consider the location of the lower and the upper grid plate as well as the upper head. The core is modelled by 18 HTSTR-components representing 50856 fuel rods contained in the 163 hexagonal fuel assemblies. Each HTSTR must be axially and radially discretized so that the heat conduction equation can be solved for this discretization considering the heat source in the pellet material. The POWER component defines the HEATSTR connected to the power source, which can be a table, predicted by the Point Kinetic model or provided by a 3D core solver such as PARCS. In this case, care must be taken when developing the TRACE model for coupled simulations with PARCS so that the feedback parameters are exchanged in a consistent manner for the core and reflector region between the thermal-hydraulic and neutronic core models. More details of the TRACE-plant model can be found in CAMIVVER-reports [33], [30]. In Figure 12, the integral TRACE plant model developed for the study of the MSLB transient is shown. There, one can find the main components of the NPP such as the RPV, the pressurizer (PZR), the steam generators (SG1 to SG4), the break modelling, the main steam header (MSH) on the secondary side, the turbine, etc.

Figure 12: Integral TRACE-model of VVER-1000 at the Kozloduy-6 site including the primary/secondary circuits, and safety systems

First of all, the stand-alone TRACE plant model is checked to be able to predict the nominal plant conditions at the time just before the MSLB is predicted with acceptable accuracy. Then, a coupled steady-state simulation using the original coupled code TRACE/PARCS or the ICoCo-based coupling scheme is performed to confirm that the nodal cross-sections are correctly generated. If this is the case, input decks for TRACE and PARCS transient scenario are developed. This TRACE-transient input (restart option) is quite small compared to the base input deck; it only includes the components, trips, control variables that have changed compared to the steady-state simulations. Then, a coupled transient simulation with TRACE/PARCS is performed, and the obtained results are discussed.

5.3 Discussion of selected results

Figure 13 shows the secondary coolant outflow through the break on the steam line of the loop-4 (initiating event) and the pressure decrease of the SG-secondary side loop-4 and loop-1 as predicted by the coupled code TRACE/PARCS. This intensive flashing and evaporation of the secondary coolant of SG-4 is caused by the pressure decrease and it leads to the enhanced heat transfer from the primary to the secondary side over the broken loop-4.

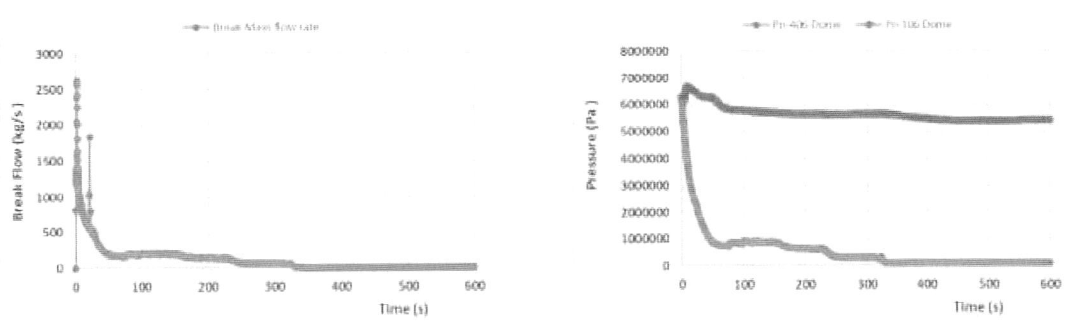

Figure 13: TRACE/PARCS prediction of the outflow through the break on the SG-4 secondary side (left) and the following secondary pressure decrease on the SG-4 dome (broken loop) compared with the intact loop SG-1

Consequently, the sector of the primary coolant in the downcomer connected to the secondary loop-4 is cooled down largely compared to the one of the other sectors until the water inventory of the SG-4 secondary side completed evaporated. During the evaporation of the water inventory of the broken SG-4 secondary side, a rapid decrease of the primary coolant temperature of the sector-4 of the downcomer by around 50 K due to the improved heat removed over the broken SG-4. Due to the mixing on the downcomer between the sector-4 with the sector-1, its coolant temperature is also reduced at the beginning moderately and later on following the trend of one of the sector-4. As assumed in this exercise in [31], the SCRAM is activated just after the break opening. Consequently, the reactor is shutdown. Figure 14 shows the respective evolution of the total reactivity (left) and the thermal power (right). In this overcooling transient, the cooldown of the primary coolant and of the fuel temperature especially in the sector-4 of the downcomer and the fuel assemblies, a positive reactivity is introduced into the core despite the core is shutdown (see left figure). But this reactivity is far away from the criticality meaning that there is no risk for re-criticality. In addition, the moderate reactivity increase during the transient does not lead to a significant increase of the power i.e., no return to power is predicted. More details are given in [31].

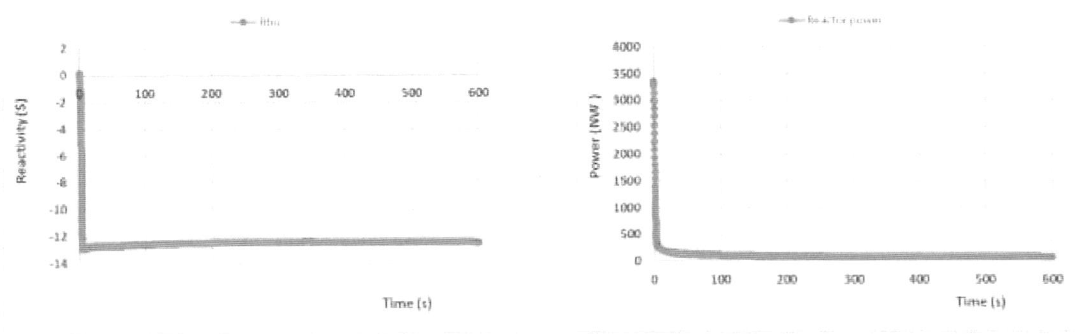

Figure 14: TRACE/PARCS prediction of evolution of the total reactivity (left) and thermal power (right) after SCRAM

6. CONCLUSIONS AND OUTLOOK

VVER-1000 and -1200 are operated and under construction in many countries. The peculiarities of the core loadings and the plant in general are challenging for the simulation tools. Hence, KIT is participating in international benchmark and projects devoted to the core/plant transient analysis with different code coupling options, where inhouse thermal hydraulic codes such as SUBCHANFLOW are being coupled to core simulations. The results of the investigations discussed here have shown the ability of the used coupled codes for the simulation of VVER-1000 transients. These investigations emphasized the role of the methods and tools for the generation of homogenized cross-sections for the 3D nodal diffusion solvers for accurate prediction of the core behavior under transients including important safety parameters. The application of the inhouse core simulator KANECS is being tested for the analysis of hexagonal cores coupled with SUBCHANFLOW based on ICoCo-approach. Further applications of the coupled codes are foreseen for VVER-1200 reactors.

ACKNOWLEDGMENT

This work has received partly funding from the H2020 CAMIVVER project, grant agreement No 945081, and from the NUSAFE program of the HGF at the Karlsruhe Institute of Technology (KIT).

REFERENCES

[1] V. Sanchez-Espinoza, F. Gabrielli, U. Imke, K. Zhang, L. Mercatali, G. Huaccho, J. Duran, A. Campos, A. Stakhanova, O. Murat, A. Mercan, J. Etcheto, Steinbrück, J. Stuckert, S. Gabriel, S.Ottenburger and a. W. T. R. Stieglit, "KIT reactor safety research for LWRs: Research lines, numerical tools, and prospects," *Nuclear Engineering and Design,* 112573 414 2023.

[2] B. Ivanov, K. Ivanov, P. Groudev, M. Pavlova and a. V.Hadjiev, "VVER-1000 Coolant Transient Benchmark Volume I: Specifications of the MCP Switching on Problem, Revision 1," NEA/OECD NEA/NSC/DOC, Paris, 2003.

[3] N. Kolev, S. Aniel, E. Royer, a. U. Bieder, D. Popov and a. T. Topalov, "VVER-1000 COOLANT TRANSIENT BENCHMARK (V1000CT), Volume II: Specifications of the VVER-1000 Vessel Mixing Problems," OECD/NEA, Paris, 2004.

[4] N. Kolev, N. Petrov, J. Donov, D. Angelova, S. Aniel, . E. Royer, K. Ivanov, E. Lukanov, Y. Dinkov, D. Popov and S. Nikonov, " VVER-1000 Coolant Transient Benchmark – Phase 2 (V1000CT-2). Vol. II: MSLB Problem – Final Specifications.," NEA/NSC/DOC(2006)6, April, 2006.

[5] K. Ivanov, V. Tereshonok, S. Nikonov, M. Lizorkin, K. Velkov and A. Pautz, "KALININ-3 coolant transient benchmark: Switching off of one of the four operating main circulation pumps at nominal reactor power," OECD report no.NEA-1848/04 zz-KALININ3, Paris, France, 2008.

[6] M. Avramova, K. Ivanov, K. Velkov, S. Nikonov, P. Gordienko, B. Shumskiy and a. O. Kavun, "Benchmark on reactivity compensation of boron dilution by stepwise insertion of control rod cluster into the VVER-1000 core. Specifications and Support Data. V1.5," NEA/EGMPEBV/DOC(2020), Paris, 2020.

[7] D. VERRIER and e. al., "Codes and methods improvements for VVER comprehensive safety assessment: the CAMIVVER H2020 Project," in *Proc. of the 28th International Conference on Nuclear Engineering (ICONE-2021).*, Online, 2021.

[8] M. Garcia, R. Vocka, R. Tuominen, A. Gommlich, J. Leppänen, V. Valtavirta, U. Imke, D. Ferraro, P. V. Uffelen, L. Milisdörferf and V. Sanchez-Espinoza, "Validation of Serpent-SUBCHANFLOW-TRANSURANUS pin-by-pin burnup calculations using experimental data from the Temelin II VVER-1000 reactor," *Nuclear Engineering and Technology.,* pp. 3133-3150, 10 53 2021.

[9] A. Stefanova, N. Zaharieva, P. Vryashkova and P. Groudev,, "The CAMIVVER Definition report with the specifications of rht NPP with VVER_1000 reactor with respect to selected transients," CAMIVVER, Paris, 2020.

[10] B. Calgaro and G. Huaccho, "Description of the core reference test cases - Part 1 + Part 2, CAMIVVER, D5.2," CAMIVVER Project, Paris, 2023.

[11] U. Imke and V. Sanchez, "Validation of the Subchannel Code SUBCHANFLOW Using the NUPEC PWR Tests (PSBT)," *Science and Technology of Nuclear Installations,* vol. ID 465059, p. 12 pp, 2012.

[12] J. Duran-Gonzalez et al, "Preliminary results for the C5G7-2D Benchmark using the PARAFISH code," in *XXXIII Congreso Anual de la Sociedad Nuclear Mexicana*, Veracruz, 2022.

[13] J. D. Gonzales, *Development of an SPN solver for Nuclear Reactor Analysis (KANECS),* Karlsruhe: KIT INR-Seminar , 15. July 2024.

[14] J. Almachi, V. Sánchez-Espinoza and U. Imke, " Extension and validation of the SubChanFlow code for the thermo-hydraulic analysis of MTR cores with plate-type fuel assemblies," *Nuclear Engineering and Design,* vol. 379, p. 111221, 2021.

[15] M. Daeubler, N. Trost, J. Jimenez, V. Sanchez, R. Stieglitz and R. Macian-Juan, "Static and transient pin-by-pin simulations of a full PWR core with the extended coupled code system DYNSUB," *Annals of Nuclear Energy,* pp. 31-44, October 2015.

[16] T. Downar, A. Ward, Y. Xu and V. Seker, "PARCS, Volume II: User's Guide. NRC - v3.3.0 Release," 2018.

[17] J. Basualdo and V. Sanchez, "PARCS-SUBCHANFLOW-TRANSURANUS multiphysics coupling for improved PWR's simulations," in *ICAPP 2017: International Congress on Advances in Nuclear Power Plants*, Fukui and Kyoto, Japan,

2017.

[18] M. Calleja, V. Sanchez and U. Imke, "Implementation of SUBCHANFLOW in the SALOME platform and coupling with the reactor dynamic code COBAYA3," Berlin, Germany, 2011.

[19] A. Ivanov, V. Sanchez, R. Stieglitz and a. K. Ivanov, "Large-scale Monte Carlo neutron transport calculations with thermal hydraulic feedback," *Annals of Nuclear Energy,* p. 204–219, 84 2015.

[20] D. Ferraro, "Monte Carlo-based multi-physics analysis for transients in Light Water Reactors," KIT. Dissertation, Karlsruhe, 2021.

[21] M. Faucher, "Coupling between Monte Carlo neutron trans-port and thermal-hydraulics for the simulation of tran-sients due to reactivity insertions," Doctoral Thesis of the Univeristy Paris-Saclay, Paris, 2019.

[22] D. Legrady, M. Pukler, B. Panka and G. Tolnai, "Coupling the GUARDYAN code to subchanflow," *EPJ Nuclear Science and Technologies,* February 2023.

[23] K. Zhang, A. C. Muñoz and Sanchez-Espinoza, " Development and Verification of the Coupled Thermal–Hydraulic Code TRACE/SCF based on the ICoCo Interface and the SALOME Platform.," *Ann. Nucl. Energy,* !%% 155 2021.

[24] V. Jauregui-Chavez, "Boiling Water Reactor Core Analysis by means of an Improved Porous Media Two-phase Flow Approach. Dissertation," KIT, Karlsruhe, 2020.

[25] A. Campos-Muñoz, V. Sanchez-Espinoza, E. Redondo-Valero and C. Queral, "Validation of Neutronic and Thermal-hydraulic Multi-physics Calculations for SMRs Rod Ejection Accident with PARCS/TWOPORFLOW," in *International Conference on Mathematics and Computational Methods Applied to Nuclear Science and Engineering*, Niagara Falls, Can, 2023.

[26] "U. NRC, TRACE V5.1051 Theory Manual, Washington, DC: Division of Safety Analysis Office of Nuclear Regulatory Research, U. S. Nuclear Regulatory Commission," 2016.

[27] P. Shoeffel, A. D. Nora, M. J. D. Eckert, D. v. d. Cron, F.-. Weyermann and A. Wielenberg, "ATHLET 3.4.0 User Manual," GRS, Garching, 2023.

[28] "MEDCoupling Developer's Guide," SALOME-platform, [Online],Available: http://docs.salome-platform.org/latest/dev/MEDCoupling/developer/index.html. [Accessed 04 02 2019]..

[29] B. Calgaro and G. Huaccho, "Description of the core reference test cases - Part 1 + Part 2," CAMIVVER Deliverable 5.2, 2023.

[30] V. Sanchez-Espinoza, K. Z. G. Huaccho, B. Calgaro, B. Vezzoni, A. Mas, T. Quenbouille, O. Bernard, A. Stefanova, N. Z. P. Vryashkova, A. H. P. Groudev and Y. Onyshchuk, "D7.4 Results of tthe Transient 2 Benchmark: Main Steam Line Break Analysis," KIT, Karlsruhe, 2023.

[31] V. H. Sanchez, G. Huaccho, K. Zhang, B. Calgaro, B. Vezzoni, A. Mas, T. Quenouille, O. Bernard, A. Stefanova, P. Vryashkova, N. Zaharieva, P. Groudev, A. Hashymov and Y. Onyshchuk, "Results of Transient 2 Benchmark: Main Steam Line Break Analysis, CAMIVVER. D7.4," CAMIVVER project, Paris, 2023.

[32] V. Sánchez-Espinoza and M. Böttcher, "Investigations of the VVER-1000 coolant transient benchmark phase - with the coupled system code RELAP5/PARCS," *Progress in Nuclear Energy 48,* vol. 48, 2006.

[33] KNPP, "Data Base for VVER1000: Safety Analysis Capability Improvement of KNPP (SACI of KNPP) in the field of Thermal Hydraulic Analysis," INRNE, Sofia, 2015.

NURER 2024
Nuclear and Renewable Energy Resources

Numerical simulation of underground hydrogen storage in a depleted gas reservoir: Effect of well configurations

Emrah Sarı
Graduate School of Natural and Applied Sciences, Gazi University, Ankara, Türkiye, emrah.sari@yahoo.com.tr
ORCID: 0009-0000-9228-4095

Erdem Çiftçi
Energy Systems Engineering, Faculty of Technology, Gazi University, Ankara, Türkiye, erdemciftci@gazi.edu.tr
ORCID: 0000-0003-2493-5962

Cite this paper as:	*Sarı, E, Çiftçi, E. Numerical simulation of underground hydrogen storage in a depleted gas reservoir: Effect of well configurations. 7th Int. Conf. on Nucl. and Ren. Energy Res. 27-30 October 2024, Antaya, Turkey*

Abstract: Producing hydrogen from excess renewable energy and subsequently converting it back to electricity when required is regarded as a viable approach. This approach also necessitates large-scale hydrogen storage. Depleted gas reservoirs are used successfully for underground natural gas storage, and these reservoirs can store large amounts of gas. However, the amount of gas that can be stored in and withdrawn from the reservoir in a given time period depends on the injection and withdrawal capacities of the well or wells. In addition, the purity of the withdrawn hydrogen emerges as a feature that needs to be taken into consideration. In this regard, a sample reservoir model was generated numerically, and seasonal hydrogen storage simulations were run to determine the effect of well perforation levels and skin factor values on underground hydrogen storage (UHS). According to simulation results, the negative skin factor is advantageous for UHS as it enhances injection and withdrawal flow rates, and the location of perforation levels, whether in the upper or lower region of the reservoir, does not significantly affect the injection and withdrawal flow rates or the purity of hydrogen.

Keywords: Underground hydrogen storage, numerical reservoir simulation, seasonal hydrogen storage

Nomenclature	
BHP	Bottom hole pressure
GIP	Gas in place
Inj.	Injection
km^2	square kilometer
m	meter
mD	milli Darcy
MRST	MATLAB Reservoir Simulation Toolbox
p	Pressure
Prod.	Production (withdrawal)
UHS	Underground hydrogen storage
Z	Gas deviation factor (Z-factor)

1. INTRODUCTION

The generation of electricity from renewable energy sources, such as solar and wind, is intermittent due to its significant dependence on atmospheric conditions. Thus, a storage medium is required to resolve this adverse situation that arises because of the nature of renewable energy sources and to ensure the balance between supply and demand [1–6]. Hydrogen is seen as a viable solution for integrating renewable energy sources into the power system and considered one of the few choices for long-term electricity storage throughout several days, weeks, or months [6, 7].

Apart from its contribution to the utilization of renewable energy sources, hydrogen can also be used by blending it with natural gas [8, 9]. Within the framework of the hydrogen plan initiated by the European

Commission in 2020, many European Union countries have established specific goals for the blending hydrogen with natural gas. The implementation of hydrogen injection into the gas network, up to a volume of around %5 - 10%, is projected to be feasible in the near future without requiring significant modifications to the transmission infrastructure and end-user installations [9]. The Ministry of Energy and Natural Resources of Türkiye has published a report about the national energy plan in 2022, and the report states that natural gas used in final sectors needed to be mixed with other clean fuels such as hydrogen and synthetic methane to reduce emissions, and the hydrogen ratio in the gas mixture was determined as 3.5% for 2035 [10]. In addition, the "Türkiye Hydrogen Technologies Strategy and Roadmap" has been published by the same Ministry, and the report states that hydrogen was declared as one of the priority areas due to its contribution to the sustainable energy future [11]. The report also emphasized that it is important to determine policies such as creating incentive mechanisms and certification programs for the use of domestic components in green hydrogen production and storage, establishing international collaborations, establishing public and private sector collaborations to encourage commercial demand and investments, training qualified human resources, and encouraging the widespread use of green hydrogen in all sectors, and that hydrogen energy and hydrogen technologies are an important necessity for Türkiye in line with the 2053 net zero emission target [11].

Hydrogen is seen as an important tool to prevent global warming and achieve net zero carbon targets both nationally and globally [11, 12]. However, since hydrogen is highly compressible, large-scale storage media are needed for hydrogen to be successfully used in the energy sector. There are different alternatives for storing hydrogen both underground and above ground, and storing hydrogen in underground geological formations is considered a solution for large-scale storage [1, 2, 13, 14]. Considering the targets for hydrogen usage in Türkiye's energy sector, it is expected that large-scale storage media will be needed for hydrogen storage in the near future. On the other hand, there are many shallow and medium-depth natural gas reservoirs in the Thrace region of Türkiye, and these natural gas reservoirs are thought to be potential solutions for large-scale hydrogen storage. Because there is a limited amount of time available for both injection and withdrawal operations, injection and withdrawal capacities are crucial when utilizing natural gas fields for seasonal gas storage. The numbers as well as the properties of the wells in the field determine both injection and withdrawal capacities. One of the notable aspects of wells is the perforation intervals, which denote the sections of the wells that permit fluid flow at the reservoir level.

Because contamination of the formation on the well wall and a reduction in permeability are commonly seen during drilling or well completion activities in natural gas wells, another notable characteristic that can substantially influence the flow rate in wells is the skin factor values of the wells. The impacted region may extend from several centimeters to several meters from the well. Conversely, on some wells, the permeability value in the vicinity of wells can be enhanced using various well stimulation techniques, including acidizing or fracturing. Therefore, the permeability surrounding the well generally differs from the permeability values in remote locations unaffected by well operations. The disparity in permeability surrounding the well is represented by the skin factor [15]. The skin factor value of a well can influence its flow rate and, consequently, the storage potential of the reservoir, either positively or negatively. This work aims to examine the impact of well numbers, perforation intervals, and skin factor values of wells on the hydrogen storage potential of a depleted natural gas reservoir with a depth of 1000 meters and relatively low capacity. The study includes several seasonal storage simulations using different well numbers and configurations.

2. METHODOLOGY

Seasonal hydrogen storage

A notable potential application of hydrogen is blending with natural gas. Natural gas demand is characterized by substantial year-round fluctuations in air temperatures and exhibits seasonal variations, especially in countries with different seasons in a year, such as Türkiye. Therefore, underground storage of hydrogen generated from renewable sources during times of diminished natural gas demand, followed by its withdrawal and blending with natural gas when natural gas demand increases, is considered a significant strategy for enhancing energy supply security and mitigating carbon emissions.

The storage strategy in this study involves storing hydrogen in the reservoir for five months annually, following a one-month shut-in period, subsequently withdrawing it from the reservoir for five months, and allowing for another one-month shut-in period. We structure the simulations to include five storage cycles, each lasting one year. The annual storage cycles are organized as detailed below.

- Hydrogen injection will be conducted over a duration of 5 months (150 days). The process initiates with a constant flow rate and transitions to maintaining a constant bottom-hole pressure (BHP) when the BHP reaches the maximum reservoir pressure.
- Wells are closed for a duration of one month (30 days).
- Withdrawal occurs over a period of 5 months (150 days), starting with a constant flow rate from the reservoir and maintaining a constant BHP when the BHP reaches the minimum reservoir pressure.
- Wells are closed for a duration of one month (30 days).

Reservoir model

This study utilized the MATLAB Reservoir Simulation Toolbox (MRST) for the purposes of reservoir modeling and simulation [16, 17]. A gas reservoir unit characterized by a surface area of 1500 m x 1000 m and a thickness of 25 m was modeled considering the rock and fluid properties of four distinct natural gas reservoirs (the Northern Marmara, Seymen, Ardic, and Karacali natural gas fields) situated in the Thrace region of Türkiye [18]. Table 1 presents the reservoir properties, while Table 2 details the fluid properties utilized in the model development. The grid dimensions of the model decrease in size as they approach the well locations, leading to a variable total number of grids depending on the number of wells in the model. The total grid number is 6190 with 3 wells in the model and 21020 when the number of wells is raised to 15. The grids exhibit uniform vertical dimensions, with a total of 10 grids arranged in the vertical direction within the model.

Table 1. Reservoir properties used in the model

Reservoir Properties	Value
Depth, m	1000
Area, km^2	1.5
Net thickness, m	25
Initial pressure, bar	106.6
Initial temperature, °C	52.8
Porosity, %	16
Permeability, mD	30
Water Saturation, %	15

Table 2. Natural gas composition used in the model

Component	Mole fraction, %
Methane	94.42
Ethane	1.73
Propane	0.62
i-Butane	0.14
n-Butane	0.16
i-Pentane	0.06
n-Pentane	0.05
Hexane	0.02
Nitrogen	2.25
Carbon dioxide	0.55
Hydrogen	0.00
Total	100

The reservoir pressure was assumed to be 50 bar in the depleted state, with the composition of the cushion gas in the reservoir as indicated in Table 2. We set the minimum reservoir pressure at 50 bar and the maximum reservoir pressure at 106.6 bar for the model. Furthermore, it is presumed that all wells within the reservoir exhibit comparable characteristics and that all wells operate simultaneously during both the injection and withdrawal periods.

Hydrogen storage capacity

Because a specific volume of gas is consistently maintained in the reservoir to ensure the required flow rate is achieved, two distinct gas capacities can be identified in depleted gas reservoirs utilized for gas storage: total gas capacity and working gas capacity [19]. Total gas capacity denotes the overall gas capacity within the reservoir, whereas working gas capacity indicates the portion of the storage gas that is utilized to meet market demand [19]. The gas that is permanently present in the reservoir and acts as a buffer is defined as cushion gas, and various studies in the literature indicate that the cushion gas ratio for utilizing depleted natural gas reservoirs for hydrogen storage ranges from 50% to 60% [18–21]. In this study, the cushion gas capacity is determined as the gas capacity that corresponds to an average reservoir pressure of 50 bar, and the cushion gas capacity constitutes 52% of the overall gas capacity.

The calculation of gas volume in the reservoir in Sm^3 varies according to the gas composition, as the gas deviation factor (Z) is dependent upon the reservoir's gas composition. Storing hydrogen in a reservoir that initially contains natural gas components results in significant alterations to both the gas composition of the reservoir and the corresponding Z factor values. Under the assumption that the hydrogen gas ratio in the reservoir is 90% over the long term, the gas in place amounts (GIP) and Z factor values were determined for different average reservoir pressure (p) values. The resulting p/Z versus GIP graph is presented in Figure 1. Under these considerations, we calculate the total gas capacity to be 468.7 million Sm^3, the cushion gas capacity to be 223.8 million Sm^3, and the working gas capacity to be 244.9 million Sm^3.

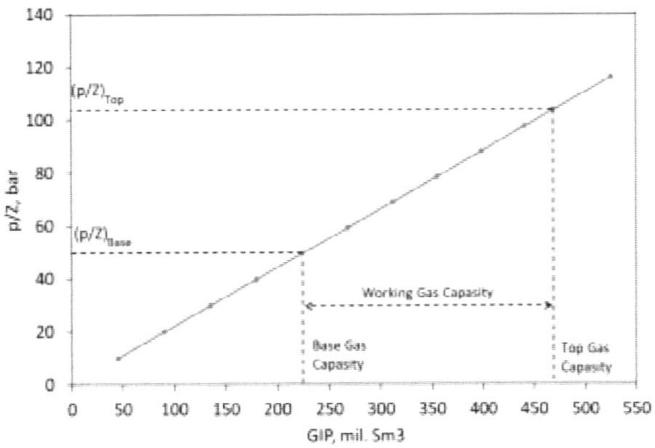

Figure 1. Plot of p/Z vs. GIP for different average reservoir pressure values

Simulation cases

Since the time that can be allocated for injecting and withdrawing the gas under seasonal storage conditions is limited, the number of wells and their corresponding injection and withdrawal capacities are critical factors in underground gas storage projects. In this study, firstly, five distinct simulations were run utilizing various well distributions to evaluate the overall seasonal injection and withdrawal capacities relative to the number of wells. Subsequently, maintaining a constant number of wells while varying the well properties, four additional simulations were run to analyze the impact of different well properties on underground hydrogen storage (UHS).

In the simulations performed for cases one to five, only the well numbers differed while the well properties were maintained as the same. In the sixth and seventh simulation cases, the perforation levels of the wells were modified, in contrast to case 1, and the simulations were performed again. In the sixth case, it is assumed that the uppermost 10 meters of the wells, beginning from the top of the reservoir, were perforated. In the seventh case, it was assumed that the lowest 10 meters of the reservoir were perforated. Figure 2 illustrates the scheme of the well perforation levels. The results obtained were compared with those from case 1 to assess the impact of well perforation levels on the volume of injected and withdrawn hydrogen, and the purity of the withdrawn hydrogen.

In the eighth and ninth simulation cases, the skin factor values of the wells were modified, in contrast to case 1, and the simulations were run again. In the eighth case, the skin factor values of the wells were set to +1, while in the ninth case, the skin factor values of the wells were set to -1. The results obtained were compared with those from case 1, in order to determine the effects of the skin factor values on the quantity of injected and withdrawal hydrogen, as well as the purity of the withdrawn hydrogen.

In all cases, the hydrogen injection period is set at 5 months (150 days), followed by a gas withdrawal phase lasting 5 months (150 days). A shut-in period of one month (30 days) is established between the injection and withdrawal phases. The total duration for each case is 5 years. The total injection and withdrawal flow rates were determined to ensure that a volume of gas equivalent to the working gas capacity could be injected into the reservoir and subsequently withdrawn within the planned injection period. Because the total working gas capacity is 244.9 million Sm^3, the total daily injection and total withdrawal flow rates are calculated as 1.63 million Sm^3/day and -1.63 million Sm^3/day, respectively. (244.9 million Sm^3/150 days = 1.63 million Sm^3/day)

In all cases, injection initiates and proceeds at a constant rate. If the BHP reaches the maximum reservoir pressure level, injection continues at a constant BHP. Similarly, withdrawal initiates and proceeds at a constant rate. In the event that the BHP decreases to the minimum reservoir pressure level, withdrawal continues at a constant BHP.

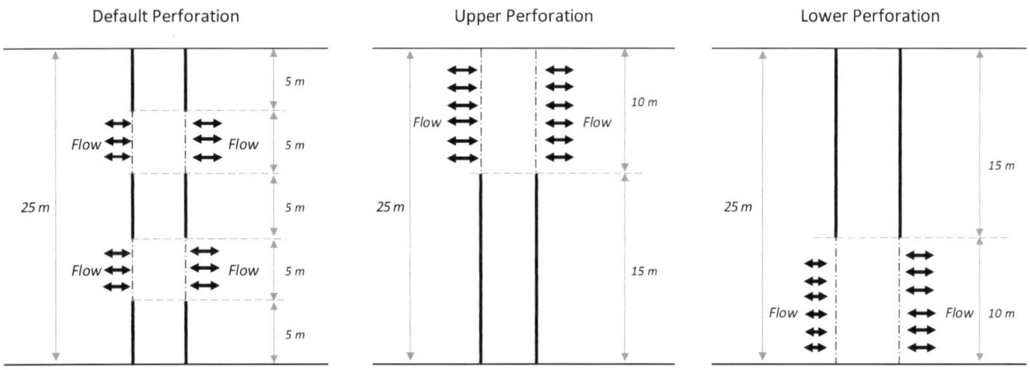

Figure 2. Schematic of well perforation levels

3. SIMULATION RESULTS

This section outlines the findings derived from the simulation results. Section 3.1 provides an analysis of the simulation results obtained from cases 1 through 5. Section 3.2 presents a comparative analysis of the simulation results from cases 1, 6, and 7, which exhibit different well perforation levels, to examine the impact of these levels on UHS. Section 3.3 analyzes the effects of varying well skin factor values on the UHS by comparing the simulation results from cases 1, 8, and 9, each defined by different well skin factor values.

Simulation results with different well numbers (Cases 1 to 5)

According to the results of simulations for cases 1 to 5, the changes in total gas flow rate and average reservoir pressure with respect to time are shown in Figure 3 (a) and (b), respectively. The results indicate that the gas flow rates remain constant for a certain period of time during both the injection and withdrawal periods, after which the gas flow rates exhibit a consistent decline. However, as the number of wells increases, the duration of the constant flow rate increases. In case 5, a total of 15 wells are utilized, and the gas flow continues with a constant flow rate throughout the majority of the injection and withdrawal periods. In addition, average reservoir pressure reaches the maximum reservoir pressure in case 5 at the end of each injection period, and the average reservoir pressure decreases to a minimum value of 53 bar, nearing the determined minimum reservoir pressure, at the end of the withdrawal periods (Figure 3 (b)).

During withdrawal periods, natural gas is also produced alongside hydrogen from the reservoir. Because natural gas exists as cushion gas within the reservoir prior to the injection of hydrogen. According to the results of cases 1 to 5, the variation in the hydrogen ratio of the gas withdrawn from the reservoir over time and the lowest hydrogen ratios recorded at the end of the related withdrawal period are illustrated in Figure 4 (a) and (b), respectively. The results indicate that the hydrogen ratio of the withdrawn gas exhibits a consistent decrease over time during withdrawn periods. However, this decline becomes less pronounced with the repetition of storage cycles. The hydrogen ratio of the gas withdrawn from the reservoir is also influenced by the number of wells, particularly during the initial cycles. Because the limited number of wells results in the accumulation of injected hydrogen gas within a confined area of the reservoir. This situation has a favorable impact on the purity of hydrogen withdrawn from the reservoir. However, the presence of a significant number of wells results in the dispersion of hydrogen gas over a broader area and more uniformly within the reservoir, and this phenomenon adversely impacts the purity of hydrogen withdrawn from the reservoir. The annual average hydrogen ratio of the withdrawn gas is 84.0% in the three-wells scenario (case 1), whereas it is 75.5% in the fifteen-wells scenario (case 4) as it is shown on Figure 5 (c). However, the annual average hydrogen ratio of the withdrawn gas during the last storage cycle is 90.2% in the case 1 and 91.7% in case 5. Thus, it is assessed that conducting hydrogen injections through a limited number of wells to a specific section of the reservoir may be suitable, especially in the initial cycles if high hydrogen purity is needed. However, as the storage cycles are repeated, there is an observed increase in the hydrogen ratio of the withdrawn gas, and the adverse impact of the high quantity of wells on hydrogen purity diminishes.

Figure 5 (a) and (b) show the amounts of hydrogen injected into the reservoir and the total amount of gas withdrawn from the reservoir, as well as the average hydrogen ratio of withdrawn gas in each storage cycle for cases 1 to 5. In the three-wells scenario, 211.4 million Sm^3 of hydrogen gas can be stored in the reservoir during the first storage cycle, whereas 154.0 million Sm^3 of gas can be withdrawn during the same cycle. On the other hand, in case 5, 244.4 million Sm^3 of hydrogen gas can be stored in the reservoir during the first cycle, whereas 231.3 million $Sm^3$of gas can be withdrawn from the reservoir during the same cycle. Considering that the volume of gas stored in the reservoir in case 5 is close to the working gas capacity, 15 wells are appropriate for the defined reservoir model under the planned seasonal storage conditions. In other terms, when a well is located inside every 0.1 km^2 region, a volume of gas approximately equivalent to the working gas capacity can be stored in the reservoir at a flow rate deemed constant.

Effects of well perforation intervals on UHS (Cases 1, 6, and 7)

This section comparatively examines the results of cases 1, 6, and 7, where the reservoir parameters and the number of wells remain constant but the well perforation levels are varied. Figure 6 illustrates the overall gas flow rate and average reservoir pressure derived from the simulations. Although comparable outcomes are seen for surface gas flow rate and average reservoir pressure in cases 6 and 7, these results exhibit partial discrepancies when contrasted with those of case 1. The gas flow rate starts to decrease earlier during the injection and withdrawal phases in cases 6 and 7 relative to case 1. According to these results, the location of perforation levels, whether in the upper or lower section of the reservoir, does not significantly affect the injection and withdrawal flow rates according to the chosen reservoir model and storage plan. However, although the total perforation interval is the same (10 m), merging the two perforation intervals into one segment of 10 meters negatively affects the flow rates in both the withdrawal and injection periods.

Upon evaluating the simulation results of cases 1, 6, and 7 concerning the hydrogen ratio of the gas withdrawn from the reservoir during the withdrawal periods, it is observed that cases 6 and 7 yield comparable results, whereas these results diverge from those of case 1, particularly during the first storage period (Figures 7 (a) and (b)). In this context, the location of well perforation, whether in the upper or lower sections of the reservoir, does not affect hydrogen purity. Nonetheless, merging the two perforation intervals into one segment of 10 meters positively impacts hydrogen purity, particularly during the first storage cycle. As storage cycles repeated, no variation in hydrogen purity was noted depending on the perforation level.

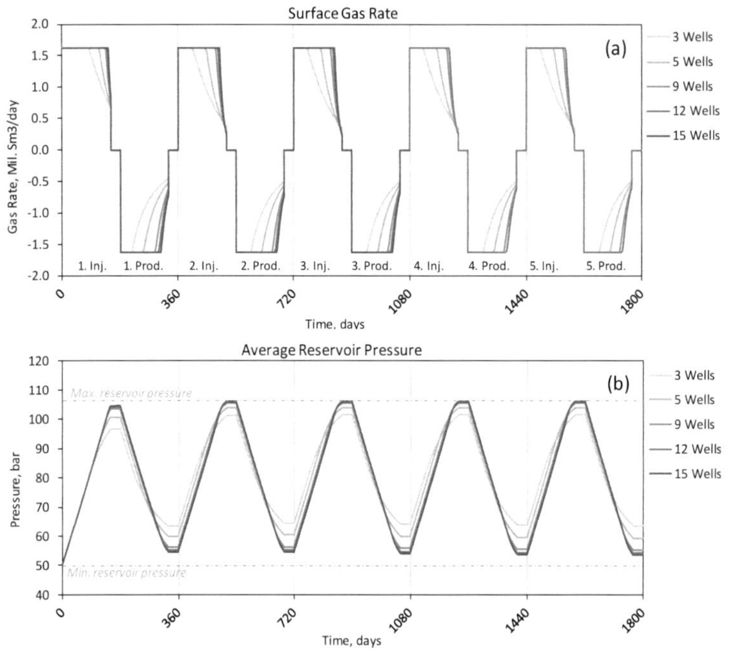

Figure 3. the variation of (a) the total gas flow rate, (b) the average reservoir pressure over time according to the simulation results of the cases 1-5

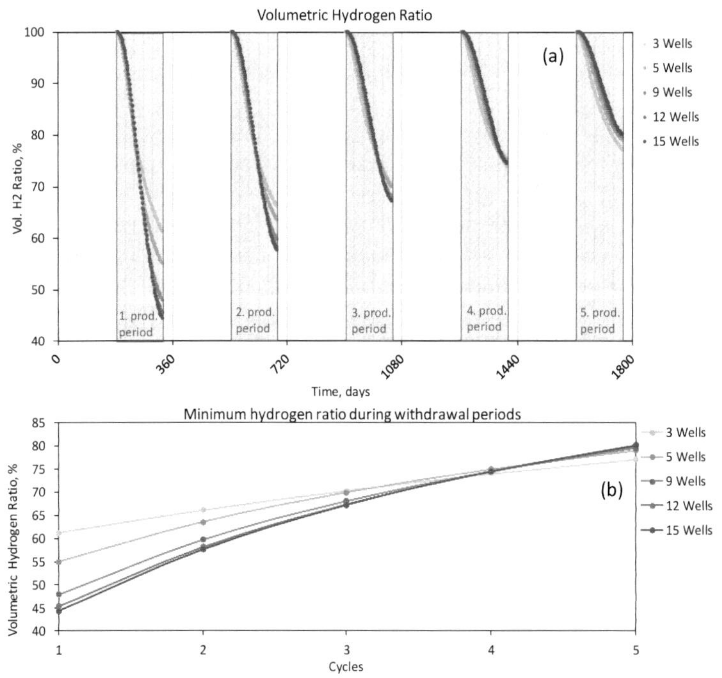

Figure 4. (a) Change in the hydrogen ratio of the gas withdrawn from the reservoir with respect to time, (b) minimum hydrogen ratio during withdrawal periods according to the simulation results of cases 1 to 5

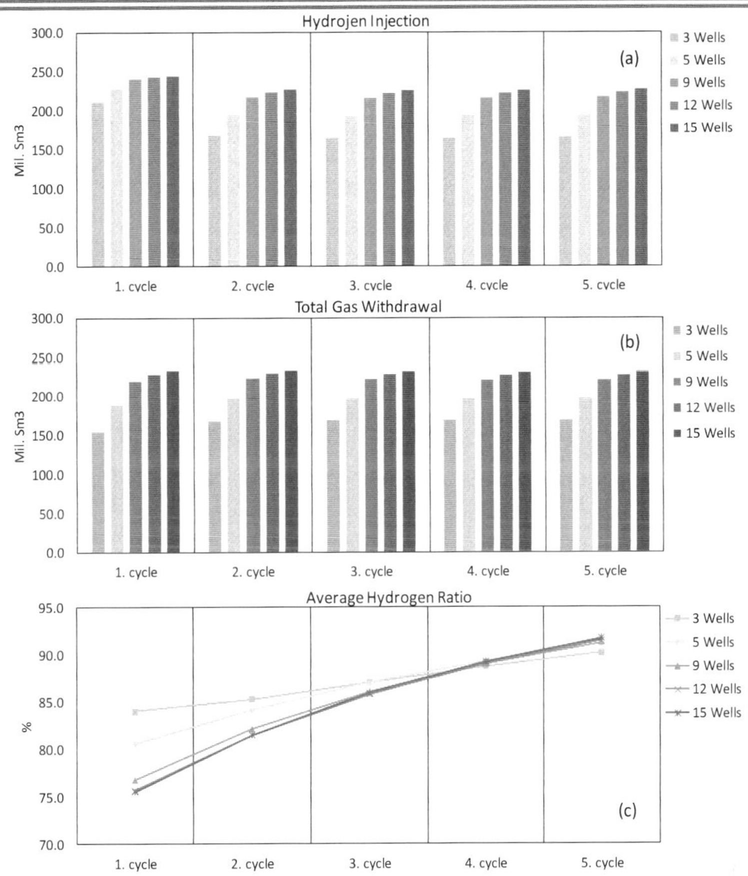

Figure 5. (a) total annual hydrogen injection, (b) total annual gas withdrawal amounts, and (c) average hydrogen ratio of the withdrawn gas according to the simulation results of the cases 1 to 5

Effects of well skin factor values on UHS (Cases 1, 8, and 9)

In this section, the simulation results of cases 1, 8, and 9, which share identical reservoir features and well numbers but differ in well skin factor values, are analyzed comparably. Figure 8 illustrates the overall gas flow rate and average reservoir pressure values derived from the simulations. The results indicate that the injection and withdrawal flow rates, as well as the maximum and minimum average reservoir pressure values at the end of related periods, vary significantly based on the skin factor value. According to these results, the value of the skin factor also influences hydrogen purity. In case 8, with a skin factor of +1, the minimum hydrogen purity is 63.5% at the end of the first storage cycle. However, in case 9, with a skin factor of -1, the minimum hydrogen purity is 58.7% at the end of the first storage cycle. Nonetheless, comparable purity levels were obtained in the third storage cycle, and the influence of the skin factor reversed from the fourth cycle onward (Figure 9 (a) and (b)).

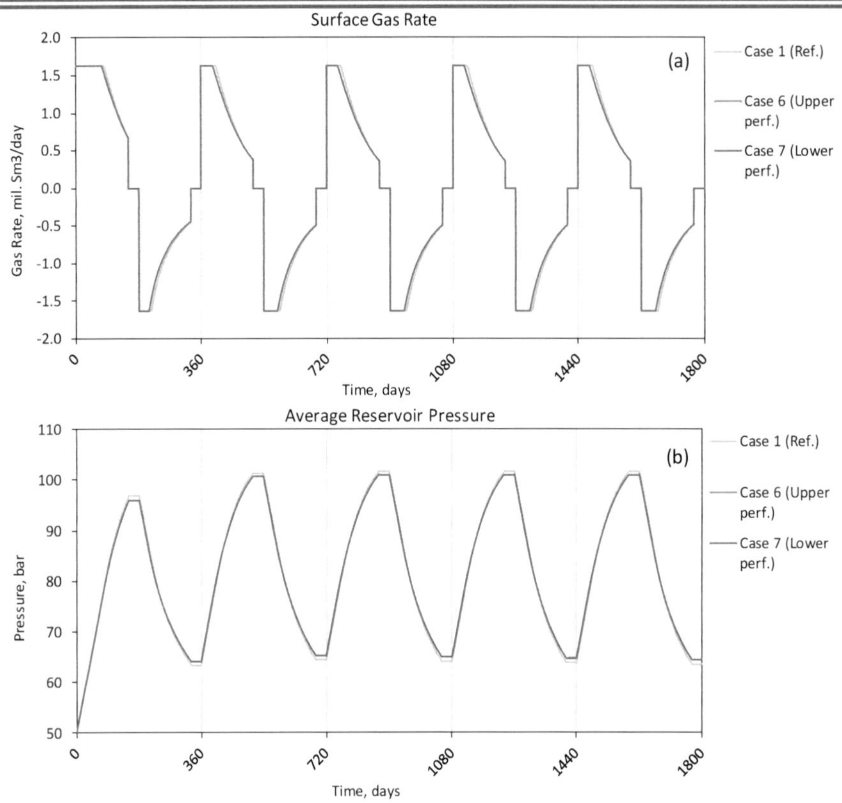

Figure 6. the variation of (a) total gas flow rates, (b) average reservoir pressure values over time according to the simulation results of the cases 1, 6, and 7

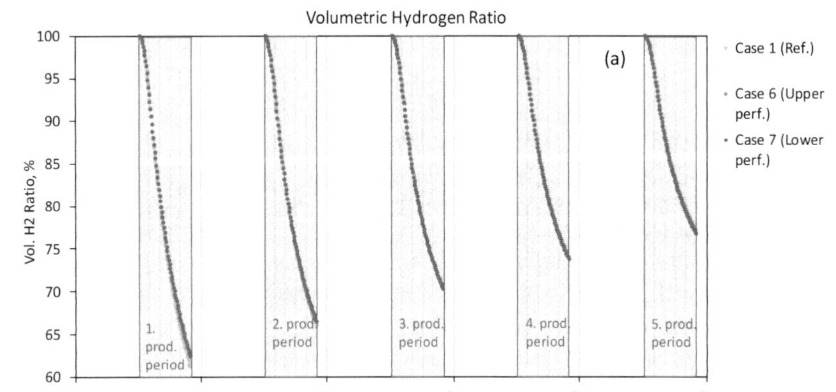

Figure 7. (a) Change in the hydrogen ratio of the gas withdrawn from the reservoir with respect to time, (b) minimum hydrogen ratio during withdrawal periods according to the simulation results of cases 1, 6, and 7

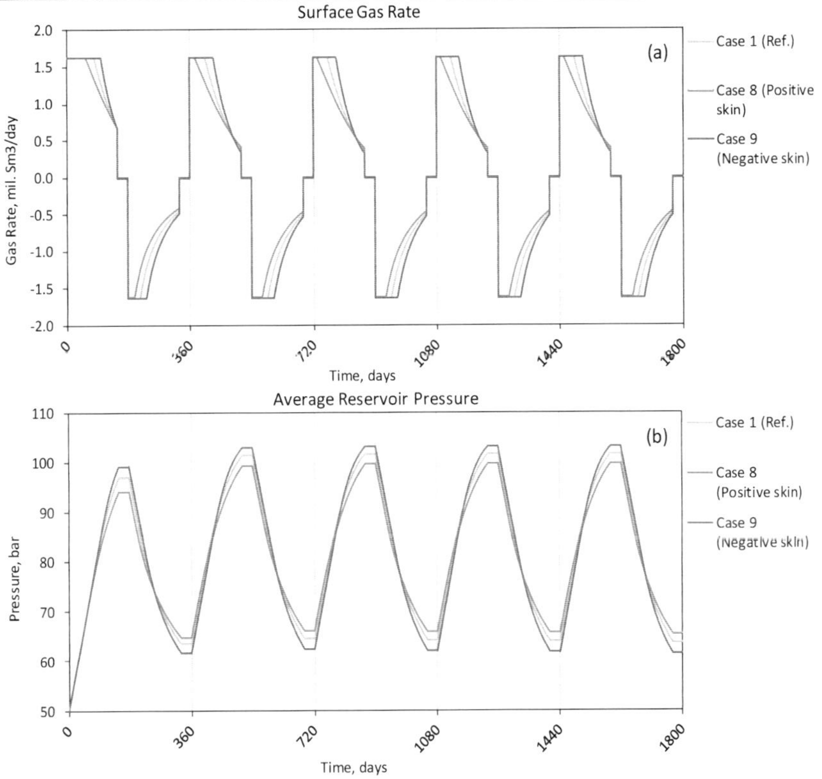

Figure 8. the variation of (a) total gas flow rates, (b) average reservoir pressure over time according to the simulation results of cases 1, 8, and 9

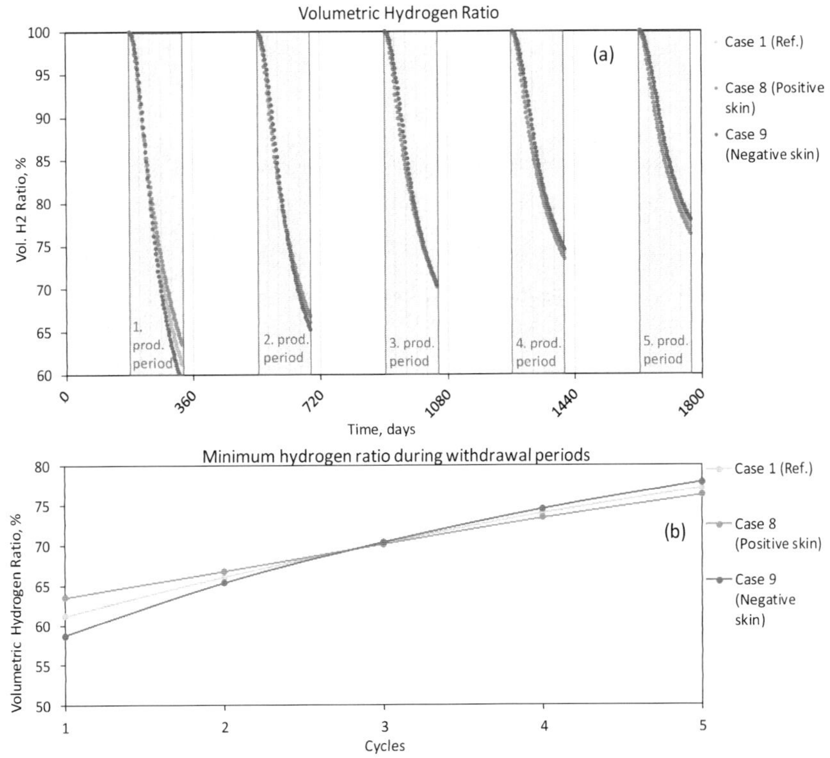

Figure 9. (a) Change in the hydrogen ratio of the gas withdrawn from the reservoir with respect to time, (b) minimum hydrogen ratio during withdrawal periods according to the simulation results of cases 1, 8, and 9

4. CONCLUSION

In this study, a sample reservoir model was developed by utilizing the average geological and fluid properties of four distinct natural gas reservoirs situated in the Thrace region of Türkiye. The total gas capacity is estimated to be 468.7 million Sm^3, the cushion gas capacity is 223.8 million Sm^3, and the working gas capacity is 244.9 million Sm^3 assuming the reservoir's minimum and maximum pressure values were 50 bar and 106.6 bar, respectively. Firstly, five distinct seasonal storage simulations were run with 3, 5, 9, 12, and 15 wells, and the results of the first five simulations with varying well counts indicate that when hydrogen is injected into the reservoir over a limited area with fewer wells, the hydrogen ratio in the gas withdrawn from the reservoir is comparatively higher until the fourth cycle. Conversely, an increased number of wells results in a comparatively lower hydrogen ratio in the gas withdrawn from the reservoir until the fourth cycle. On the other hand, the hydrogen ratio of the withdrawn gas increases as storage cycles are repeated in all cases.

Upon comparison analysis of the results from cases 1, 6, and 7 to assess the impact of well perforation levels on hydrogen storage;

- The positioning of perforation levels, whether in the upper or lower region of the reservoir, does not substantially influence the injection and withdrawal flow rates based on the selected reservoir model and storage plan. Nonetheless, despite the overall perforation interval remaining constant at 10 meters, consolidating the two perforation intervals of 5-meters into a single 10-meters segment adversely impacts the flow rates during both the withdrawal and injection phases.
- Similarly, the positioning of perforation levels, whether in the upper or lower region of the reservoir, does not affect the hydrogen ratio of withdrawn gas. However, the consolidation of the two perforation intervals into a single 10-meter length enhances hydrogen purity during the first storage cycle.

Upon comparison analysis of the results from scenarios 1, 8, and 9, to assess the influence of the skin factor values of the wells on hydrogen storage;

- The findings indicate the injection and withdrawal flow rates, along with the maximum and minimum average reservoir pressure values at the end of relevant periods, vary considerably depending on the skin factor value.
- The skin factor values also affect hydrogen purity. In scenario 8, with a skin factor of +1, the minimum hydrogen purity is 63.5% at the end of the first storage cycle. In scenario 9, with a skin factor of -1, the minimum hydrogen purity is 58.7% at the end of the first storage cycle. However, similar purity levels were achieved in the third storage cycle, and the effect of the skin factor reversed from the fourth cycle onward.

It has been concluded that substantial quantities of hydrogen can be stored in relatively low-capacity gas reservoirs, with the majority being retrievable under seasonal storage conditions. Furthermore, hydrogen can be withdrawn with a purity above 80%, and the purity of the hydrogen enhances with each subsequent cycle. The negative skin factor value is seen as beneficial for UHS since it positively influences injection and withdrawal flow rates. On the other hand, the position of perforation levels, whether in the upper or lower region of the reservoir, does not substantially influence the injection and withdrawal flow rates or hydrogen purity.

REFERENCES

[1] Zivar, D., Kumar, S., & Foroozesh, J. (2021). Underground hydrogen storage: A comprehensive review. International Journal of Hydrogen Energy, 46(45), 23436–23462. https://doi.org/10.1016/J.IJHYDENE.2020.08.138

[2] Muhammed, N. S., Haq, B., Al Shehri, D., Al-Ahmed, A., Rahman, M. M., & Zaman, E. (2022). A review on underground hydrogen storage: Insight into geological sites, influencing factors and future outlook. Energy Reports, 8, 461–499. https://doi.org/10.1016/J.EGYR.2021.12.002

[3] Pfeiffer, W. T., & Bauer, S. (2015). Subsurface Porous Media Hydrogen Storage – Scenario Development and Simulation. Energy Procedia, 76, 565–572. https://doi.org/10.1016/J.EGYPRO.2015.07.872

[4] Amid, A., Mignard, D., & Wilkinson, M. (2016). Seasonal storage of hydrogen in a depleted natural gas reservoir. International Journal of Hydrogen Energy, 41(12), 5549–5558. https://doi.org/10.1016/J.IJHYDENE.2016.02.036

[5] Elberry, A. M., Thakur, J., Santasalo-Aarnio, A., & Larmi, M. (2021). Large-scale compressed hydrogen storage as part of renewable electricity storage systems. International Journal of Hydrogen Energy, 46(29), 15671–15690. https://doi.org/10.1016/J.IJHYDENE.2021.02.080

[6] Hagemann, B., Rasoulzadeh, M., Panfilov, M., Ganzer, L., & Reitenbach, V. (2015). Mathematical modeling of unstable transport in underground hydrogen storage. Environmental Earth Sciences, 73(11), 6891–6898. https://doi.org/10.1007/S12665-015-4414-7

[7] Hydrogen - Fuels & Technologies - IEA. (n.d.). Retrieved July 5, 2023, from https://www.iea.org/fuels-and-technologies/hydrogen

[8] HyBlend: Opportunities for Hydrogen Blending in Natural Gas Pipelines | Department of Energy. (n.d.). Retrieved April 22, 2023, from https://www.energy.gov/eere/fuelcells/hyblend-opportunities-hydrogen-blending-natural-gas-pipelines

[9] Blending hydrogen into the EU gas system. (n.d.). Retrieved May 22, 2023, from https://joint-research-centre.ec.europa.eu/jrc-news/blending-hydrogen-eu-gas-system-2022-01-19_en

[10] Türkiye National Energy Plan 2022. (n.d.). Retrieved April 25, 2023, from https://enerji .gov.tr/Media/Dizin/EIGM/tr/Raporlar/TUEP/T%C3%BCrkiye_National_Energy_Plan.pdf

[11] Türkiye Hydrogen Technologies Strategy and Roadmap. (n.d.). Retrieved July 9, 2023, from https://enerji.gov.tr/Media/Dizin/SGB/tr/Kurumsal_Politikalar/HSP/ETKB_Hidrojen_Stratejik_Plan2023.pdf

[12] Hydrogen for Net-Zero. (2021). Retrieved December 21, 2022, from www.hydrogencouncil.com

[13] Tarkowski, R. (2019). Underground hydrogen storage: Characteristics and prospects. Renewable and Sustainable Energy Reviews, 105, 86–94. https://doi.org/10.1016/j.rser.2019.01.051

[14] Pan, B., Yin, X., Ju, Y., & Iglauer, S. (2021). Underground hydrogen storage: Influencing parameters and future outlook. Advances in Colloid and Interface Science, 294, 102473. https://doi.org/10.1016/J.CIS.2021.102473

[15] Tarek Ahmed. Reservoir engineering handbook (Second Edition). Texas:Gulf Professional Publishing, 2000, pp. 420.

[16] Lie, K.-A. (2019). An Introduction to Reservoir Simulation Using MATLAB/GNU Octave. Cambridge University Press. https://doi.org/10.1017/9781108591416

[17] Lie, K.-A., & Møyner, O. (Eds.). (2021). Advanced Modeling with the MATLAB Reservoir Simulation Toolbox. Cambridge University Press. https://doi.org/10.1017/9781009019781

[18] Çınar, G. The Natural Gas Evaluation of The Thrace Basin of Turkey. MSc, Istanbul Technical University, Istanbul, 1996.

[19] Türeyen, Öİ. Effect of wellbore conditions on the performance of underground gas-storage reservoirs. MSc, Istanbul Technical University, Istanbul, 2000.

NURER 2024
Nuclear and Renewable Energy Resources

Reactor Physics Analysis of ATF cladding system for 24-month cycle APR-1400

Khurrum Saleem Chaudri
Department of Mechanical and Nuclear Engineering, Khalifa University of Science and Technology, Abu Dhabi P.O. Box 127788, United Arab Emirates.
Emirates Nuclear Technology Center (ENTC), Khalifa University of Science and Technology, P.O. Box 127788, Abu Dhabi, United Arab Emirates.
khurrum.chaudri@ku.ac.ae
ORCID: 0000-0002-4283-6207

Mohammad Alrwashdeh
Department of Mechanical and Nuclear Engineering, Khalifa University of Science and Technology, Abu Dhabi P.O. Box 127788, United Arab Emirates.
Emirates Nuclear Technology Center (ENTC), Khalifa University of Science and Technology, P.O. Box 127788, Abu Dhabi, United Arab Emirates.
mohammad.alrwashdeh@ku.ac.ae
ORCID: 0000-0002-0149-7704

Saeed A. Alameri
Department of Mechanical and Nuclear Engineering, Khalifa University of Science and Technology, Abu Dhabi P.O. Box 127788, United Arab Emirates.
Emirates Nuclear Technology Center (ENTC), Khalifa University of Science and Technology, P.O. Box 127788, Abu Dhabi, United Arab Emirates.
saeed.alameri@ku.ac.ae
ORCID: 0000-0001-8926-2004

Cite this paper as: *Chaudri, Khurrum Saleem, Alrwashdeh, Mohammad, Alameri, Saeed A,. Reactor Physics Analysis of ATF cladding system for 24-month cycle of APR-1400. 7th Int. Conf. on Nucl. and Ren. Energy Res. 27-30 October 2024, Antalya, Turkey*

Abstract: Longer cycle lengths for nuclear power plant operations can provide economic benefits regarding availability and capacity factors over its lifetime. This study aims to analyze ATF fuel-clad combinations for the 24-month cycle of APR-1400. Higher-density fuels like U_3Si_2 and UB_2 are analyzed with Cr-coated Zircalloy, SiC, and SS claddings. UO_2 fuel for a 24-month cycle requires LEU+ enrichment, i.e.,≥ 5%. Higher-density fuels can potentially eliminate the usage of LEU+ fuel and still provide a 24-month cycle length. Higher thermal conductivity leading to lower centerline temperature and potentially higher linear heat generation rate are additional benefits of such fuels. ATF cladding candidates such as Cr-coated Zircalloy, SiC, and SS can provide longer operations without producing hydrogen in the worst-case scenario of cooling loss in a severe accident. This work aims to study the neutron physics characteristics of different ATF systems and analyze the potential pros and cons.

Keywords: *Accident Tolerant Fuel, APR-1400, Reactor Physics, 24-month cycle*

Nomenclature	
ATF	Accident Tolerant Fuel
PWR/LWR	Pressurized Water Reactor/Light Water Reactor
BoC/EoC	Beginning of Cycle/End of Cycle
EFPD	Equivalent Full Power Days
LEU+	Low Enriched Uranium Plus (referring to fuel enrichment >5% and <10%)
HM	Heavy Metal
CS	Chemical Shim

1. INTRODUCTION

In the era of renewable energy, nuclear energy must stay competitive in economic terms. One option considered by different operators is to shift to longer cycle lengths e.g. an 18-month to 24-month cycle. Studies performed show that to achieve the required cycle length, LEU+ enrichment is required (Khalefih et al., 2023). Fuel manufacturers are moving in the direction of usage of LEU+ fuel and getting necessary regulatory approvals (*Global Nuclear Fuel Gets Approval for Fuel Enriched up to 8 Percent U-235 -- ANS / Nuclear Newswire*, n.d.). Still, there are problems of uranium dioxide fuel such as low thermal conductivity, pellet cracking at higher burnups etc. In the aftermath of Fukushima nuclear accident, fuel materials in addition to cladding options are being studied for ATF concept. The fuel materials considered, specifically high-density ones, are particularly of interest from an economic standpoint. The examples of such fuels are UN, UC, U_3Si_2, UB_2 etc. These fuels have high heavy metal density which can support longer cycle lengths (Ahmed et al., 2016). Active research is being carried out to solve the problems these materials have. For example, composite fuels (U_3Si_2+UB_2, UN+U_3Si_2, UN+U_3Si_5) (Zakova & Wallenius, 2012), mixing with additives (UN+ZrO_2) (Zakova & Wallenius, 2012), coating the nuclear fuel to add an additional barrier (Chaudri et al., 2013) etc. have been proposed by researchers. Composite fuels like U_3Si_2+UB_2 offer advantages like high heavy metal density. Addition of UB_2 in U_3Si_2 can help in increasing the onset of steam fuel reaction (Turner & Abram, 2020). New burnable absorber concepts like Highly Intensive and Discrete Gadolinia/Alumina Burnable Absorber (HIGA) (Kim et al., 2024) can be used for the excess reactivity control of these longer cycle length concepts.

Different cladding materials such as chromium coated zircalloy, stainless steel 304 and silicon carbide are being considered to rectify the problem of high temperature steam zircalloy reaction which leads to hydrogen production. The issue with the development of introducing new materials in reactor core is that it takes a long regulatory process to get the necessary approval to use these materials in actual nuclear power plants. Hence silicon carbide, which shows good properties for use as cladding, will only be available in a comparatively long time (Alrwashdeh & Alameri, 2022b) . Stainless steel or other iron based cladding materials such as FeCrAl or APMT show good characteristics according to strength and corrosion resistance but are marred by its very high neutron absorption cross-section but result in a severe cycle-length penalty (Alrwashdeh & Alameri, 2021). Chromium-coated zircalloy cladding provides a middle-term option as main cladding material is still zircalloy but it is now shielded by a layer of chromium coating to avoid direct contact with high temperature steam in case of an accident (Alrwashdeh & Alameri, 2022a). Studies have shown that there is some diffusion taking place between the zircalloy and chromium coating which might affect the desired properties of materials negatively. To avoid this, a thin layer of Niobium or Molybdenum can be placed between zircalloy and the chromium coating. From reactor physics standpoint, Nb is found to be better choice than Mo due to smaller absorption cross-section. The thickness of the Nb-layer and Cr-coating, to minimize the affect on neutron economy, were fond to be in the range of 1-5 µm and 10-20 µm respectively.

In this work, neutronics feasibility of U_3Si_2+UB_2 composite fuel with different cladding materials is studied for APR-1400 nuclear reactor to obtain 24-month cycle length. The criteria on which the suitability of studied fuel/clad system is judged are cycle length of 650+ days (corresponding to 24-month cycle with 90% capacity factor) along with reasonable radial power peaking factor. CASMO4E/SIMULATE3 code system is used to perform the required analysis. Article is organized into sections. Section 2 gives the methodology, section 3 gives the results and discussion, while section 4 presents the conclusion and future direction of this work.

2. Methodology

To carry out the relevant analysis for candidate fuel cladding systems, CASMO4E/SIMULATE3 (Zimmerman et al., 1999) two-step code system is used. Relevant fuel assemblies making the APR-1400 core are modeled in CASMO4E (Rhodes et al., 2004). Geometry, composition of constituting materials, and operating conditions serve as input for CASMO4E code. Several branching calculations for fuel temperature, coolant temperature, and chemical shim concentration are devised by the code to cover the design space. Two-dimensional transport theory calculations are performed by CASMO4E to generate the homogenized cross-section for all the assemblies. Auxiliary code CMSLINK reads the outputs generated by CASMO4E code and writes them in a format suitable for SIMULATE3 (Smith, 1989) in a file with .lib extension. Two-dimensional core is modeled in SIMULATE-3 code with the provision of axial buckling to incorporate the effect of axial leakage and give an equivalent solution to three-dimensional modeling. The flowchart of calculation scheme used by the code is shown in Figure 1. The cross-section library used by CASMO4E is based on ENDF/B-VI.8 database (Dunford, 1992). The CASMO version 4E is educational version, hence doesn't have access to the latest cross-section libraries.

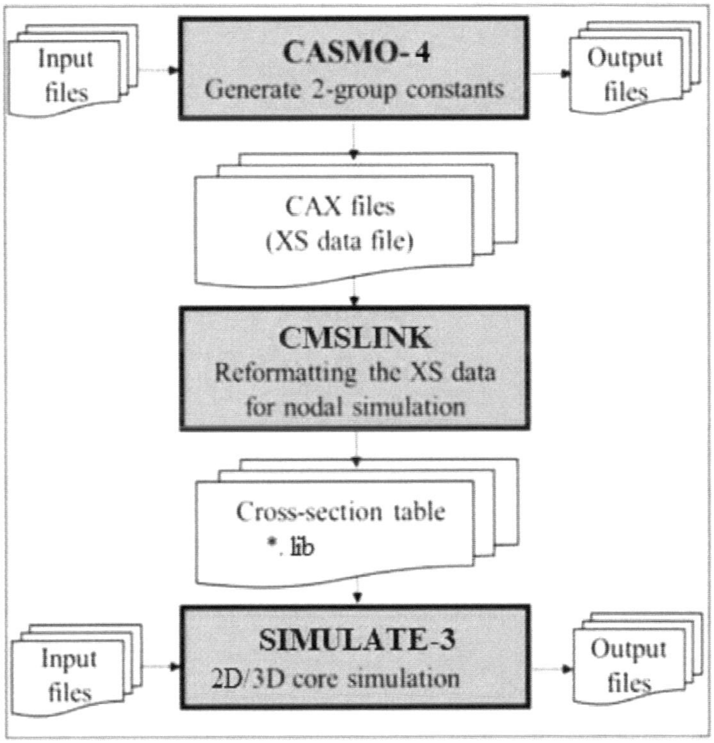

Figure 1: CASMO4E/SIMULATE-3 code calculation scheme

CASMO4E/SIMULATE3 code system is well verified for the conventional UO_2/Zr fuel clad system of PWRs. Two-step code systems like CASMO4E/SIMULATE3 are limited in the sense of the cross-section library as that library uses representative neutron spectrum of system under consideration for weighting and hence development of multi-group cross-section library. To establish the suitability of the code system for the proposed accident tolerant fuel clad system, assembly level calculations for the proposed designs are verified against Monte-Carlo code calculations. OpenMC (Romano et al., 2015) code is used for this purpose. OpenMC is a community developed open-source code based on Monte-Carlo theory for neutronics calculations of nuclear reactors. The use of continuous energy neutron interaction cross-sections for relevant nuclides with full heterogenous geometry modeling makes it suitable for all kinds of nuclear reactors, irrespective of neutron spectrum. OpenMC utilized ENDF/B-VIII.0 cross-section library (Brown et al., 2018).

For the case of U_3Si_2, and U_3Si_2-UB_2 fuel, fuel assembly with provision of 16 burnable absorber locations is analyzed by CASMO4E and OpenMC codes. The mentioned fuel assembly cases are burnt to 70 MWd/tHM and multiplication factors are compared. Fuel temperature, coolant temperature, and chemical shim concentration values used to perform the analysis are 900K, 580.4K, and 540ppm respectively. The fuel enrichment for higher enrichment and lower enrichment pins is 4.9% and 4.2% respectively. The burnable absorber pin locations are filled with the moderator. The cladding is Zircalloy for the verification studies. The multiplication factor and relative difference in percentage is shown in Figure 2 and Figure 3. The difference is less than one percent with underpredicted value by CASMO4E at BoC which switches to overprediction by the EoC. Looking at the trend and magnitude, it can be assumed that the difference is small enough to safely use CASMO4E for cross-section generation for the ATF candidates under study in this work. The biggest con of Monte-Carlo theory reactor physics codes is their calculation time. For the case being studied here where different cases are studied for multiple cycles (equilibrium core), the calculation time becomes unreasonable. Hence to get an insight with reasonable calculation times, CASMO4E/SIMULATE will be used to perform the studies presented here.

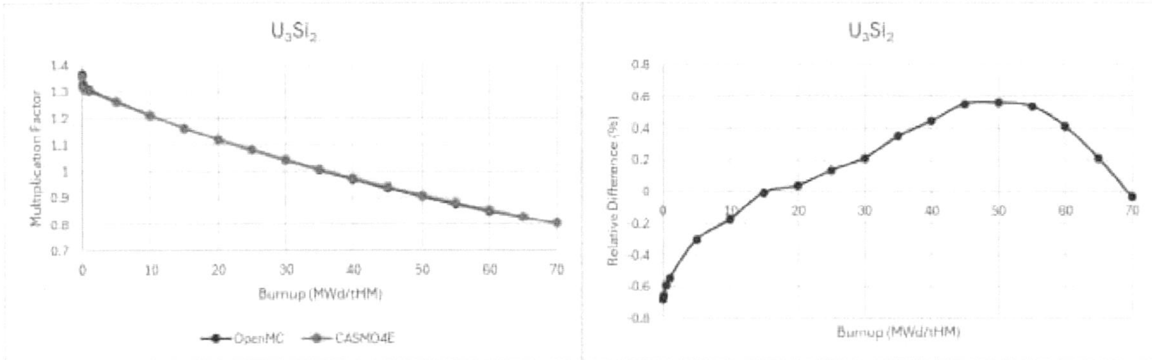

Figure 2 (a): Multiplication factor comparison for U_3Si_2 case (b): Relative difference (%) for the two codes

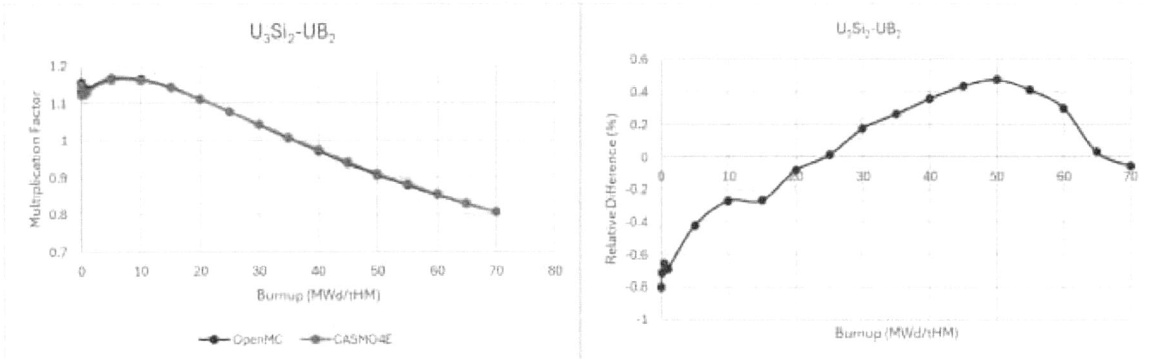

Figure 3 (a): Multiplication factor comparison for U_3Si_2-UB_2 case (b): Relative difference (%) for the two codes

The equilibrium core loading pattern is shown in Figure 4. In this figure, J1 and J2 represent the once-burnt fuel assemblies while H1 and H2 represent the fresh fuel assemblies.

J	H	G	F	E	D	C	B	A	
H2	J2	H2	H2	J2	H2	J1	J2	H1	09
J2	H2	J2	J1	J2	H2	J2	H2	H1	10
H2	J2	H2	J2	H2	J1	J1	J1	H1	11
H2	J1	J2	H2	J2	H2	H2	J1	H1	12
J2	J2	H2	J2	J1	J2	H1	J1		13
H2	H2	J1	H2	J2	J2	H1	H1		14
J1	J2	J1	H2	H1	H1	H1			15
J2	H2	J1	J1	J1	H1				16
H1	H1	H1	H1						17

Legend: Fresh; Once Burnt; Water

Figure 4: Equilibrium core reloading pattern for APR-1400

The assembly description for J1/H1 and J2/H2 is shown in Figure 5. The pins shown in yellow indicate the lower enrichment fuel pins, the purple color denotes higher enrichment fuel pins, while the orange color shows the location of burnable absorber pins. The light blue color in the figure represents water which is acting as coolant/moderator.

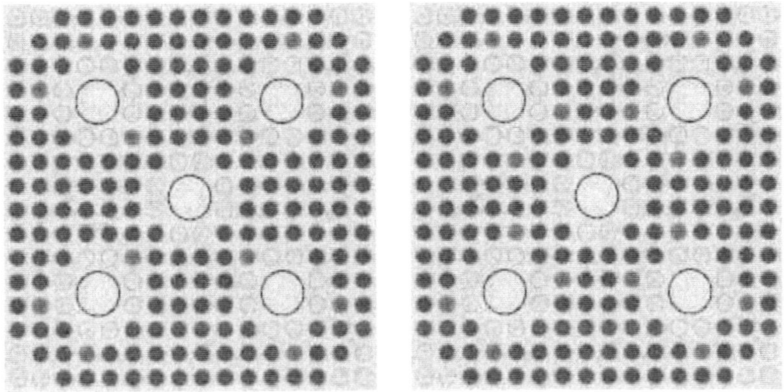

Figure 5(a) J1/H1 fuel assembly *(b) J2/H2 fuel assembly*

3. Results and Discussion

The target cycle length for equilibrium core configuration is 650+ EFPDs. This corresponds to a 24-month cycle with 90% capacity factor. As discussed in the introductions section, LEU+ (>5% enrichment) fuel is required to achieve 650+ EFPD goal with UO_2 fuel, owing to its relatively smaller heavy metal density. The higher heavy metal density of U_3Si_2, UB_2 and UC can help lower the enrichment which was seen during this study. For U_3Si_2 and U_3Si_2+UB_2 case, the density came out to be 4.9 and 4.2 wt% and for UC, even lower enrichment values of 4.2 and 3.5 wt% provided the required cycle length.

U_3Si_2 Fuel Option:

For Uranium silicide fuel option (12.2 g/cm^3), the heavy metal density is 11.3 g/cm^3. This value is greater than 9.2 g/cm^3 for Uranium dioxide fuel (10.44 g/cm^3) currently prevalent in nuclear industry. This helps in loading more uranium in nuclear reactor core with reasonable enrichment levels. In case of uranium silicide fuel, it was seen that a cycle length of 723.9 EFPDs was obtained with enrichment of 4.9 and 4.2 wt%. Similar EFPDs goal with Uranium dioxide fuel required 5.9 and 5.2 wt%. The multiplication factor trend over the course of cycle is shown in Figure 6. Initial multiplication factor is quite high (1.23001). The reason for this high value is the absence of any control mechanism i.e. no BA or CS is considered. To control the excess reactivity, Highly Intensive and Discrete Gadolinia/Alumina Burnable Absorber (HIGA) absorber rods are used at 12/16 BA locations in fuel assemblies. In Figure 6, the curve for uranium silicide with 12/16 HIGA rods show reasonable suppression of excess reactivity. The initial multiplication factor drops down to 1.4112 with a cycle length of 700.9 days, which satisfies the criteria.

U_3Si_2+UB_2 Fuel Option:

To find a better suppression of excess reactivity, which can help in reducing the CS in coolant/moderator, the addition of the Uranium diboride fuel in Uranium silicide fuel is considered. Material science studies have shown that the addition of uranium diboride to uranium silicide can help increase the onset of oxidation as compared to uranium silicide only. Additionally, uranium diboride has higher uranium density (11.7 g/cm3) as compared to uranium silicide. Presence of boron allows it to suppress the excess reactivity also. Different amounts of uranium diboride are mixed with uranium silicide to see the impact on multiplication factor. This mixing of uranium diboride is done uniformly in all the fuel rods. Concentrations of 0.5 and 1.0 wt% uranium diboride mixed homogenously with uranium silicide are studied. Figure 7 shows the variation of multiplication factor the 3 studied cases. Cycle length of 723.9, 702.9 and 678.2 days are obtained form U_3Si_2, U_3Si_2+UB_2 (0.5 wt%), and U_3Si_2+UB_2 (1.0 wt%) cases. The case with 1.0 wt% uranium diboride suppresses the excess reactivity very efficiently.

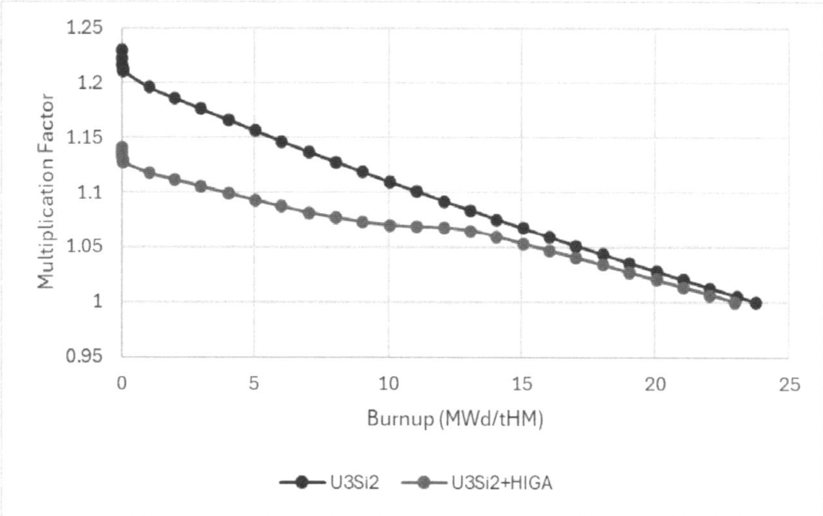

Figure 6: Equilibrium core multiplication factor evolution over the cycle length with and without BA

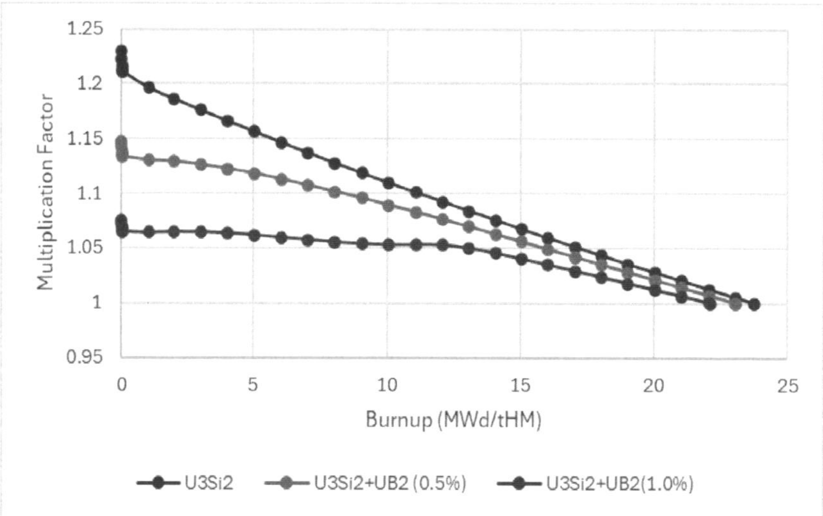

Figure 7: Equilibrium core multiplication factor evolution over the cycle length for different wt% of UB_2

The results shown in Figure 7 don't incorporate any heterogenous burnable absorber at 12/16 locations in each assembly. In order to explore the option of further suppression of the excess reactivity, HIGA rods are considered at those BA rod locations. Different loading of gadolinia is considered for the smallest reactivity swing over the cycle length. Another option studied is addition of different loading of uranium diboride in 2wt% enriched uranium silicide rods placed at the BA locations. This kind of configuration is being used currently in APR-1400. The cycle length obtained is 723.9, 674.2, 656.6, and 675.1 EFPD for uranium silicide, uranium silicide + 0.5 wt% uranium diboride + 5 wt% HIGA BA rods, uranium silicide + 1.0 wt% uranium diboride + 2.5 wt% HIGA BA rods, and uranium silicide + 1.0 wt% uranium diboride + (uranium silicide + 7.0 wt% uranium diboride) BA rods respectively. The results are shown in Figure 8.

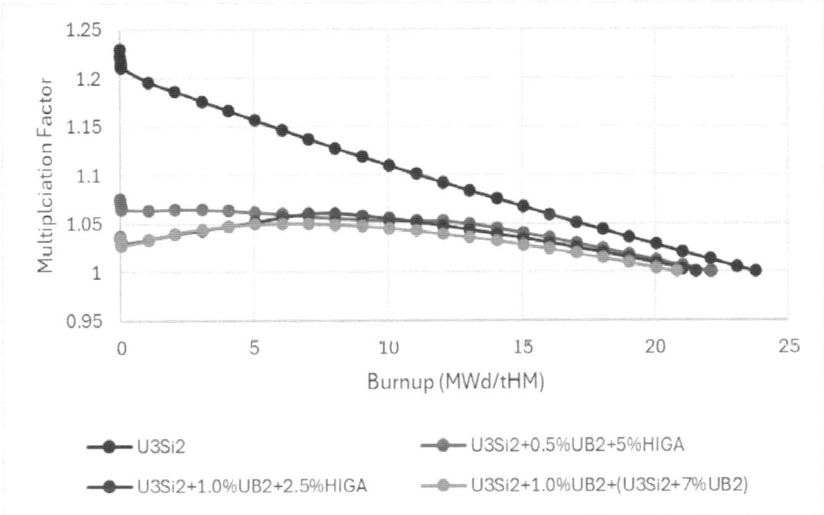

Figure 8: Equilibrium core multiplication factor evolution over the cycle length for different BA configurations

Another important parameter besides the cycle length is the radial power peaking factor (PPF). To choose a suitable configuration, its PPF should also abide by some limit. The overall PPF value for APR-1400 is 2.43. Considering the 2D analysis being performed here, a fixed axial PPF value of 1.3 along with engineering PPF value of 1.05 is assumed. This leads to a maximum radial PPF value of 1.78. Figure 9 displays the radial PPF value of the studied cases.

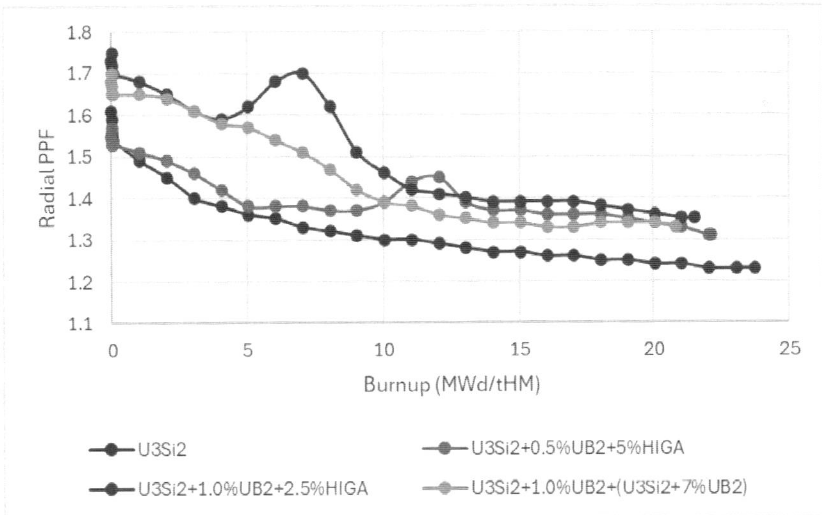

Figure 9: Equilibrium core radial PPF value over the cycle length for different BA configurations

The radial PPF shows higher values with the inclusion of burnable absorbers. As per Figure 8, the best reactivity suppression was obtained by uranium silicide + 1.0 wt% uranium diboride cases. Here in Figure 9, the same two cases show the highest value of radial PPF. These higher values still satisfy the criteria of <1.78.

Cladding Neutron Physics Impact: To study different cladding concepts and their neutron physics impact on APR-1400 equilibrium core configuration, uranium silicide + 1.0 wt% uranium diboride + (uranium silicide + 7.0 wt% uranium diboride) BA rods configuration studied previously was selected. Cladding options such as Chromium coated Zircalloy, SiC, and Stainless Steel 304 are analyzed here. Figure 10, shows the multiplication factor variation over the cycle length for the studied cases. The cycle length for Zircalloy, Cr-coated Zircalloy, SiC, and SS-304 came out to be 675.1, 670.2, 704.7, and 443.8 days respectively. The major outcome of the result is the severe reduction of cycle length when using SS-304. This is due to its very high absorption cross-section. For SS-304 to be used as a cladding, enrichment need to be increased to obtain the required cycle length. In comparison Cr-coated impact is minimal i.e. 5 EFPDs. On the other hand, SiC can have a positive impact on cycle length, an increase of ~30 EFPDs, as compared to Zircalloy. The reason for this being its absorption cross-section value which is even smaller than Zircalloy. Cr-coated zircalloy case considers 2.2 μm thick Nb as liner while thickness of Cr-coating is 10 μm. The Nb-liner helps stop the diffusion of Cr-coating into Zircalloy cladding. This option serves as a viable near-term solution to tackle steam + Zr reaction, as happened in Fukushima nuclear accident.

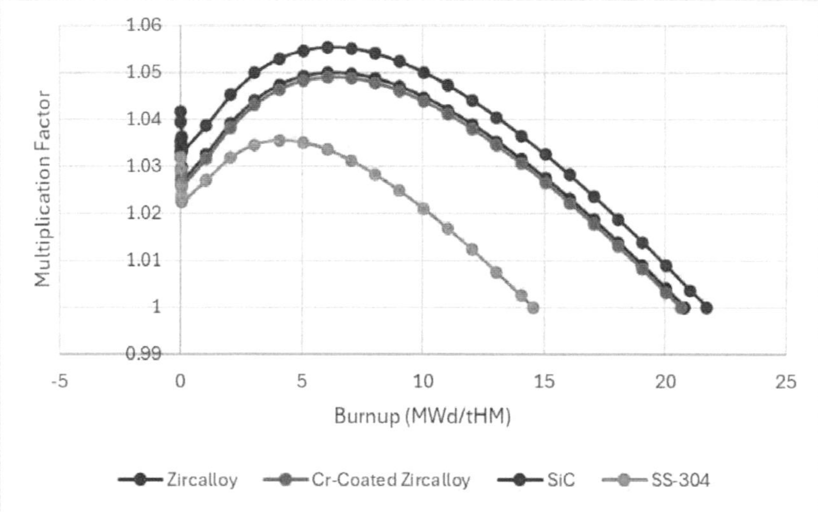

Figure 10: Equilibrium core multiplication factor value over the cycle length for different cladding materials

4. Conclusion and Future Work

The feasibility of ATF fuel/clad system for implementation of 24-month cycle in APR-1400 is studied. Uranium silicide and uranium diboride composite fuel provided very good cycle length and excess reactivity management for the required cycle length. To further suppress the excess reactivity, HIGA and 2 % enriched uranium diboride + uranium silicide fuel mixed in 7-93 wt% composition was studied. In terms of reactivity suppression and radial PPF, 2 % enriched uranium diboride + uranium silicide fuel mixed in 7-93 wt% composition option provide better results. Using this as the final fuel choice, neutron physics studies for Cr-coated zircalloy, SiC, and SS-304 claddings were performed. The analyses showed that SS-304 cause a severe cycle length penalty due to parasitic absorption of neutrons due to its very high absorption cross-section as compared to other cladding options being studied. SiC cladding can have a positive effect on the cycle length due to its smaller absorption cross-section as compared to zircalloy cladding. Cr-coated zircalloy showed similar performance in terms of cycle length to zircalloy with a minimal negative impact. In future, burnable absorber design studies will be performed to find an optimal solution for best reactivity suppression. The two-dimensional core model will be converted to a full three-dimensional model. Temperature reactivity coefficients for the ATF options will also be studied.

ACKNOWLEDGMENT

This research was supported by the UAE Ministry of Education (CRPG-2019, Grant No.: 1570604539) and the Emirates Nuclear Technology Center (ENTC), Khalifa University of Science and Technology, UAE. This work was partly supported by the National Research Foundation of Korea (NRF) Grant funded by the Korean Government (MSIT) (2022M2E9A304619011).

REFERENCES

Ahmed, H., Chaudri, K. S., & Mirza, S. M. (2016). Comparative analyses of coated and composite UN fuel – Monte Carlo based full core LWR study. *Progress in Nuclear Energy*, *93*, 260–266. https://doi.org/10.1016/J.PNUCENE.2016.08.014

Alrwashdeh, M., & Alameri, S. A. (2021). Preliminary neutronic analysis of alternative cladding materials for APR-1400 fuel assembly. *Nuclear Engineering and Design*, *384*, 111486. https://doi.org/10.1016/J.NUCENGDES.2021.111486

Alrwashdeh, M., & Alameri, S. A. (2022a). Chromium-Coated Zirconium Cladding Neutronics Impact for APR-1400 Reactor Core. *Energies, 15*(21), 8008.

Alrwashdeh, M., & Alameri, S. A. (2022b). SiC and FeCrAl as potential cladding materials for APR-1400 neutronic analysis. *Energies, 15*(10), 3772.

Brown, D. A., Chadwick, M. B., Capote, R., Kahler, A. C., Trkov, A., Herman, M. W., Sonzogni, A. A., Danon, Y., Carlson, A. D., Dunn, M., Smith, D. L., Hale, G. M., Arbanas, G., Arcilla, R., Bates, C. R., Beck, B., Becker, B., Brown, F., Casperson, R. J., ... Zhu, Y. (2018). ENDF/B-VIII.0: The 8th Major Release of the Nuclear Reaction Data Library with CIELO-project Cross Sections, New Standards and Thermal Scattering Data. *Nuclear Data Sheets, 148*, 1–142. https://doi.org/https://doi.org/10.1016/j.nds.2018.02.001

Chaudri, K. S., Tian, W., Su, Y., Zhao, H., Zhu, D., Su, G., & Qiu, S. (2013). Coupled analysis for new fuel design using UN and UC for SCWR. *Progress in Nuclear Energy, 63*, 57–65. https://doi.org/10.1016/j.pnucene.2012.11.001

Dunford, C. L. (1992). Evaluated nuclear data file, ENDF/B-VI. *Nuclear Data for Science and Technology: Proceedings of an International Conference, Held at the Forschungszentrum Jülich, Fed. Rep. of Germany, 13–17 May 1991*, 788–792.

Global Nuclear Fuel gets approval for fuel enriched up to 8 percent U-235 -- ANS / Nuclear Newswire. (n.d.). Retrieved July 29, 2024, from https://www.ans.org/news/article-5770/global-nuclear-fuel-gets-approval-for-fuel-enriched-up-to-8-percent-u235/

Khalefih, H., Oh, T., Jeong, Y., & Kim, Y. (2023). LEU+ loaded APR1400 using accident tolerant fuel cladding for 24-month two-batch fuel management scheme. *Nuclear Engineering and Technology*. https://doi.org/10.1016/J.NET.2023.04.009

Kim, J. S., Jung, T. S., & Yoon, J. (2024). Reactor core design with practical gadolinia burnable absorbers for soluble boron-free operation in the innovative SMR. *Nuclear Engineering and Technology*. https://doi.org/https://doi.org/10.1016/j.net.2024.03.015

Rhodes, J., Smith, K., & Edenius, M. (2004). *CASMO-4E: Extended Capability CASMO-4, User's Manual. Studsvik Scandpower*. Report SSP-09/443 University Release.

Romano, P. K., Horelik, N. E., Herman, B. R., Nelson, A. G., Forget, B., & Smith, K. (2015). OpenMC: A state-of-the-art Monte Carlo code for research and development. *Annals of Nuclear Energy, 82*, 90–97. https://doi.org/https://doi.org/10.1016/j.anucene.2014.07.048

Smith, K. S. (1989). SIMULATE-3: Advanced Three-Dimensional Two-Group Reactor Analysis Code. *STUDSVIK/SOA-8904*.

Turner, J., & Abram, T. (2020). Steam performance of UB2/U3Si2 composite fuel pellets, compared to U3Si2 reference behaviour. *Journal of Nuclear Materials, 529*, 151919. https://doi.org/10.1016/j.jnucmat.2019.151919

Zakova, J., & Wallenius, J. (2012). Fuel residence time in BWRs with nitride fuels. *Annals of Nuclear Energy, 47*, 182–191. https://doi.org/https://doi.org/10.1016/j.anucene.2012.03.033

Zimmerman, S. G., Brittingham, J. C., Reed, M. L., Bandera, R. P., & Crawley, P. F. (1999). PWR Reactor Physics Methodology Using CASMO-4/SIMULATE-3. *United States: Arizona Public Service Company*.

Prediction of fuel variability impacts on the combustion characteristics of $NH_3/H_2/N_2$ fuel mixtures using machine learning techniques

Israfil Soyler
School of Engineering and Materials Science, Queen Mary University of London, London, E1 4NS, United Kingdom,
i.soyler@qmul.ac.uk
ORCID: 0009-0001-9483-9995
Cihat Emre Ustun
School of Engineering and Materials Science, Queen Mary University of London, London, E1 4NS, United Kingdom,
c.e.ustun@qmul.ac.uk
ORCID: 0000-0003-3598-5973
Amin Paykani
School of Engineering and Materials Science, Queen Mary University of London, London, E1 4NS, United Kingdom,
a.paykani@qmul.ac.uk
ORCID: 0000-0002-1344-5549
Nader Karimi
School of Engineering and Materials Science, Queen Mary University of London, London, E1 4NS, United Kingdom,
n.karimi@qmul.ac.uk
ORCID: 0000-0002-4559-6245

Cite this paper as: *Soyler, I, Ustun, E.C, Paykani, A, Karimi, N. Prediction of fuel variability impacts on the combustion characteristics of $NH_3/H_2/N_2$ fuel mixtures using machine learning techniques. 7th Int. Conf. on Nucl. and Ren. Energy Res. 27-30 October 2024, Antalya, Turkey*

Abstract: As a promising hydrogen (H_2) carrier, pure ammonia (NH_3) faces low reactivity problems compared to conventional fuels. Products of catalytic NH_3 cracking ($NH_3/H_2/N_2$) can overcome the low reactivity issue. However, such blended fuels have high compositional uncertainties due to the operational variations, leading to significant issues such as flame instability and unexpected emissions. This underlines the importance of uncertainty quantification (UQ) and minimizing unexpected outcomes. UQ methods require many parameters, although surrogate models can be used to reduce the computational cost of the UQ approach. In the present work, artificial neural networks (ANNs) were used as a surrogate model to predict and quantify the uncertainty effects on $NH_3/H_2/N_2$ mixtures. The training dataset was collected on laminar flame speed, NO formation and adiabatic flame temperature from 216 different fuel compounds. This dataset was gathered at three different initial pressures and temperatures across 11 equivalence ratios. Training dataset consists of 21,384 data points which was divided into train (80%), validation (10%) and test sets (10%). The ANN architecture was optimised to minimise the mean absolute error while maximising coefficient of determination (R^2). The results suggest that the ANN model achieves R^2_{test} of >0.99 for all quantities of interest. Finally, ANN achieves speed-up ratios up to 8,000 compared to Cantera 1D simulations.

Keywords: *Uncertainty quantification, Data-driven modelling, Ammonia cracking, Hydrogen*

Numerical simulation of thermal-hydraulic performance of staggered tube bank with longitudinal triangular fins

Mustafa Soyler
Department of Energy Systems Engineering, Osmaniye Korkut Ata University, Osmaniye, Türkiye,
mustafasoyler@osmaniye.edu.tr,
ORCID: 0000-0003-4767-5825

Cite this paper as:	*Soyler, M. Numerical simulation of thermal-hydraulic performance of staggered tube bank with longitudinal triangular fins. 7th Int. Conf. on Nucl. and Ren. Energy Res. 27-30 October 2024, Antaya, Turkey*

Abstract: The use of fins for heat transfer enhancement is widely employed due to their effectiveness in increasing surface area and improving thermal performance. Triangular fins can enhance heat and flow performance by increasing the heat transfer surface area and elevating the turbulence of the flow. In this study, the flow and heat transfer characteristics of a tube bank with added longitudinal triangular fins were numerically investigated. The thermal-hydraulic performance of a triangular-finned tube bank was compared to that of a bare tube bank. To assess the thermal efficiency of the tube bank, longitudinal triangular fins with constant thickness (t/D=0.08) and different lengths (L/D = 0.50, 1.0 and 1.5) were used. The Reynolds number is within the range of 5000 ≤ Re ≤ 25000. At the Reynolds number of 10000, the L/D =1.5 has a Nusselt number that is about 25-30% higher than that of the L/D = 0.5. The result show that employing triangular fins leads to a reduction in friction factor and an increase in the Nusselt number, significantly enhancing the thermal-hydraulic performance of the tube bank.

Keywords: *Tube bank, Triangular fins, Nusselt number, friction factor*

Characterization of Microstructure and Corrosion Properties of Powder Metallurgy Ti-Al-Sn-Zr-Mo Alloy for Nuclear Applications

Eda Ergün Songül
İstanbul University-Cerrahpaşa, İstanbul, Turkey, eda.ergunsongul@iuc.edu.tr,
ORCID: 0000-0002-1123-502X

İlven Mutlu
İstanbul University-Cerrahpaşa, İstanbul, Turkey, imutlu@iuc.edu.tr,
ORCID: 0000-0002-4989-9219

Cite this paper as: *Ergün Songül, Mutlu, İ. Characterization of Microstructure and Corrosion Properties of Powder Metallurgy Ti-Al-Sn-Zr-Mo Alloy for Nuclear Applications. 7th Int. Conf. on Nucl. and Ren. Energy Res. 27-30 October 2024, Antaya, Turkey*

Abstract: In this study, a Ti-20Al-5Sn-8Zr-15Mo alloy was manufactured using powder injection moulding for use as structural (cladding) material for fuel rods and fuel assemblies in pressurized light water nuclear power reactors. Al, Zr, Mo, and Sn were added to Ti in order to decrease its thermal neutron absorption cross-section for neutron economy. The corrosion behaviour of the alloy was investigated at both room temperature and high temperatures in NaCl and Na2SO4 environments. Electrochemical corrosion tests were performed to determine the corrosion rate, and cyclic polarization was used to evaluate localized corrosion. In addition, an electrochemical potentiokinetic reactivation test was conducted to investigate the degree of sensitization to intergranular corrosion, and the critical pitting potential test characterized the critical repassivation potential. It was found that resistance against localized corrosion was enhanced with Mo, while the corrosion rate was reduced with Zr and Mo. Furthermore, the addition of Sn improved the sinterability of the alloy, and Al enhanced its resistance at high temperatures. Thermal neutron irradiation was carried out using a 241Am-Be neutron irradiation cell, and neutron irradiation was also modelled using the Monte Carlo code. Non-destructive tests were used to characterize the microstructure and mechanical properties.

Keywords: *Ti-Al-Sn-Zr-Mo, Corrosion, Neutron irridation, Nondestructive test, Pitting*

Numerical Analysis of Air-Channel Photovoltaic Thermal Collectors and Evaluation of Hydrogen Production

Murat ÖZTÜRK
National Defence University, Ankara, Turkey, muratzr610@gmail.com,
ORCID:0000-0002-0668-8075

Erdem ÇİFTÇİ
Gazi University, Ankara, Turkey, erdemciftci@gazi.edu.tr,
ORCID:0000-0003-2493-5962

Cite this paper as: *Öztürk, M. Çiftçi, E. Numerical Analysis of Air-Channel Photovoltaic Thermal Collectors and Evaluation of Hydrogen Production. 7th Int. Conf. on Nucl. and Ren. Energy Res. 27-30 October 2024, Antaya, Turkey*

Abstract: In this study, PVT (photovoltaic thermal collector) containing various air channels was designed and numerical analysis was carried out. Then heat was transferred to the water fluid with a classical type of heat exchanger and the obtained temperature values were examined. Finally, PEM electrolysers, which are available in the literature, were discussed and hydrogen production performance was evaluated with current models. Ansys Fluent software was used for numerical analysis of the developed models and the Computational Fluid Dynamics (CFD) approach was benefited. In this study, which examined 3 different PVT models, it was seen increase in the air outlet temperature that directly affecting hydrogen production, with adding air channels. In the results, the air output temperature was obtained as 34.50 °C for conventional PVT, 36.133 °C for 10-channel PVT and 45.25 °C for 18-channel PVT. In addition, the output temperatures of the water transmitting to the PEM electrolyser in the heat exchanger was also calculated as 31.60 °C for the conventional model, 35.54 °C for 10-channel model and 43.23 °C for 18 channels model. In this sense, the increase in water temperature in current models increases the amount of reaction occurring in the PEM electrolyser and affects the rate of hydrogen production. Depending on the current studies evaluated, the best production rate in is predicted to occur in the 18-channel model.

***Keywords**:* *Solar energy, energy storage, hydrogen production, computational fluid dynamics (CFD), PEM electrolyser*

1. INTRODUCTION

The growing world population directly affects energy demand. In this sense, the use of renewable energy sources, which can be alternative to basic energy sources such as fossil fuels, is increasing day by day. At the beginning of these energy sources, which can provide clean and continuous, solar energy systems are also coming. Photovoltaic thermal collectors that obtain thermal energy from solar energy are one of the most common solar energy systems.
Ural (2019) experimentally examined a solar air collector with absorbent plates from textile fabric. In his results, he stated that energy efficiency the model with the factory fabric was about 25% better and confirmed its availability [1]. Dawood et al. (2022) studied the performance of 7 circular hole window type solar air collectors that can move. In their results, they calculated the maximum thermal efficiency as 71% [2]. Ishak et al. (2023) performed exergy analysis of a PVT system with reverse circular flow jet multiplication. In their results, they stated the maximum exergy under 900W/m2 solar radiation, at 47.2 W [3]. Çiftçi and Ozturk (2023) have experimentally analyzed the thermal performance of a flexible aluminum air duct PVT and have indicated its thermal efficiency at around 74% [4]. On the other hand, in literature, many new collector models have been developed and their analysis performed in order for the collectors to reach a higher thermal performance [5-7].
On the other hand, many methods have been developed to store the energy produced. One of them is the production and storage of hydrogen by electrolysis. In this sense, a PVT supported hydrogen production model is developed and analyzed in many studies. Ma et al. (2024) Aimed for hydrogen production with a PVT-supported PEM electrolyzer and performed their analysis. In this experimental and numerical study, they determined that system efficiency increased with ambient temperature and that the PVT was added to improve the working efficiency [8]. Soyturk et al.

(2023) carried out analyses for hydrogen production with a small-scale PVT unit. In this system based on the Rankine cycle, they specified the rate of hydrogen production as 16 g/s [9]. Similarly, with different designs, the PVT-electrolyser model has been developed and evaluated in many studies [10-12].

In this study, a new model of air-based PVT-PEM electrolyzer was proposed. In the developed model, PVT and heat exchanger were used and numerical analysis of these elements was carried out. Subsequently, existing studies were examined and the potential for hydrogen production was evaluated. The analysis was carried out with the Computational Fluid Dynamics (CFD) approach and the fluid output temperatures occurring in each element were considered.

2. MATERIAL AND METOD

In this study, an PVT integrated hydrogen production system was modeled and its performance was evaluated. This model includes air-based PVT collector, conventional heat exchanger and PEM electrolyzer. Figure 1 shows the scheme of the model that has been developed. Initially, an air-channel PVT collector was developed and numerical analysis was carried out. Then a heat exchanger containing the air and water fluids was modeled and similarly numerical analysis was carried out. Finally, in line with the results obtained, the existing literature was evaluated and the hydrogen production performance of the developed system was examined. In this context, Ansys Fluent software was used for numerical analysis and Computational Fluid Dynamics (CFD) approach was used.

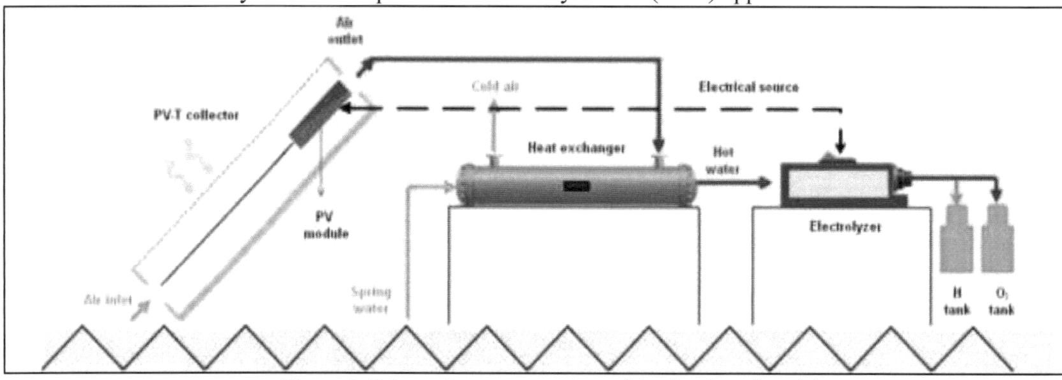

Figure 1. Schematic representation of the developed model

On the other hand, the dimensions of the PVT designed are 980x300x75 mm. The air inlet and outlet points are performed of 75x75 mm channels and the absorber plate is designed 500x300 mm. Figure 2(a) shows the PVT model designed and figure 2(b) shows the absorber plate designed. As basis, it's modeled 3 different collectors as the conventional type, 18 air channels and 10 air channels, and for each of them, the thermal analysis was carried out. The number of mesh applied in numerical analysis is 8590630 for the conventional type, 10 air ducts for 8637550 for 18 air ducts and 8675086 for 8 channels. In addition, the collector material was applied aluminum and was preferred copper material for the absorbent surface. The air intake speed and temperature were set at 2.8 m/s and 20 °C respectively, and analyses were completed under 1300 W/m^2 solar radiation.

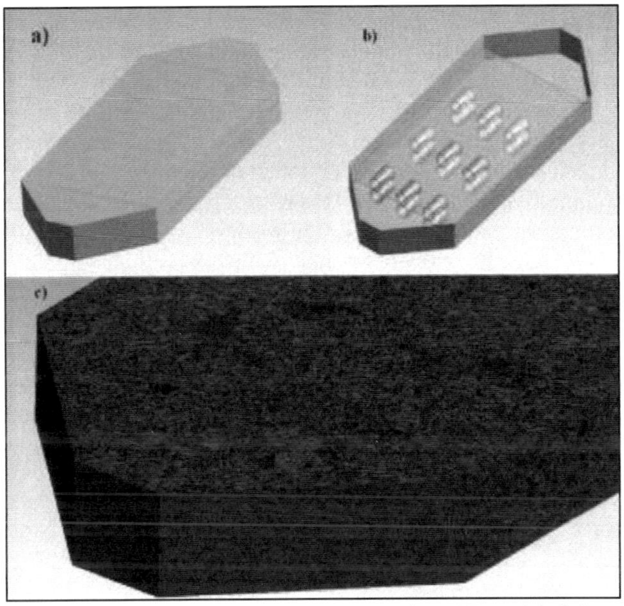

Figure 2(a). PVT collector model, figure 2(b). Absorbent plate, located in PVT modeled, figure 2(c). PVT collector mesh structure

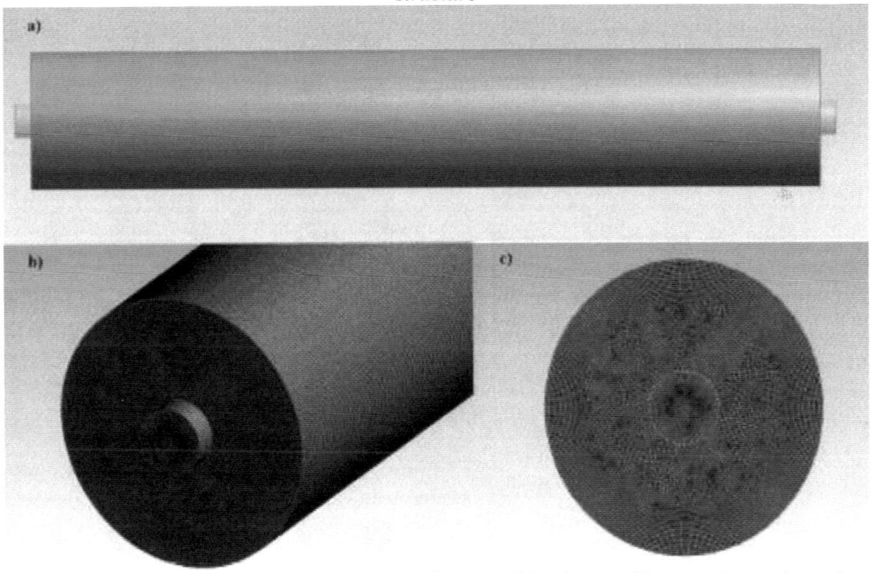

Figure 3(a). Model of designed heat exchanger, figure 3(b). Side view of the designed heat exchanger's mesh model, figure 3(c). Front view of the designed heat exchanger's mesh model

The hot air obtained in the PVT collector is intended to be transferred to the water fluid, so a numerical analysis of the water temperature that can be obtained in a conventional heat exchanger is performed. The modeled heat exchanger is a conventional type heat exchanger with shell-tube. The inner and outer pipe diameters are 25 mm and 100 mm, respectively. In addition, the pipe length is determined as 500 mm and the pipe material is preferred as copper. The number of mesh for the analysis to be performed is determined as 6802791, and the inlet temperature and speed of the cold fluid to the heat exchanger are determined as 10 °C and 0.5 m/s. In addition, the hot air is also the conditions at the PVT output. Figure 3(a), figure 3(b), and figure 3(c) respectively show the heat exchanger model, the side view of the mesh structure, and the front view of the mesh structure. Depending on the water fluid temperature obtained, hydrogen production performance was evaluated and shared in the results.

3. RESULTS AND DISCUSSION

In this section, the results obtained in PVT and heat exchanger, which were analyzed numerically, were shared and the potential of hydrogen production was evaluated. Figure 4 shows the air outlet temperatures in the PVT collector as a result of the analysis. In the results obtained air outlet temperature for conventional type PVT is 34.50 °C, the air outlet temperature for the 10-channel PVT is 36.133 °C and the air outlet temperature for the 18-channel PVT is 45.25 °C. The increase in air ducts directly affects air temperature and increases collector performance. The surface area where the air comes into contact is increasing, thus reaching higher fluid temperatures. On the other hand, figure 5 shows the water output temperatures obtained in the modeled heat exchanger. The water output temperature is obtained as 31.60 °C, 35.54 °C and 43.23 °C, respectively, for conventional type PVT integrated heat exchanger, 10-channel PVT integrated heat exchanger and 18-channel PVT integrated heat exchanger. The air heated in the collector directly affects the water temperature in the heat exchanger. In this sense, the highest temperature is derived from the 18-channel PVT integrated heat exchanger.

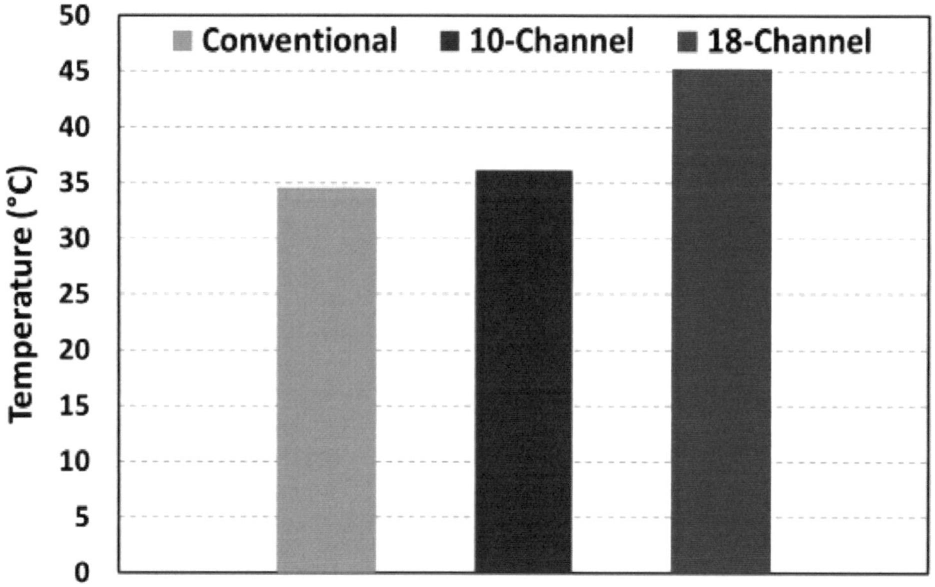

Figure 4. Air outlet temperature of PVT collectors

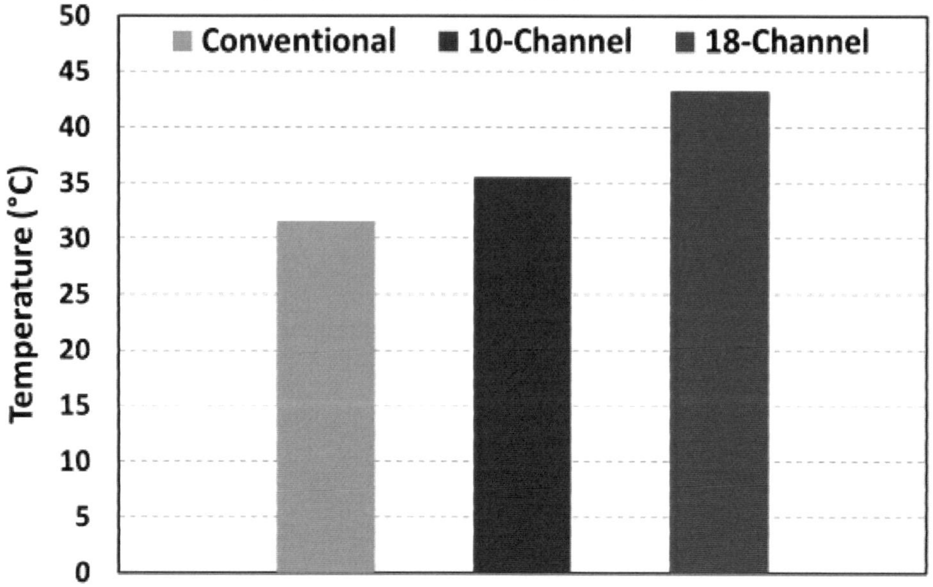

Figure 5. Water outlet temperature in heat exchangers

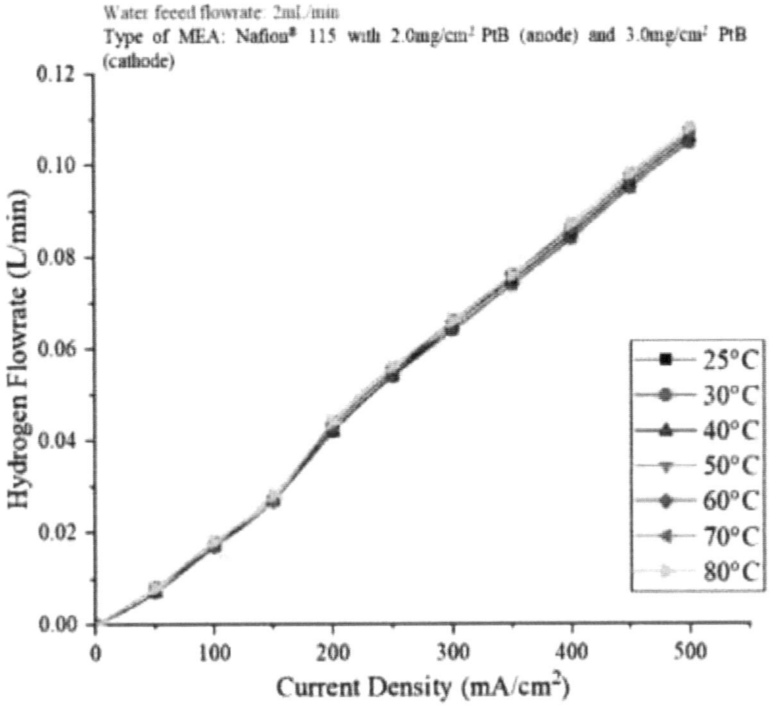

Figure 6. Hydrogen production at different operating temperatures [13]

Figure 6 shows hydrogen production at different operating temperatures. In this study, the maximum water output temperature was obtained from the 18-channel PVT collector integrated heat exchanger. In this sense, it is estimated that maximum hydrogen production can be obtained from 18-channel PVT collector. Azam et al. (2023) stated that the maximum hydrogen flow at figure 6 is at 80 °C operating temperature. In this sense, the increase in the temperature of the fluid entering the electrolyser and the increase in hydrogen production are changing at the right rate.

4. CONCLUSION

In this study, a hydrogen production system supported by PVT collectors containing a variety of air channels is examined. 10 and 18 air ducts were added to a conventional PVT collector, later, and numerical analysis of both the PVT collectors and the heat exchanger used to transfer heat to the water fluid was performed with Ansys Fluent software, and the hydrogen production performance was evaluated. As a result of the analysis, the following results were obtained:

•With the addition of air ducts, increased in the air outlet temperatures were seen. In addition, the increase in air temperatures directly affects the water temperature obtained from the heat exchanger.

•The increase in fluid temperatures directly affects hydrogen production.

•The air-output temperatures from conventional PVT, 10-air channel PVT and 18-air channel PVT were achieved as 34.50 °C, 36.133 °C and 45.25 °C respectively.

•In addition, the water temperatures calculated from the conventional PVT integrated heat exchanger, 10-channel PVT integrated heat exchanger, 18-channel PVT integrated heat exchanger were obtained as 31.60 °C, 35.54 °C and 43.23 °C respectively.

•Maximum hydrogen production can be estimated to be obtained from 18 air-channel PVT integrated models.

REFERENCES

[1] Ural, T.. Experimental performance assessment of a new flat-plate solar air collector having textile fabric as absorber using energy and exergy analyses, *Energy* 2019; 188, 116116.

[2] Dawood, N. I., Jalil, J. M., & Ahmed, M. K.. Investigation of a novel window solar air collector with 7-moveable absorber plates. *Energy* 2022; 257, 124829.

[3] Ishak, MAAB, Ibrahim, A., Fazlizan, A., Fauzan, MF, Sopian, K. ve Rahmat, AA. Ters dairesel akışlı jet çarpmalı çift taraflı fotovoltaik termal (PVT) güneş kollektörünün ekserji performansı. *Isı Mühendisliğinde Vaka Çalışmaları* 2023 ; 49 , 103322.

[4] Öztürk, M., & Çiftçi, E. Upgrading the performance of a solar air collector with flexible aluminum air ducts and graphene nanoplatelet-enhanced absorber coating. *Thermal Science and Engineering Progress* 2023 ; 40, 101760.

[5] Öztürk, M., Yüksel, C., & Çiftçi, E. Energy, exergy and sustainability analysis of a photovoltaic-thermal solar system with nano-enhancement and thermal energy storage integration. *Process Safety and Environmental Protection* 2024; 187, 593-604.

[6] Li, S., Zeng, F., & Wei, X. A Study on the Performances of Solar Air Collectors Having a Hemispherical Dimple on the Absorber Plate. *Fluid Dynamics & Materials Processing* 2024; 20(5).

[7] Ul-Abdin, Z., Zeman, M., Isabella, O., & Santbergen, R. Investigating the annual performance of air-based collectors and novel bi-fluid based PVThermal system. *Solar Energy* 2024; 276, 112687.

[8] Ma, Y., Zhao, M., Bai, F., Yu, R., Liu, L., & Wang, J. Numerical simulation and experimental verification of solar PVT coupled PEM electrolyzer system for hydrogen production. *Fuel* 2024; 365, 131323.

[9] Soyturk, G., Kizilkan, O., Ezan, M. A., & Colpan, C. O. PVT integrated hydrogen production with small-scale transcritical power cycle. *Process Safety and Environmental Protection* 2023; 180, 351-360.

[10] Nasser, M., Awad, M. M., & Hassan, A. A. 4E assessment of all-day clean electricity generation systems based on green hydrogen integrated system using PV and PVT solar collectors and wind turbines. *International Journal of Hydrogen Energy* 2024.

[11] Akrami, E., Nemati, A., Nami, H., & Ranjbar, F. Exergy and exergoeconomic assessment of hydrogen and cooling production from concentrated PVT equipped with PEM electrolyzer and LiBr-H2O absorption chiller. *International Journal of Hydrogen Energy* 2018; 43(2), 622-633.

[12] Senthilraja, S., Gangadevi, R., Marimuthu, R., & Baskaran, M. J. I. J. O. H. E. Performance evaluation of water and air based PVT solar collector for hydrogen production application. *International journal of hydrogen energy* 2020; 45(13), 7498-7507.

[13] Noor Azam, A. M. I., Li, N. K., Zulkefli, N. N., Masdar, M. S., Majlan, E. H., Baharuddin, N. A., ... & Shaffee, S. N. A. Parametric study and electrocatalyst of polymer electrolyte membrane (PEM) electrolysis performance. *Polymers* 2023; 15(3), 560.

NURER 2024
Nuclear and Renewable Energy Resources

TsunamiFloodProtection- System

Jens Wieneke
INNOMECOM AG, Weekendweg 5, Einigen, CH-3646 Einigen, Switzerland, jens .wieneke@innomecom.com
ORCID: 0009-0008-2286-0729

Cite this paper as: *Wieneke, Jens. TsumaniFloodProtection- System. 7th Int. Conf. on Nucl. and Ren. Energy Res. 27-30 October 2024, Antaya, Turkey*

Abstract:

Part of the Fukushima problem

The Fukushima disaster in 2011 illustrates the devastating effects that tsunamis can have on nuclear power plants and represents an alarming turning point in the debate about the safety of these plants. Although the Fukushima Daiichi nuclear power plant already had a flood wall of 4 meters in height, this was raised to 5.7 meters after the Indian Ocean tsunami in 2004 - a response to the lessons learned from this event.

It has become clear that the construction of flood walls alone is not sufficient to cope with the potential risks of tsunamis for nuclear power plants. A multidisciplinary approach and, in the best case, a complementary passive system is needed crease the safety of these plants and thus prevent similar disasters in the future.

Development of the TFP-System

The TFP- System protects critical building openings from emergency diesel buildings against Tsunmis and heavy flooding – target is to save the diesel buildings and to save the electrical power supply after the external hazard.

In order to meet the high nuclear safety standards, the closing function of the TsunamiFloodProtection (TFP) system was reduced to the essentials.

The system is fully redundant thanks to design features such as the dual float system design and the option of installing two modular installation frames with the modular TFP modules on top of each other.

The passive closing and passive opening of the TFP system is achieved by the buoyancy forces (hydrostatic buoyancy) of the polluted water caused by a tsunami or severe flooding, which act on the redundant float systems.

It can withstand large masses of mud and floating debris, extreme weather conditions as well as high earthquake loads.

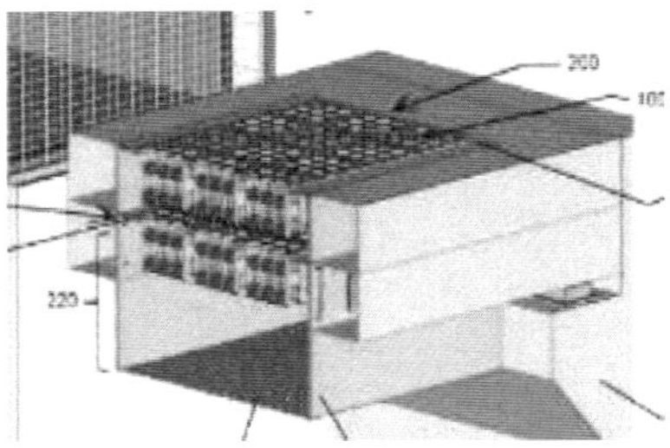

Keywords: *Tsunami, Severe flooding, Nuclear Safety, Investment Protection, TFP- System*

Mitigating Recirculation Zones in Molten Salt Fast Reactor Models

Ismail Yagci
Sinop University, Sinop, Türkiye, iy.nuclear@gmail.com
ORCID: 0009-0004-7936-9261

Zekeriya Parlak
Sakarya University, Sakarya, Türkiye, zparlak@sakarya.edu.tr
ORCID: 0000-0002-2487-0065

Huseyin Sahiner
Sinop University, Sinop, Türkiye, huseyinshnr@gmail.com
ORCID: 0000-0002-3191-1590

Cite this paper as: *Yagci, I, Parlak, Z, Sahiner, H. Mitigating Recirculation Zones in Molten Salt Fast Reactor Models. 7th Int. Conf. on Nucl. and Ren. Energy Res. 27-30 October 2024, Antaya, Turkey*

Abstract: The Molten Salt Fast Reactor (MSFR) is a liquid-fueled nuclear reactor that uses molten salt as both fuel and coolant. It works at atmospheric pressure with no solid components in the reactor core. As part of the Generation IV reactor concepts, it can utilize the thorium fuel cycle (Th232/U233). The reactor emphasizes passive safety, with a liquid fuel design that eliminates the risk of meltdown. In case of overheating, the fuel automatically drains into subcritical tanks for a safe shutdown. However, the thermal-hydraulic behavior of molten salts remains poorly understood, particularly the formation of recirculation zones near the blanket. Recirculation zones can lead to temperature hot spots and material stress. This study aims to minimize recirculation issues in MSFR reactors using ANSYS Fluent CFD simulations. The investigation covered multiple models, including a 16-inlet/outlet MSFR and a 6-inlet Small Modular MSR (SM-MSR). The 6-inlet SM-MSR model showed fluid behavior and more stable flow, indicating better performance. The obtained geometry allows spiral motion rather than recirculation zones, even under varying physical conditions.

Keywords: *Molten Salt Reactor, Probabilistic Risk Analysis, Fault Tree, Event Tree*

NURER 2024
Nuclear and Renewable Energy Resources

Experimental Studies of a Double-Circuit Solar Installation with Thermosiphon Circulation with A Heat Pump

Gulnur Tyulepberdinova
Al-Farabi Kazakh National university, Almaty, Kazakhstan, tyulepberdinova@gmail.com
ORCID: 0000-0002-4322-8983

Murat Kunelbayev
Al-Farabi Kazakh National university, Almaty, Kazakhstan, Institute of Information and Computational Technologies CS MES RK, Almaty, Kazakhstan, murat7508@yandex.kz
ORCID: 0000-0002-5648-4476

Gulshat Amirkhanova
Al-Farabi Kazakh National university, Almaty, Kazakhstan, gulshat.aa@gmail.com
ORCID: 0000-0003-3933-5476

Ardak Tolepberdinova
Al-Farabi Kazakh National university, Almaty, Kazakhstan, ardak.9t@mail.ru
ORCID: 0009-0000-6692-858X

Cite this paper as:	*Gulnur Tyulepberdinova, Murat Kunelbayev, Gulshat Amirkhanova, Ardak Tolepberdinova Experimental Studies of a Double-Circuit Solar Installation with Thermosiphon Circulation with A Heat Pump. 7th Int. Conf. on Nucl. and Ren. Energy Res. 27-30 October 2024, Antaya, Turkey*

Abstract: This paper examines thermosiphon circulating flat solar collectors paired with heat pumps for heating and hot water supply on livestock farms. A mathematical analysis divided the two-circuit solar installation into three components: 1) flat solar collector and heat pump evaporator; 2) heat pump condenser and storage tank; 3) heat pump itself.

The study proposed a new heat pump system that includes a flat solar collector, fan model, and circulation pumps. Key parameters analyzed include the temperature of the supplied medium, cooling capacity, and power consumption of the heat pump compressor. Significant findings indicate that increasing the collector area and absorption coefficient enhances system performance. The heat balance equation allowed calculations of heat transfer rate, energy consumption, and collector efficiency. The proposed system is advantageous due to its extended operational time and low capital costs. Experimental results revealed energy conversion efficiency ranging from 2.5 to 4.5 units.Using 50 liters of water, heating from 200°C to 450°C was achieved in 60 minutes, with collector efficiency between 40% and 50%. Power generation from flat solar collectors ranged from 1.6 to 3 kW/m², linked to reduced optical losses and heat flow resistance. The study suggests future solar heat supply systems could integrate neural networks and predictive machine learning for further optimization.

***Keywords**:* *flat solar collector, dual-circuit solar system with thermosiphon circulation, mathematical model, heat pump performance, exergy analysis.*

1. INTRODUCTION

When using traditional energy, which requires the production of fossil fuels or coal, it leads to pollution and harms the environment. Such reasons have prompted many researchers to turn to renewable energy sources and use them. In [1] describes the early attempts to develop solar energy, the use of passive solar energy in architecture before the 2nd

World War, and the subsequent growth of the modern photovoltaic industry. It explores the role of the supposed long-term potential of solar energy in combination with the capital intensity and cyclical nature of the industry. In [2] offers current coverage of the mathematical theory of solar heat supply systems, the design of systems and applications in various market sectors, along with an emphasis on the design and analysis of solar systems using simulations. In [3], a mathematical and numerical study of computational fluid dynamics was developed for comparison between conventional and spiral models to increase thermal characteristics using a spiral standing pipe, which increases the performance and temperature of the heat exchanger. In the article [4], a mathematical analysis of the heat transfer process of flat plate solar collectors with a ribbed tube was developed and investigated, which showed an increase in heat transfer. In [5], a solar water heater with a cross-section of an absorption pipe was developed. It was recorded that the absorption tubes absorb a large amount of solar radiation compared to another absorption tube, in which the efficiency of the absorption tube increased by 10-12%. In [6], a flat solar collector was made in which there is an absorbent tube, an absorbent plate, and a transparent cover. In the study, various mathematical modeling methods were used to improve the increase in the heat transfer characteristics of a flat plate collector by inserting (twisted tapes, and wire coil) into a copper tube and by changing the surface area of the copper tube in accordance with the design. The article is used to improve the heat transfer characteristics of a solar flat plate collector by using an internal corrugated copper tube instead of the existing copper tube. In [7], a new mathematical model was presented of the effect of the geometry of the cross-section of a heat pipe without a wick and the filling coefficient of its working fluid on the performance of flat plate solar collectors. Three groups of non-contact heat pipes having three different cross-sectional geometries (namely circular, elliptical, and semicircular cross-sections) were designed and manufactured. It has been experimentally proved that flat plate solar collectors with heat pipes without a wick of elliptical cross-section have better performance than solar collectors of circular cross-section, with low water filling coefficients. The paper [8] presents experimental results of using a water-based nanofluid Al2O3 as a working fluid for a flat tube in a plate-type solar collector. A mathematical model of thermodynamics and energy parameters in a solar collector was also developed. An experimental test facility was developed, which includes a solar collector, a closed system of the working fluid, and measuring instruments (flow meter, thermocouples, temperature meter, and digital solar energy meter). In the basic version, experiments were carried out with deionized water at a flow rate of 1 l/min. In the second case, Al2O3 nanoparticles are mixed with deionized water to obtain a nanofluid with a volume fraction concentration of 0.5%. The maximum difference between the temperature at the outlet and at the inlet of the solar collector was 14.4 °C with solar illumination of about 788 W/m, while in the case of water, the maximum temperature difference was 10.7 °C with solar illumination of about 781 W/m. In [9], the characteristics of heat transfer and consumption of a device for collecting solar energy using CFD modeling were studied. In the article [10], the concept was studied in the example of a sheep farm. The newly-made design performs these functions; it recycles heat, organizes its movement and accumulation, and also equalizes the unstable distribution. The key components of the system are a solar thermal power plant (SES), a milk cooler, an air conditioning unit, a thermal pump (HP), a heating battery, a mechanical control design, as well as heating and hot water devices. The main goal, a lower price of the energy produced and the elimination of unevenness of SE, in comparison with the familiar SPP, is achieved due to the flow of energy from the sources listed above The purpose of this work is experimental studies of a two-circuit solar installation with thermosiphon circulation with a heat pump. For the experimental study, the work was carried out in five stages. In the first stage, a new design of a flat solar collector was developed. In the second stage, a heat pump was developed. In the third stage, a two-circuit solar installation with a thermosiphon was developed. In the fourth stage, the theoretical part was developed. At the fifth stage, studies were experimentally tested to measure the temperature of the coolant in the evaporator, and condenser, changes in the efficiency of the heat pump depending on the date and time, as well as energy and exergy indicators from hourly changes in meteorological conditions in Almaty (Kazakhstan).

2. SECTIONS

The operation of the proposed installation is as follows. In a flat solar collector 1, in accordance with the invention: instead of a single transparent glass, a heat-insulating transparent double-glazed window 2 with double glass and with reduced pressure, as well as a perimeter frame 1, the bottom 4 is made of plywood with a thickness of 8 mm. and a heat-insulating film 6 with foil is glued to them. In the gap formed between the double-glazed window and the bottom of the frame in accordance with the invention: instead of a flat pipe (copper, aluminum), a flexible thin-walled stainless corrugated tube 4 is laid. in the form of a coil 3. The ends of the tube attached to the inlet and outlet protruding pipes of the siphon 5. There is a siphon of the dispenser tank 7, a pipeline with a valve for cold water 8, as well as a dispenser tank 9. A dispenser tank 9 with a solar collector 1 is connected using a circulation pipe 10. The heat pump 11 consists of an evaporator 12, a compressor 13, a throttle valve 14, a condenser 15. The accumulator tank 16 includes a condenser heat exchanger 17, an automatically controlled backup electric heater 18, a heating system heat exchanger 19. Installation equipped with automatically controlled circulation pumps of the first 20 and second circuits 21.

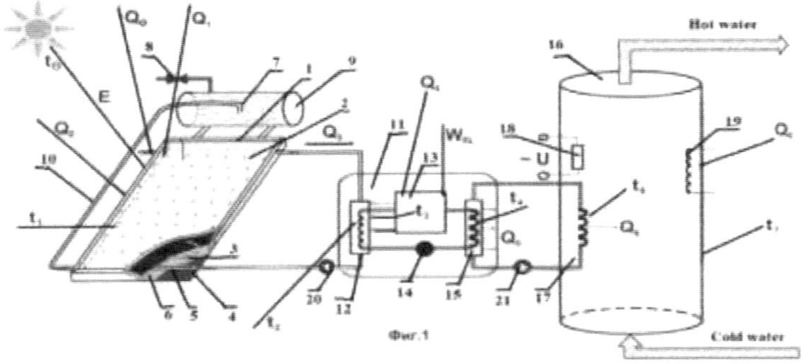

Figure 1. Principal diagram of double circuit soalr installation with thermosiphon circulation

Figure 1 shows a schematic diagram of a dual-circuit solar system with thermosiphon circulation. The system, in this study, is used for hot water supply and space heating during the winter period and mainly consists of a flat solar collector, a siphon metering tank, a liquid heat pump, and a hot water storage tank.

2.1. DOUBLE-CIRCUIT SOLAT INSTALLATION WITH THERMOSIPHON CIRCULATION

As mentioned above, a double-circuit solar system with thermosiphon circulation can absorb heat from a solar source. The bottom of the wooden frame 7 is made of 8 mm thickness plywood. and heat-insulating film 5 with foil glued to it.

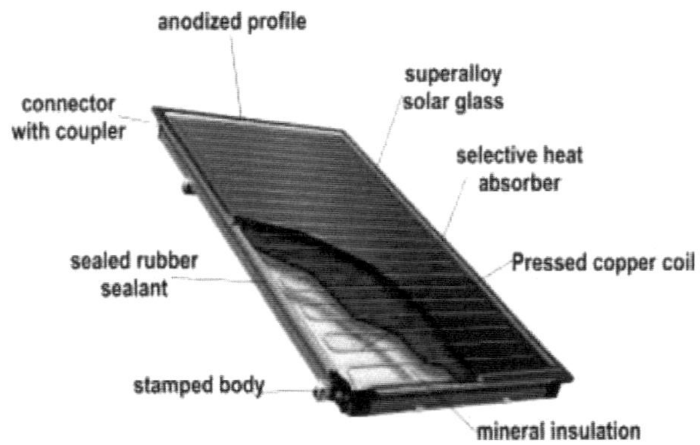

Figure 2. Principal diagram of flat solar collector

Figure 2 shows a model of a flat solar collector. The essence and novelty of the proposed one are that, unlike the well-known design principle, the collector contains a transparent double-glazed window 2 with double glass and with reduced pressure, as well as a perimeter frame 1.

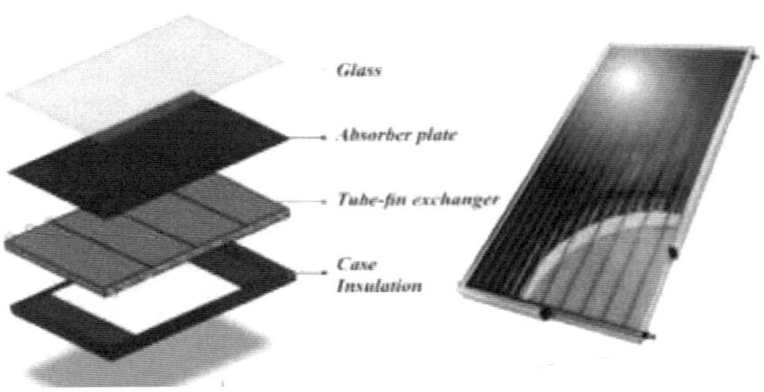

Figure 3. Principal diagram of flat solar collector in parts

Figure 3 shows in the gap, formed between the double-glazed window and the bottom of the frame, a flexible thin-walled stainless corrugated tube of 4-16 mm is laid in the coil form. The edges of the tube are attached to the inlet and outlet protruding pipes 6.
As shown in Figure 4, the solar energy passes through the glass and hits the absorber plate, which heats up, converting solar energy into thermal energy. Heat is transferred to the working fluid, which passes through tubes, attached to the absorber plate.

Figure 4. Flat solar collector mockup

Figure 4 shows a full-scale model of a flat solar collector. The solar collector is the main heat generating unit of the solar installation. To achieve this goal, we have developed a fundamentally new flat solar collector, on the basis of which various types of solar systems will be created, both according to sizes and design, applied to water heating and premises heating. Ethylene glycol was used in the working fluid. Ethylene glycol ($C_2H_6O_2$) is a transparent colorless liquid with elements of oily consistency.

2.2 Theoretical analysis

Based on the purpose of the work, the maximum heating capacity is taken as the target function:

$$Q = f(E; W_{EQ}; Q_1 ... Q_n; t_0 ... t_n; S ... S_n) \rightarrow \max \tag{1}$$

Provided that the heat supply to the consumer is uninterrupted:

$$Q + Q_{EQ} \geq Qconsumer$$

Where Q_1 ... Q_n- the totality of heat flows [Joule], $t_0 ... t_n$- temperature characteristics of the device, ^0C; $S ... S_n$- the surface area of the main elements (solar collectors, capacitor surfaces, battery tank), m^2.
To perform a theoretical analysis, the installation is divided into three conditional blocks:1. Formed by a flat solar collector and a heat pump evaporator; 2. Formed by a heat pump condenser and a storage tank; 3. In the center formed by the heat pump itself.
In view of the fact that the circuits form a sequential chain of blocks, the maximum heat of the performance of each individual circuit, and inside the circuit-at the maximum heat of the performance of each individual element.
Analysis of the first contour. The totality of the processes of the first circuit is expressed by the heat balance equation:

$$(E - Q_0 - Q_1 + 2Q_2)S - Q_3 = C_1 \frac{dt_1}{d\tau} \tag{2}$$

where C_1- the total heat capacity of the primary circuit (includes the heat capacity of the coolant), Joule/K; t_1-the current temperature of the coolant,^0C; τ- time, hour.
In this study, "Dimensionless Time" was used - this is just a way of scaling the time coordinate with respect to the mass of the hole. Some sources also use a dimensionless radial coordinate, which is constructed basically in the same way as the ratio of the usual radial coordinate to the time-constant mass of the hole.
We can also say that "Dimensionless Time" is when we can express the known time interval from sunrise to sunrise as 1 day, 24 hours, 1440 minutes, or 86,400 seconds. The numbers change, but we are talking about the same time interval. A quantity is called dimensionless if its value remains unchanged with any choice of the system of units.

It follows from (2) that the value of the derivative can serve as a measure of the estimation of heating capacity, that is, the higher it is, the greater the heating capacity:

$$\frac{dt_1}{d\tau} \rightarrow \max \tag{3}$$

Based on (2), a detailed heat balance equation is compiled:

$$[E - k_0E - k_1 * E(1-k_0) + U_L(t_0 - t_1)]S - W_{EL}[a - b(t_4 - t_1) - 1] = C_1\frac{dt_1}{d\tau} \tag{4}$$

The solution of equations (4) is made:

$$\int_0^{\tau} d\tau = C_1 \int_{t_1=t_0}^{t_1} \frac{dt_1}{[E - k_0E - k_1 * E * (1-k_0) + U_L(t_0 - t_1)]S - W_{EL}(a-1) + W_{EL} * b(t_4 - t_1)} \tag{5}$$

The dependence of the current temperature of the primary circuit coolant on a set of related factors is obtained:

$$t_1 = -\frac{[W_{EL} * b(t_4 - t_0)]\exp[-\frac{(W_{EL} * b + U_L)\tau}{C_1}}{(W_{EL} * b + U_L S)} + \frac{[E(1-k_0-k_1+k_1 * k_0)S - W_{EL}(a-1)]\{1-\exp[-\frac{(W_{EL}b+U_L)\tau}{C_1}]\} + W_{EL}b * t_4 + U_L S * t_0}{(W_{El}b + U_L S)} \tag{6}$$

Verification was performed (6). It was received when τ=0, t_1=t_0 which meets the condition of the task. At $\tau=\infty$ the same time, we obtain a formula for calculating the temperature of the coolant at the outlet of the first circuit with steady-state operation.

$$t_1 = \frac{[E - k_0E - k_1 * E(1-k_0) - U_l * t_0]S - W_{EL}(a + b * t_4 - 1)}{(W_{EL}b + U_L S)} \tag{7}$$

The analysis showed that when the installation works with a heat pump, the direction of the heat flow between the gel collector and the environment changes.
Analysis of the heat pump circuit. For solving the analysis of the thermal pump, we will formulate the aggregate of thermal pump contour processes by means of the thermal balance equation:

$$Q_3 + W_{EL} * \eta_{EL} * \eta_k + Q_4 - Q_5 - Q_6 = C_2 d\tau\frac{dt_5}{d\tau} \tag{8}$$

where: $W_{EL} \cdot \eta_{EL} \cdot \eta_K$ – the energy entering the contour from operation of a compressor, Joule; Q_4 – utilizable heat, emitted with a compressor during the working process, Joule; Q_5 and Q_6 – accordingly, heat losses through the condenser heat exchanger surface and the energy, supplied by the thermal pump to absorber tank, Joule; C_2 – heat capacity of condenser contour transfer medium, Joule/K; t_5 – transfer medium current temperature at the output from a contour,^{0}C.
The main element of the circuit under study is the heat pump. Figure 5 shows a diagram of a new technical solution of the thermal pump in section.
The proposed device allows you utilize the heat, generated by the compressor during operation and at the same time cool it.

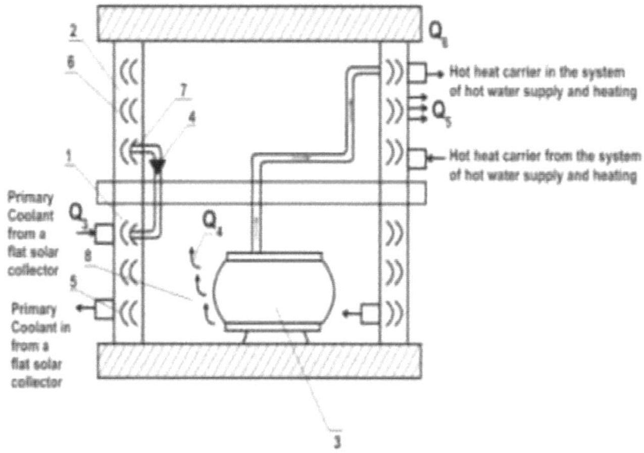

Figure 5. Scheme of new technical solution of TP in section

Figure 6 shows hereby 1 is the heat exchanger of the evaporator; 2 - condenser heat exchanger; 3 - compressor; 4 - throttle valve; 5 and 6 - tubes for the coolant of the evaporator and condenser; 7- hole for introducing tubes into the capacitor housing; 8 - mesh shell-divider of air flow.

The heat exchangers of evaporator 1 and condenser 2 are made in the form of ring-shaped containers, formed by the inner and outer cylindrical shells with radii R1 and R2 and are mounted coaxially above each other, at the bottom of the evaporator, above the condenser, forming an internal cylindrical air cavity. Compressor 3 is installed in the cavity of the evaporator heat exchanger. To ensure optimal heat transfer from the compressor to the evaporator, a cylindrical mesh shell-divider 8 is placed in the gap between them. The coolants of the 1st and 2nd circuits circulate inside the ring tanks, removing heat from tubes 5 and 6. Thus, the compressor is practically located inside the "cold pan", the walls of which are cooled by the coolant of the tubes 5. As a result, the heat, generated by the compressor is absorbed by the HP evaporator, increasing its performance, and the compressor is simultaneously cooled without the use of a fan.

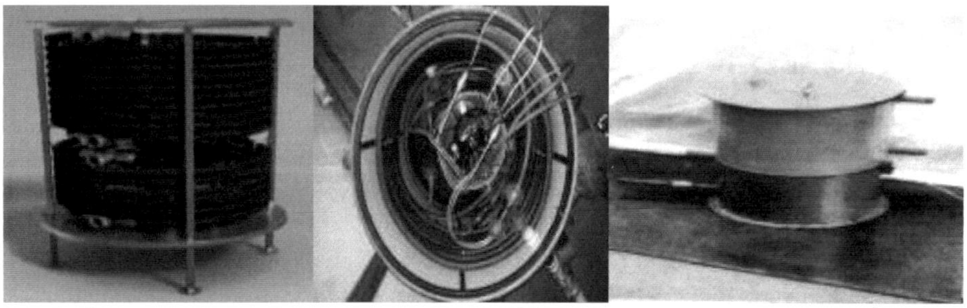

a) b) c)

Figure 6. Heat pump

Figure 6 shows a diagram of the new technological conclusion of the heat pump in the section. The proposed installation allows you to use the heat generated by the compressor during operation and simultaneously cool it.

Table 2. Technical characteristics of the new heat pump

#	Design parameters	Dimensions
1	Evaporator and condenser external diameters (D_2), mm	375,0
2	Internal diameter (D_1), mm	325 and 307
3	Heat exchangers jackets height, mm	355 and 382
4	Heat exchangers jackets width, mm	20 and 30
5	Heat exchangers jackets volume, mm	0.52 and 0,8
6	Shell-divider diameter, mm	238
7	Surpassing the evaporator height over compressor height, mm	80,0

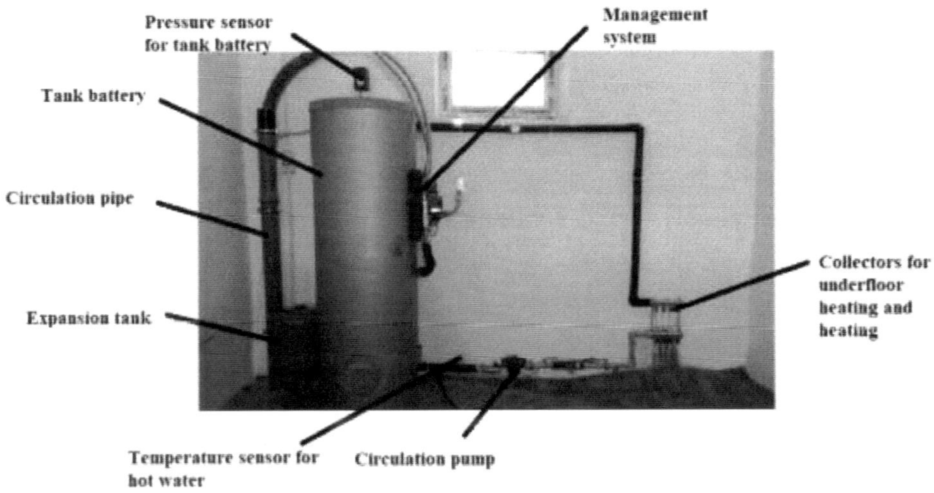

Figure 7. Solar heating system with thermosiphon

Figure 7 shows an industrial design and a general view of a two-circuit solar installation, which consists of a heat pump, an accumulator tank, a circulation pump, and an expansion tank.

The novelty of the research herein is the development of a double-circuit solar system with thermosiphon circulation, which has a flat solar collector, representing the heat-insulating transparent double-glazing unit with reduced pressure, a heat carrier is made of the thin-walled corrugated stainless tube. Heat, obtained from solar flow, heats the liquid in coils, which is removed from the collector, and its place is occupied with cold liquid from the siphon there happens constant heat circulation, which upgrades heat transmission effectiveness, eliminating additional webs between the panel and heat insulation. There is, as well, a thermal pump, where a condenser and evaporator are made in the form of the heat exchanger of "spiral in spiral" type, heat exchanger pipelines are located one over another, which increases the square and heat exchange intensity.

The scientific novelty of this study is a new mathematical model of the joint functioning of a flat solar collector with a heat pump, which allows us to deduce the basic calculations and prove the feasibility of using an adsorber with a high total effect of absorbing solar energy and atmospheric air heat. Also, in the new mathematical model, the hypothesis of heating capacity is formulated and used as the sum of solar radiation energy, heat from the environment and heat released during the operation of a heat pump compressor. The analysis showed that the new solution, without taking into account the energy costs of the cooling fan drive, contributes to an increase in the heating capacity of a dual-circuit solar installation.

Using equation (8) there is compiled a detailed balance equation:

$$Q_3 + W_{EL} * \eta_{EL} * \eta_k + Q_3(1-\eta_k) + W_{EL}(1-\eta_{EL}) - k_{T2} * S_2(t_5 - t_8) - W_{EL}[a - b(t_5 - t_1) = C_2 d\tau \frac{dt_5}{d\tau} \tag{9}$$

where: t_8 – environmental temperature,^{0}C; κ_{T2} –heat exchange factor of thermal pimp condenser heat exchanger heat insulation, Kcal/degree.

The dependence of the coolant temperature is derived:

$$t_5 = \frac{k_{T2}S_2t_8 + [Q_3 + W_{EL}\eta_k + Q_3(1-\eta_k) + W_{EL}(1-\eta_{EL}) - W_{EL}a - W_{EL} * b * t_t] * \{1 - \exp(\frac{(k_{t2}S_2 + W_{EL}(a-b)) * \tau}{C_2})\}}{W_{EL} * b * \exp(\frac{(k_{T2}S_2 + W_{EL}(a-b)) * \tau}{C_2} + k_{T2}S_2 - W_{EL} * b} \tag{10}$$

There has been carried out the formula checking for $\tau = 0$, where from obtained, that $t_4 = t_8$.

At $\tau = \infty$, we receive the temperature formula under steady state operation:

$$t_5 = \frac{k_{T2}S_2 * t_8 + [Q_3 + W_{EL}\eta_k + Q_3(1-\eta_k) + W_{EL}(1-\eta_{EL}) - W_{EL}a - W_{EL}b * t_1]}{k_{T2}S_2 - W_{EL} * b} \quad (11)$$

Analysis has shown, that the new solution, without taking into account the energy costs of the cooling fan drive, contributes into an increase in heat output.

Analysis of the second circuit. The totality of the processes of the second circuit is expressed by the equation of the thermal balance of the circuit:

$$Q_6 + Q_{EL} - Q_7 - Q_8 - Q_9 = C_3 \frac{dt_6}{d\tau} \quad (12)$$

where C_3- heat capacity of the second circuit coolant, kcal/deg; t_6- the current temperature of the water in the battery tank; Q_{el}- heat flow from a backup source, Joule; Q_7 and Q_8- accordingly, the thermal energy used for hot water supply and heating, Joule; Q_9- heat losses of the battery tank, Joule.

By analyzing equation (12) is reduced to the expanded form:

$$Q_6 + P_3 * \tau - k_{var} * \sum n_{7i}v_{7i} - V_8 * n_8 - k_{T3} * \delta_3 * S_3(t_7 - t_8) = C_3 \frac{dt_7}{d\tau} \quad (13)$$

where k_{var}- the coefficient of simultaneity of the hot water consumer, is accepted on average 0.7...0.8; n_{7i} and v_{7i}- quantity and rate of consumption of the i-th consumer of hot water; V_8- the volume of the heated room, m^3; n_8- heating norm for the climate zone.

The result of solution (13) was the dependence of the current temperature of the coolant in the storage tank on a combination of factors:

$$t_7 = t_8 + \frac{Q_6 + P_3 * \tau - k_{var} * \sum n_{7i}v_{7i} - V_8 * n_8}{k_{T3}\delta * S_3}[1 - \exp(-\frac{k_{T3}\delta * S_3}{C_3})\tau] \quad (14)$$

Checking the correctness of the formula are the values t_7. at characteristic points. For example , when $\tau=0$ we get the expected value $t_7=t_8$.

At $\tau=\infty$, we obtain a formula for calculating the temperature of the coolant when entering the steady-state operation mode:

$$\frac{Q_6 + P_3 * \tau - k_{var} * \sum n_{7i}v_{7i} - V_8 * n_8}{k_{T3}\delta * S_3} \quad (15)$$

The analysis showed that the constant heat consumption from the accumulator tank for hot water supply and heating is the most effective way to increase heating capacity. This contributes to the maximum heat transfer Q6, therefore maintaining the heat pump's efficient operation by providing the heat pump's leading conversion coefficient. To increase the heating capacity, it is also necessary to provide thermal insulation for the battery tank.

The practical significance of such research at this time is very wide. In residential and administrative buildings, solar energy is mainly used in the form of heat to meet the needs of hot water supply, heating, cooling, ventilation, drying, etc. The use of solar heat from an economic point of view is most advantageous when creating hot water supply systems and in installations for heating water close to them in terms of technical implementation (in swimming pools, industrial devices). Hot water supply is necessary in every residential building, and since the needs for hot water change relatively little throughout the year, the efficiency of such installations is high and they pay off quickly. As for solar heating systems, the period of their use during the year is short, during the heating period the intensity of solar radiation is low and, accordingly, the collector area is much larger than in hot water systems, and the economic efficiency is lower. Usually, when designing, a solar heating and hot water supply system are combined. In solar cooling systems, the operating period is even lower (three summer months), which leads to prolonged downtime of equipment and a very low coefficient of their use.

For experimental studies, a self-recording potentiometer of the brand KSP-4 manufactured in Russia was used. Designed for measuring DC power and voltage, as well as non-electrical quantities converted into electrical signals and active resistance. They work with primary converters whose resistance, including the resistance of the communication line, does not exceed 200 ohms. The maximum supply voltage is 10V. Designed for switching DC and AC 1A electrical circuits with a voltage of 220V. If the current exceeds 1A, an intermediate relay must be used.

The M80 pyranometer with GSA galvanometer is used to measure solar radiation falling on the surface. The M80 pyranometer with GSA galvanometer has a spectral characteristic from 400 to 1100 nm. The pyranometer can operate at ambient temperatures from -50 to +50°C and a relative humidity of 100% at +25 °C. The measurement range of solar illumination is from 1 W/m2 to 1300 W/m2, as well as the measurement range of energy effects: from 1 W/m2 to 500 kW/h/m2

The laser thermometer "Center-350" is designed to measure the surface temperature of an object, it is applicable to various hot, dangerous or hard-to-reach objects, quickly and safely contactless. The laser thermometer has a temperature range from -20 to 500 ^{0}C. The radiation coefficient for various objects is 0.95. The maximum supply voltage is 9V. The laser thermometer is configured to measure objects in 500 ms, it also has an optical resolution of 8:1.

The automatic wireless weather station Vantage Vue was used, which provided accurate and reliable weather monitoring, as well as continuously recorded meteorological data at regular intervals (every 60 s) using the Weather-Link program controlled by a PC-based system. The ambient temperature recorded by the weather station had an accuracy of 0.5 °C. The industry standard pyranometer measures incident solar radiation with a nominal accuracy of 5 W/m^2.

The wind speed during the collector tests was not to exceed 5 m/s. Wind speed measurements were to be carried out in the immediate vicinity of the collector at a height corresponding to half the height of the collector.

The tests were carried out at a water flow rate through the collector of 25 kg/(m^2·h) and a water temperature at the inlet to the collector of 20, 30, 40, and 50 °C. The water temperature was changed from test to test. Thus, at least 4 tests were to be carried out. The duration of the tests was at least 2 h. Five thermocouples were connected to a data monitoring system that measured the temperature at various points within the system.

3.Results

At the beginning of the study, a model of a flat solar collector was developed, as well as a two-circuit solar installation with thermosiphon circulation. We made a theoretical analysis for two circuits and for the heat pump circuit.

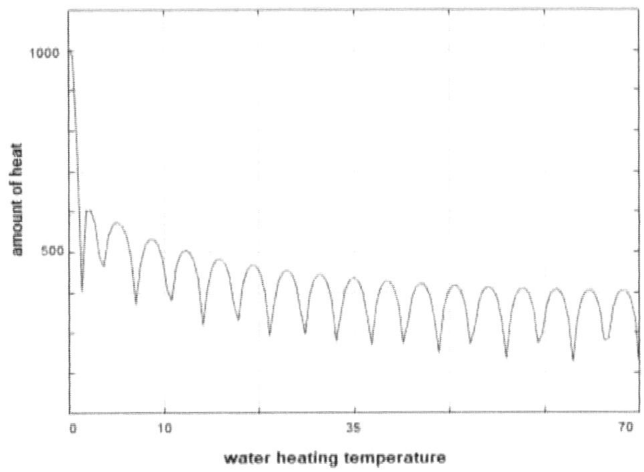

Figure 8. Dependence of thermal pump heat amount on water heating temperature

Figure 8 shows the dependence of a heat pump heat amount on the temperature of water heating. It can be seen from the graph, that in the winter period the heat pump's heat amount drops according to weather conditions, and also depends on the specific heat of the substance and the temperature change in the evaporator and condenser. With temperature change, the amount of heat decreases according to the first and second thermodynamics laws.

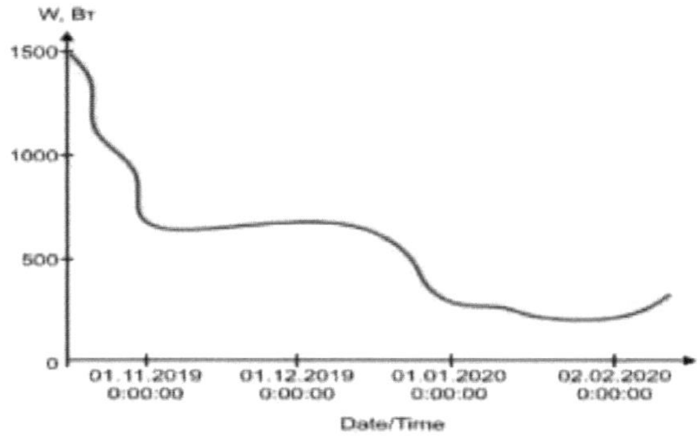

Figure 9. Dependence of thermal pump capacity on time and date

As can be seen from Figure 9, the thermal power was calculated in the winter period from 01.11.2019 to 02.02.2021. It can be seen from the graph, that at a power of 1500 Wt. during the winter period (November, December, January and February), the power drops due to the climatic conditions of the given region.

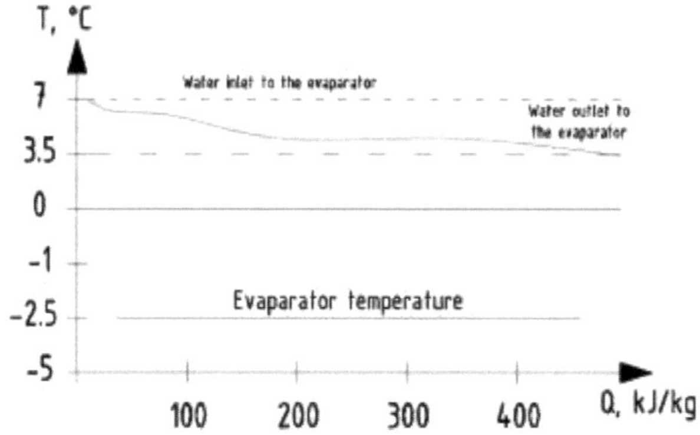

Figure 10. Diagram of transfer media temperature changes in evaporator

Figure 10 shows a diagram of the temperature changes in the coolant in the evaporator and condenser of the heat pump. In order to increase the COP, it is necessary to increase the temperature of the water at the inlet (outlet) of the evaporator. An increase in temperature is only possible to a certain limit. The maximum approach to this temperature causes an increase in the surface of the tubes, which must be lowered into the tank (increase the heat exchange surface). A decrease in this temperature leads both to a drop in the COP and to the heat pump danger at the minimum temperature of the water, leaving the evaporator.

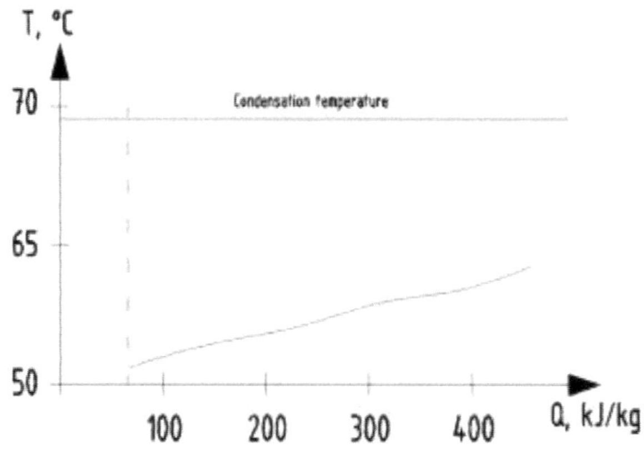

Figure 11. Diagram of transfer media temperatures changes in condenser

Figure 11 demonstrates the diagram of transfer medium temperatures changes in the thermal pump condenser. As a result of the analysis of the obtained data, it was found, that at an initial water temperature of 50 ° C in the evaporator, more intense heating of the water in the condenser is observed in the time interval of 100 minutes. This can be explained by a higher heat transfer coefficient. Further, the intensity of the temperature increase lowers, since the temperature in the evaporator decreases and the transformation coefficient lowers accordingly.

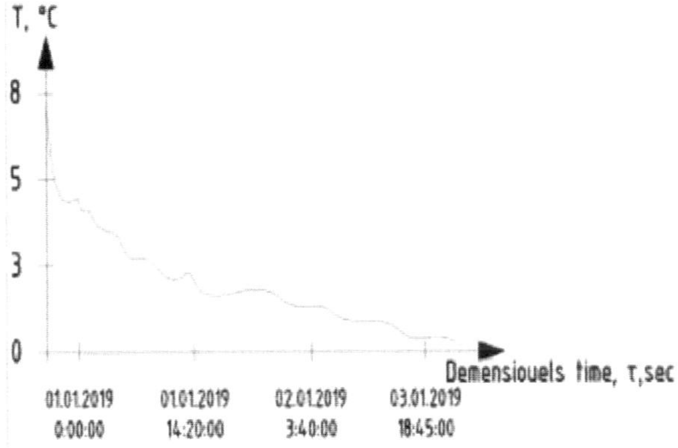

Figure 12. Graph of thermal pump COP change according to date and rime

As can be seen from Figure 12, the average coefficient COP of heat pump transformation from date and time affects the decrease in water temperature at the inlet and outlet of the condenser, lowers condensation temperatures (in the condenser), increases the evaporation temperature (in the evaporator) and increases the water temperature at inlet (outlet) to the evaporator.

4.Discussion

The heat pump was running continuously from 10 am to 7 pm, so the collection of solar energy took place during this period of time. When the heat pump was running, the water temperature at the outlet of the flat solar collectors gradually increased in the morning, and then gradually in the afternoon, the water temperature at the inlet continuously increased and decreased when the heat pump was turned off. When the temperature of the hot water in the storage tank increased, the temperature of the water at the entrance to the flat collectors increased. Another reason is that the ambient temperature and solar radiation also increased in the morning, and then gradually decreased in the afternoon. The water temperature at the outlet of the flat collectors reached its maximum value of 74.5 °C at 14:05. When the heat pump was started, the variable mass flow in the circuit increased greatly, and then gradually decreased.

5. CONCLUSIONS

The experimental results of the study show, that the energy and exergy model of solar collectors with thermosiphon circulation with heat pumps efficiently works in heating mode with accurately measured outdoor temperature, typical of the climatic zone of Almaty (Kazakhstan). The analysis of the first circuit showed that when the installation works with a heat pump, the direction of the heat flow between the gel trap and the environment changes. The analysis of the heat pump showed that the new solution, without taking into account the energy consumption for the cooling fan drive, contributes to an increase in thermal power. The analysis of the second circuit showed that the constant consumption of heat from the storage tank for hot water supply and heating is the most effective way to increase heating capacity. This contributes to the maximum heat transfer Q6, thereby maintaining the efficient operation of the heat pump by ensuring the best conversion coefficient of the heat pump. To increase the heating capacity, it is also necessary to provide thermal insulation of the battery tank. Energy analysis shows, that with the new flat solar collectors, the average annual values were 2.5 kW and 8 kW, respectively. The efficiency of the flat solar collector was also high, the COP system in winter from November to February was improved by 4%.

ACKNOWLEDGMENT

Ministry of Education and Science of the Republic of Kazakhstan, Grant/Award Number: AP23488439.

REFERENCES

[1] Jones GG, Bouamane L. Power from sunshine: a business history of solar energy. *Harvard Business School Operating Paper Series*; 2012

[2] Duffie JA, Beckman WA. *Solar engineering of thermal processes*. Wiley; 1991.

[3] Freegah B, Asim T, Albarzenji D, Pradhan S, Mishra R. Effect of the shape of connecting pipes on the performance output of a closed-loop hot water solar Thermo-syphon. In *3rd International Workshop and Congress on eMaintenance*; 2014.

[4] Manjunath MS, Karanth KV, Sharma NY. Three dimensional numerical analysis of conjugate heat transfer for enhancement of thermal performance using finned tubes in an economical unglazed solar flat plate collector. In *Proceedings of World Congress on Engineering* (Vol. 3, pp. 2245- 2249); 2011.

[5] Shrirao PN, Pente SS, Mahure AN. Comparative thermal analysis of a flat plate solar collector using aerofoil absorber tube with conventional circular absorber tubes. *Int J Basic Appl Res*. 7(12): 98- 107.

[6] Balachandar M, Narendran A. Experimental investigation of solar flat plate collector with inner grooved copper tube. *Int J Eng Res*. 2016; 5(9): 695- 700.

[7] Hussein HMS, El-Ghetany HH, Nada SA. Performance of wickless heat pipe flat plate solar collectors having different pipes cross sections geometries and filling ratios. *Energy Convers Manag*. 2006; 47(11–12): 1539- 1549.

[8] Bute JV, Kongre SC. Experimental investigation of a solar flat plate collector. *Int Eng J Res Dev*. 2015; 2(5): 36- 46.

[9] Yao K, Li T, Tao H, Wei J, Feng K. Performance evaluation of all-glass evacuated tube solar water heater with twist tape inserts using CFD. *Energy Procedia*. 2015; 70: 332- 339.

[10] R.Omarov, S.Abdygaliyeva, D.Omar,M.Kunelbayev,"Integrated system for the use of solar energy in the animal farm,:Scientia Iranica, Volume.24,Issue.6, 2018,pp.3213-3222.

Utilizing gamma-irradiated Iranian propolis (bee glue) extract for the green synthesis of silver nanoparticles

Farimehr Ghanizadeh Geraili
Department of Medical Radiation Engineering, Faculty of Nuclear Engineering, Shahid Beheshti University, Tehran, Iran. ghanifarmehr@gmail.com. 0000-0002-6210-9891

Seyed Mahmoud Reza Aghamiri*
Department of Medical Radiation Engineering, Faculty of Nuclear Engineering, Shahid Beheshti University, Tehran, smr-aghamiri@sbu.ac.ir. 0000-0002-5080-0482

Marzieh Heidarieh*
Nuclear Science and Technology Research Institute, Tehran, Iran, mheidarieh@aeoi.org.ir. 0000-0003-2097-960X

Cite this paper as: *Ghanizadeh Geraili, F, Aghamiri, S.M.R, Heidarieh, M. Utilizing gamma-irradiated Iranian propolis (bee glue) extract for the green synthesis of silver nanoparticles. 7th Int. Conf. on Nucl. and Ren. Energy Res. 27-30 October 2024, Ankara, Turkey*

Abstract: Green synthesis uses natural compounds to produce nanoparticles without toxic substances, suitable for medical and pharmaceutical applications. It offers an environmentally friendly alternative to conventional processes, addressing global environmental concerns. Propolis, a natural beehive product, contains bioactive compounds with broad biological and pharmacological properties. The antioxidant property of the alcoholic extract gamma irradiated with a dose of 10 kGy has shown a significant increase compared to the non-irradiated state. The biosynthesis of nanoparticles involves the regeneration of their salt ions and the neutralization of the electric charge. These changes occur rapidly and at room temperature, indicating the high speed of this method and its lack of high temperatures for the formation of silver nanoparticles. The synthesis was monitored by UV–vis spectrometry, and the biosynthesized silver nanoparticles showed high antioxidant activity. Scanning electron microscope (SEM) analysis revealed sizes of approximately 50-90 nm in the gamma-irradiated group. In conclusion, the use of the green synthesis method and the absence of the production of additional toxic substances in the synthesis of nanoparticles from Iranian propolis make the produced silver nanoparticles safe for medical applications.

***Keywords**:* *Green synthesis, Gamma rays, propolis, DPPH, Nanoparticles*

1. INTRODUCTION

Various chemical, biological, or physical techniques can be employed in order to establish nanoparticles[1].The traditionalist strategy utilized for the purpose of synthesizing Ag nanoparticles is the application of chemical and physical techniques so as to generate NPs[2, 3]. Natural substances extracted from microorganisms or plants are used in green synthesis[4, 5]. Silver nanoparticles (AgNPs) are biosynthesized by fungi, bacteria, actinomycetes, plants, and yeast. As a result, avoiding the use of toxic materials is important to encourage their use in the pharmaceutical and medical fields[2, 6]. Given the problems that our world is dealing with today and also the limited chance of exploring applicable solutions, green chemistry has emerged in order to avoid the utilization of ecologically destructive items and materials[7-10].In its generalized version, biodiversity studies serve as a solution promoting the use of nature's capacity and the use of non-harmful components. Therefore, different procedures have been sought for the application of nanomaterials so as to contribute to the same field by using green forms, e.g., the nanoparticles (NPs) produced using plant extracts[10-13].
The most common technique is the application of the extracts obtained from plants. This is attributable to the fact that this approach is the simplest extracellular synthesis in order to obtain metallic NPs on a large scale. As a result, scholars have utilized varied types of plant extracts in order to develop NPs[14-17]. Given its cost-effectiveness and straightforward forms, both industrial and academic sections have focused on the applications of AgNPs produced via

green techniques in a variety of fields, such as the environment, industries, and biomedicine[18, 19]. The Ag nanoparticles have their own unique optical, chemical, magnetic, mechanical, and electrical characteristics[1]. One may characterize AgNPs by desirable antibacterial characteristics, and as a result, Ag has been broadly utilized as an antibacterial and healing agent for many years[20]. In addition, AgNPs are characterized by anti-inflammatory, antifungal, antiplatelet, and antiviral characteristics[21, 22]. According to recent investigations, biosynthesized silver nanoparticles have strong antioxidant, anti-cancer, and antimicrobial characteristics[18, 19, 23, 24]. Nanotechnology serves as a developing system of drug delivery against a great number of serious diseases, and remedy for tumors and physiological disorders[21, 25, 26].

As a natural resinous product prepared by honey bees, propolis is collected by them from exudates and leaf buds of a variety of flowers and herbs and is mixed with pollen, wax, and enzymes of bees. Furthermore, the complex bioactive compounds of propolis are dependent on the sources of plant available in local greenery and present a wide range of pharmacological and natural characteristics[27, 28]. A number of studies have shown that bee propolis can be used to synthesize NPs, which is attributable to its wide range of biomolecules, e.g., flavonoids, alkaloids, steroids, terpenoids, phenols, etc.[28-30]. Propolis went under gamma irradiation with doses equal to 10, 20, 30, 40, and 50 KGy. According to the acquired results, when irradiated up to 10 KGy, the thermal characteristics of propolis are not affected by the irradiation process[31, 32]. In addition, it was found that the alcoholic extract of Iranian propolis (EEP) gamma irradiated at a 10 KGy dose was rich in phenolic compounds, and as a result, the gamma irradiated EEP (10 KGy) may be regarded as the most optimal dose amongst the investigated ones. The present investigation shows that one can apply the gamma-irradiated EPP as an enriched natural antioxidant agent and characterized by its applicability for therapeutic and pharmaceutical purposes[31].As a result, the present Investigation compared the characteristics of silver nanoparticles biosynthesized using non-irradiated and irradiated (10 KGy gamma-ray dose) Iranian propolis.

2. MATERIAL AND METHODS

The whole chemicals utilized in the current investigation were procured with maximum purity. The silver nitrate salt ($AgNO_3$, MAY & BAKER Ltd. Dagenham, England). It contains $AgNO_3$ (purity of >99.9%), Cu, Pb, chloride, and propolis prepared from a local apiary (Marvdasht, Fars, Iran).

Extraction, collection, and gamma irradiation of propolis specimens:

For the purpose of preparing the alcoholic extract of propolis, liquid nitrogen was initially used to crash the propolis into small pieces. Subsequently, 25 grams of it was mixed with a volume of 250ml of ethanol solution (80%). After shaking the prepared mixture at room temperature for 48 h at the horizontal level (150 rpm), a filter paper (Whatman grade 4) was used to filter the obtained alcoholic extract twice. Then, a rotary device was utilized to evaporate the alcohol in order to obtain a pure extract with alcoholic nature. The acquired pure extract was then weighed. Then, its 10 % (weight by volume) solution was prepared in 80 % alcohol and subsequently stored at a temperature of 4 °C within a dark glass container until use[33].

In accordance with the results of Heidarieh and colleagues in 2021, the alcoholic extract underwent irradiation at Iranian Nuclear Science and Technology Research Institute by applying a dose of 10 KGy of γ ray as the optimum dose by using a Cobalt-60 source (manufactured by PX-30 Issledovapel, Russia) in a 0.02 Gray/s dose range[31]. The specimens were stored at a 4°C temperature for the purpose of further testing after irradiation.

Synthesis and characterization:

Preparing AgNO3 solution

Initially, silver nitrate salt (molecular mass=169.87 M) was prepared. In a George balloon (volume=100 ml), 0.17 g/l of $AgNO_3$ featuring a purity of 99.9% was mixed with 100 milliliters of dehydrogenated water so as to prepare 1 mM silver nitrate solution.

Preparation of AgNPs using EEP

Various ratios of silver nitrate and EEP were used to find the optimal ratio for combining them. Eventually, the effective ratio was determined on the basis of the investigation conducted by Cardoso-Avila et al. (2021)[17]. Simply put, 5 ml of the reducing agent (EEP) was mixed with silver nitrate solution (1 mM) in the amount of 5 milliliters in both non-irradiated/irradiated scenarios. It is noteworthy that the whole steps were taken in the dark. The above solutions were shaken for 1 hour away from light, and finally, the presence of silver nanoparticles was detected by observing the color change of the samples.

Recording the color changes

After 15 min, the specimens' color changed to yellow, and then, after 1h, they changed to a brownish color. Thus, after the formation of the NPs, the color of the concerned solutions turned into a dark brown one, which is a sign of the formation of AgNPs[34].

Analysis of the UV-Visible spectrophotometry:

Analysis of the UV-Visible spectrophotometry

With regard to the study conducted by Kischkel et al. (2020), the absorption peak observed at the wavelength of 430 nm confirms the synthesis of AgNPs. Therefore, AgNPs were analyzed using ultraviolet-visible spectroscopy (UV2601, Rayleigh, China)[35].
Therefore, the light absorption by the concerned solutions was investigated using VU-visible spectrophotometry (wavelength= 300-700 nanometers)[34]. The generated spectrum reveals that the light has been absorbed by the studied solutions, and by examining the resulting spectrum, one can find out if AgNPs are synthesized.

Morphological Analyses

Through centrifugation (8000×g for 20 min, T= 25°C) and membrane filtration (0.45 μm), the nanoparticles were separated from the initial solution[35]. The specimens were then dried within an oven (T=45 °C) for 24 hours. Subsequently, the remaining samples deposit underwent the conductivity process and surface preparation by magnetron sputter coater (CY-VTC-3R, China) so as to prepare them for scanning electron microscope (SEM) analysis, and finally, the Field Emission Scanning Electron Microscope (FE-SEM) device (JSM-7610F, Japan) was utilized so as to decide the size and morphology of biosynthesized AgNPs (in both scenarios), and for this purpose, the resulting deposition underwent SEM.

DPPH free radical scavenging technique

Based on the technique presented by Brand-Williams et al., the antioxidant activity using the method of the free radical scavenging performance of DPPH (2,2-diphenyl-1-picryl-hydrazyl-hydrate) was studied for alcoholic extract and biosynthesized AgNPs specimens in both scenarios of non-irradiated state and the specimens irradiated using a 10 KGy γ ray dose. For the purpose of DPPH assay, each extracted sample (1 mg/L) was mixed with DPPH ethanol solution (100 millimolar, vol= 0.1 ml). After that, a spectrophotometric ((UV/VIS Spectrophotometer) microplate reader was used to determine absorption at the desired wavelength in the dark at 37 °C for 30 minutes. Then, the DPPH inhibition percentage was measured for each sample in accordance with the rate of fixed DPPH in the solution as soon as the reaction completed. Measurements were conducted in triplicate. In addition, one may estimate the rate of inhibition effect as Equation (1):

$$\text{Percentage of the inhibition rate} = [1 - (A_2 - A_1)/A_0] \times 100. \quad (1)$$

Where A_1 stands for the absorption of light in the presence of specimen, A_2 stands for the light absorption of the specimen with no DPPH radical, and A_0 represents the absorption of the control[31].
In the present investigation, the blank sample is composed of a mixture prepared using 2 milliliters of ethanol (CH3CH2OH) and two milliliters of extract/nanoparticle suspension in both scenarios, while a specimen that contained two milliliters of the extract and DPPH (2 ml) was used as a negative control.

Determining the overall content of polyphenol

The Folin-Ciocalteu (F-C) colorimetric technique was utilized in order to determine the overall phenolic content of non-irradiated/ irradiated samples. The Folin–Ciocalteu reaction is regarded as an antioxidant assay serving on the basis of electron transfers in which the reductive capacity of antioxidant is measured. This technique, which is presented by Singleton and Rossi with slight modifications[36, 37], has been commonly used in measuring the total polyphenol/phenol content of biological samples and plant-based foods. The specimens underwent analysis using flame atomic absorption spectrometry. For this purpose, the readings were carried out using an atomic absorption spectrometry device within the 300-700 nanometers range. Then, the calibration curve was prepared using the dilutions of a pure Ag solution (1 mM). The considered wavelength was equal to 328.1 nanometers. This is because that with regard to the guidelines provided by the manufacturer, it represents the silver radiation absorbance wavelength[35].

Statistical analysis:

The statistical software SPSS ver.19 was employed to analyze the obtained data. One-way ANOVA was utilized in the present investigation to analyze the data statistically. The treatments were also compared using Tukey's test. The considered significance levels of these tests were <0.05.

3. RESULTS and DISCUSSION

The biosynthesis of NPs refers to the regeneration of their salt ions and, in fact, neutralizing their electric charge. The occurrence of these changes at T=37 °C within a short period of time illustrates the high pace of the same technique and its non-dependence on high temperature for the generation of the AgNPs[34]. The synthesis/characterization of biosynthesized NPs are the first steps of the present investigation. The biosynthesis was carried out on the basis of reducing Ag+ to silver atoms (Ag0), which was followed by atoms' agglomeration and nanoparticles' stabilization. One could characterize these steps by the observed solution color changes, which paved the way for monitoring the synthesis using UV–vis spectrometry[35, 38]. According to the results of this investigation, EEP, in particular, the irradiated scenario, is of a great potential to regenerate the metal ions of $AgNO_3$ salt and transform them into atoms of metal in nanometer dimensions[17].
After the preparation of the alcoholic extract of propolis in both non-irradiated/irradiated scenarios and its combination with a 1 mM $AgNO_3$ solution and a period of 30 minutes, the resulting mixture was kept in the dark. Subsequently, the color of the specimens changed to a yellowish one. After a period of 1h, the samples' color changed to brown, which reflects the generation of Ag nanoparticles in the solutions. The synthesis of Ag nanoparticles leads to the generation of solutions with various colors, e.g., golden yellow, brownish yellow, and pale yellow[39]. In the present study, as Figure 1 shows, the color of the specimens changed from yellowish color to brown. This is in agreement with the earlier investigations and represents the formation of colloidal suspension (Hydrosol) of AgNPs. As a result, UV-vis spectrophotometry was employed so as to approve the presence of AgNPs in the samples.

Figure 1. Specimen color change after the creation of AgNPs

Spectrophotometry analysis:

The surface plasmon resonance (SPR) is behind the unique optical characteristics of metal nanoparticles that are changed subject to factors, e.g., the refractive index of the ambient, the size and shape of the NPs, and the distance between the nanoparticles[40]. Given that AgNPs absorb light within the 300-500 nanometers range, according to the spectrophotometry results, the absorption band characteristic of SPR was recorded at 345-350 nanometers for AgNPs. The specification of the absorption band of SPR for the biosynthesized AgNPs at 345-350 nm have been shown in Figures 2 and 3. As the graphs show, non-irradiated EEP has a better absorption performance for the reduction of the $AgNO_3$ salt in comparison with the irradiated case, and it leads to the synthesis of more NPs. However, due to the instability of the biosynthesized AgNPs in the non-irradiated scenario in comparison with the irradiated case, the major peak observed in the curves caused by the discharge of the stable NPs in both durations of 1 h and 24 h following the exposure of the $AgNO_3$ salt to EEP in the two tested scenarios decreased significantly.

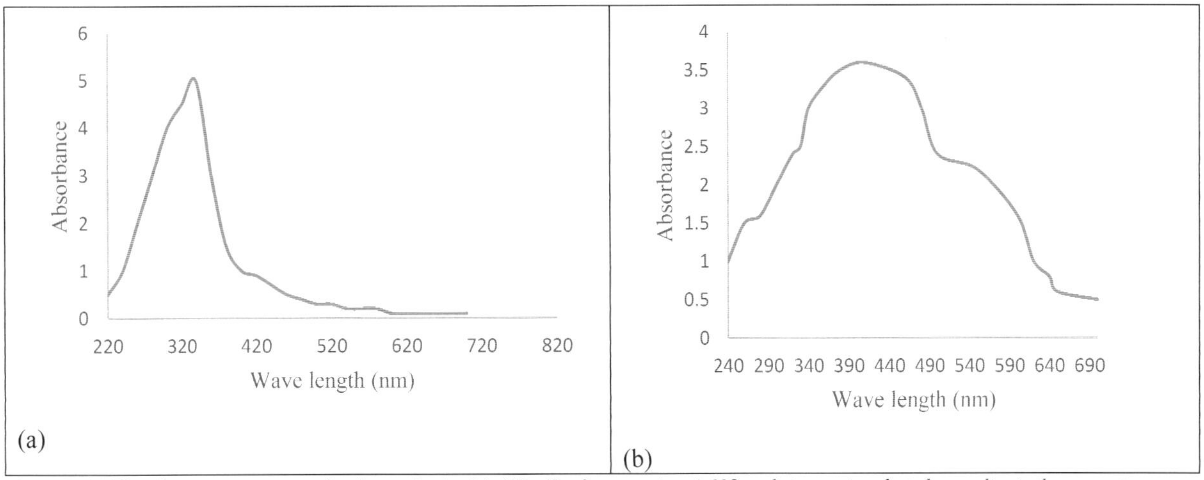

Figure 2. (a) The absorption spectrum for the synthesized AgNPs 1h after exposing $AgNO_3$ salt to non-irradiated propolis- in the present case, due to the destabilization of the colloidal solution of NPs, the major peak is greatly reduced due to the discharge of stable nanoparticles.; (b) The absorption spectrum recorded by the spectroscopy device for the synthesized AgNPs 1h after exposing $AgNO_3$ salt to irradiated EEP.

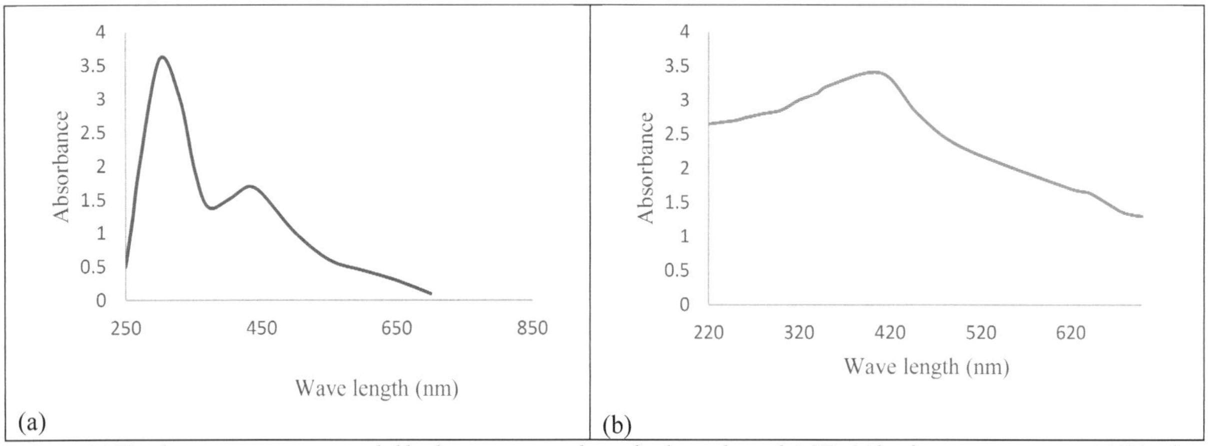

Figure 3. (a) The absorption spectrum recorded by the spectroscopy device for the synthesized AgNPs 24 h subsequent to exposing AgNO3 salt to non-irradiated propolis- in the present scenario, as a result of the destabilization of the colloidal solution of NPs, the major peak is greatly reduced due to the discharge of stable nanoparticles; (b) The absorption spectrum recorded by the spectroscopy device for the synthesized AgNPs 24 h subsequent to exposing AgNO3 salt to irradiated propolis.

Given the light absorption by the AgNPs within the 300-500 nm range and the characteristic absorption band of surface plasmon resonance recorded for the AgNPs at 345-350 nm, we concluded that AgNPs were present during the process of synthesis.

Results of the SEM analysis:

Based on the results of SEM analysis, we conclude that, in terms of size, the AgNPs biosynthesized by employing the irradiated EEP are within the 50-90 nm range. Moreover, the size of the Ag nanoparticles generated by utilizing the non-irradiated extract was within the 100-580 nm range (Figures 4-9).
The reason behind the uncontrolled agglomeration of nanoparticles in a liquid suspension is the high surface energy of them. The same tendency of the NPs for agglomeration makes their application difficult. As a result, further homogenization as well as the disintegration of agglomerates are regarded as crucial steps in material processing most of the time[41]. As a result, in the same SEM analysis technique, it should be mentioned that NPs are characterized by a potential tendency for aggregation, which is ascribable to their high level of surface energy. Also, electron microscopic image in Figure 9 shows the adhesion of nanoparticles.

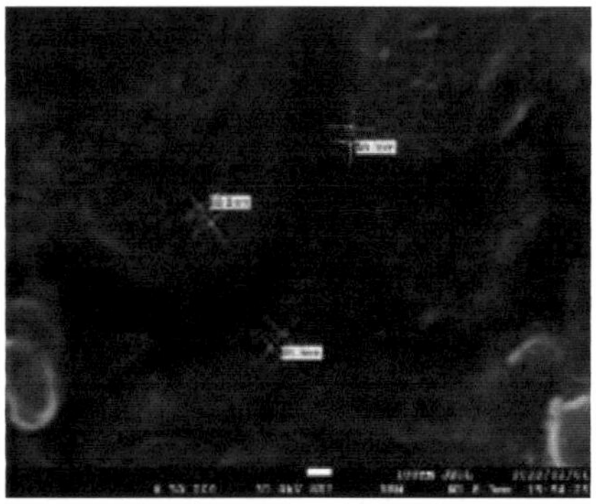

Figure 4. Biosynthesized AgNPs as soon as being exposed to the irradiated EEP after one hour. As seen, a small number of NPs was observed characterized by very small sizes.

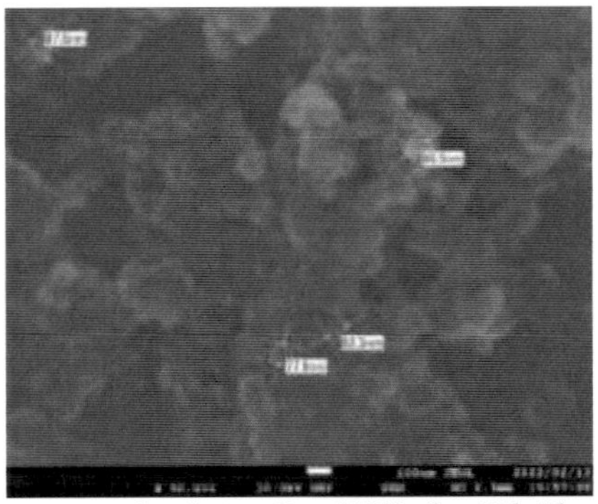

Figure 5. Biosynthesized AgNPs as soon as being exposed to the irradiated EEP after 12 h. The observed NPs were at their maximum number.

Figure 6. Biosynthesized AgNPs following being exposed to the irradiated EEP after a period of 72 h. The passage of time resulted in removing a number of particles featuring smaller sizes and led to increased particle sizes.

Figure 7. Biosynthesized AgNPs by exposing non-irradiated propolis to the alcoholic extract after 1 hour.

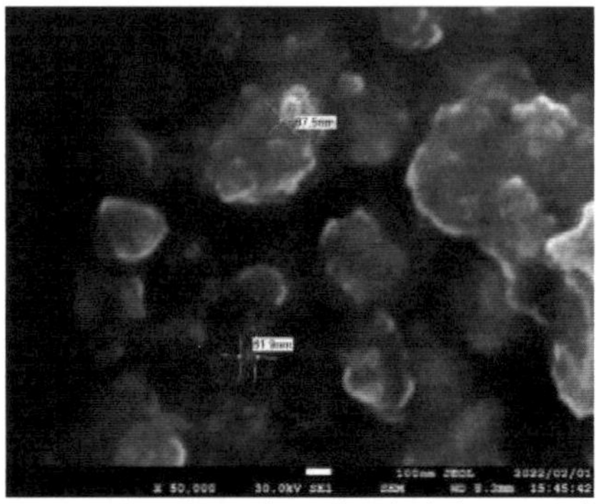

Figure 8. Biosynthesized AgNPs after exposing non-irradiated propolis to the alcoholic extract for 12 hours. Even though the sizes of NPs have decreased, in this case, much less uniformity is observed in comparison with the scenario with irradiated propolis.

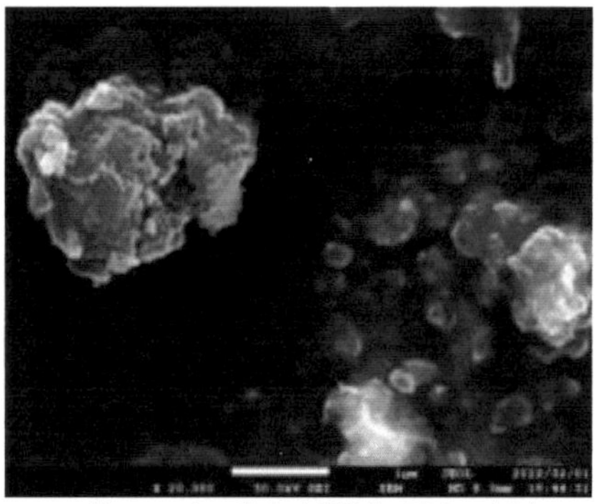

Figure 9. Biosynthesized AgNPs after being exposed to non-irradiated EEP for 72 h, which resulted in the removal of a number of particles featuring smaller sizes and increased the particle sizes. This phenomenon was observed with higher frequency in non-irradiated propolis.

According to the obtained results, the finally obtained AgNPs in the irradiated group are homogeneous featuring a smaller size and a spherical shape.

Results of DPPH analysis:

A comparison made between the two specimens of non-irradiated/irradiated EEP indicated that the antioxidant characteristics in the irradiated EEP increased at a 10 KGy γ ray dose in comparison with the non-irradiated EEP (Table 1), and this is attributable to the tremendous positive effect occurring when the irradiated propolis extract is used in comparison with the non-irradiated scenario[31]. The antioxidant activity of the silver nanoparticles biosynthesized at a 150 µg/ml concentration in the A' group indicated a noticeable increase in comparison with other groups and tested concentrations ($P < 0.05$). Moreover, the antioxidant properties showed a significant increase by expanding the concentration of alcoholic extract and biosynthesized nanoparticles in both irradiated and non-irradiated scenarios ($P < 0.05$). The antioxidant activity of EEP in all concentrations utilized in B' and B conditions in comparison with the silver nanoparticles biosynthesized in A' and A groups and also the glutathione standard indicated a noticeable decline ($P < 0.05$). The difference between all investigated concentrations of biosynthesized nanoparticles in A and A' conditions was significant ($P < 0.05$).

Table 1. Antioxidant activity of biosynthesized AgNPs and EEP in both states using DPPH method. Glutathione (GSH) is used as a standard in this test and data are reported based on the mean ± standard deviation.

concentration(µcg/ml)	GSH	A'	A	B'	B
50	81.22±2.15[a]	75.22±1.31 [b]	64.59±1.35[cd]	66.08±1.74 [c]	61.47±1.89 [d]
100	87.64±3.23[a]	80.36±3.94 [b]	69.41±2.81 [b]	70.49±2.11 [b]	63.98±2.51 [c]
150	98.51±2.29[a]	94.83±2.67 [a]	88.35±2.54 [b]	84.15±2.21 [c]	83.62±1.67 [c]

A': Ag nanoparticles biosynthesized by the irradiated EEP
A: Silver nanoparticles biosynthesized by employing non-irradiated EEP
B': Irradiated EEP
B: Non-irradiated EEP
GSH: Glutathione

4. CONCLUSION

Besides the fact that the use of irradiation is safe and harmless, it causes the controlled decrease of metal ions without the need for extra chemical reducing agents or the formation of unfavorable products of oxidation-reduction. It is noteworthy that the antioxidant characteristic of the case irradiated using a 10 KGy gamma-ray dose was increased significantly in comparison with that of the non-irradiated scenario[31]. According to the results of the present investigation, the biosynthesized AgNPs, in particular, when employing the regenerative characteristic of the irradiated EEP, showed significant antioxidant activity. The alcoholic extract of propolis subject to irradiation by γ ray with a dose of 10 KGy is rich in phenolic compounds.

In addition, given the application of the biosynthesis technique and the lack of extra toxic materials produced in the course of the synthesis of Ag nanoparticles, it can be concluded that the generated NPs can be used safely for medical purposes.

It is recommended to test the biosynthesized AgNPs based on in vivo conditions so as to determine a variety of the drug mechanisms, such as release, delivery, and interaction between the human body and the drug delivery system, so as to decide the potential side effects in the near future.

REFERENCES

[1] Keshari, AK, Srivastava, R, Singh, P, Yadav, VB, and Nath, G.Antioxidant and antibacterial activity of silver nanoparticles synthesized by Cestrum nocturnum. *Journal of Ayurveda and integrative medicine* 2020; 11(1):37-44. https://doi.org/10.1016/j.jaim.2017.11.003

[2] Urnukhsaikhan, E, Bold, B-E, Gunbileg, A, Sukhbaatar, N, and Mishig-Ochir, T.Antibacterial activity and characteristics of silver nanoparticles biosynthesized from Carduus crispus. *Scientific Reports* 2021; 11(1):21047.

[3] Mounika, K, Venakatesh, J, Manogna, K, and Paul, AD.Biosynthesis and characterization of silver nanoparticles from tuber extract of Cyperus rotundus and study of its antibacterial activity. *RESEARCH JOURNAL OF PHARMACEUTICAL BIOLOGICAL AND CHEMICAL SCIENCES* 2017; 8(5):378-387.

[4] Botteon, C, Silva, L, Ccana-Ccapatinta, G, Silva, T, Ambrosio, S, Veneziani, R, Bastos, J, and Marcato, P.Biosynthesis and characterization of gold nanoparticles using Brazilian red propolis and evaluation of its antimicrobial and anticancer activities. *Scientific Reports* 2021; 11(1):1974. https://doi.org/10.1038/s41598-021-81281-w

[5] Benedec, D, Oniga, I, Cuibus, F, Sevastre, B, Stiufiuc, G, Duma, M, Hanganu, D, Iacovita, C, Stiufiuc, R, and Lucaciu, CM.Origanum vulgare mediated green synthesis of biocompatible gold nanoparticles simultaneously possessing plasmonic, antioxidant and antimicrobial properties. *International Journal of Nanomedicine* 2018; 13(1041-1058. https://doi.org/10.2147/IJN.S149819

[6] Francis, S, Joseph, S, Koshy, EP, and Mathew, B.Green synthesis and characterization of gold and silver nanoparticles using Mussaenda glabrata leaf extract and their environmental applications to dye degradation. *Environmental Science and Pollution Research* 2017; 24(17347-17357. https://doi.org/https://doi.org/10.1007/s11356-017-9329-2

[7] de Marco, BA, Rechelo, BS, Tótoli, EG, Kogawa, AC, and Salgado, HRN.Evolution of green chemistry and its multidimensional impacts: A review. *Saudi pharmaceutical journal* 2019; 27(1):1-8. https://doi.org/10.1016/j.jsps.2018.07.011

[8] Hurst, GA.Systems thinking approaches for international green chemistry education. *Current Opinion in Green and Sustainable Chemistry* 2020; 21(93-97. https://doi.org/https://doi.org/10.1016/j.cogsc.2020.02.004

[9] Falcone, PM, and Hiete, M.Exploring green and sustainable chemistry in the context of sustainability transition: The role of visions and policy. *Current Opinion in Green and Sustainable Chemistry* 2019; 19(66-75. https://doi.org/https://doi.org/10.1016/j.cogsc.2019.08.002

[10] Castillo-Henríquez, L, Alfaro-Aguilar, K, Ugalde-Álvarez, J, Vega-Fernández, L, Montes de Oca-Vásquez, G, and Vega-Baudrit, JR.Green synthesis of gold and silver nanoparticles from plant extracts and their possible applications as antimicrobial agents in the agricultural area. *Nanomaterials* 2020; 10(9):1763.

[11] Gamboa, SM, Rojas, E, Martínez, V, and Vega-Baudrit, J.Synthesis and characterization of silver nanoparticles and their application as an antibacterial agent. *Int. J. Biosen. Bioelectron* 2019; 5(5):166-173.

[12] Vaid, P, Raizada, P, Saini, AK, and Saini, RV.Biogenic silver, gold and copper nanoparticles-A sustainable green chemistry approach for cancer therapy. *Sustainable Chemistry and Pharmacy* 2020; 16(100247. https://doi.org/https://doi.org/10.1016/j.scp.2020.100247

[13] Mondal, P, Anweshan, A, and Purkait, MK.Green synthesis and environmental application of iron-based nanomaterials and nanocomposite: a review. *Chemosphere* 2020; 259(127509. https://doi.org/10.1016/j.chemosphere.2020.127509

[14] Scampicchio, M, Wang, J, Blasco, AJ, Sanchez Arribas, A, Mannino, S, and Escarpa, A.Nanoparticle-based assays of antioxidant activity. *Analytical chemistry* 2006; 78(6):2060-2063. https://doi.org/10.1021/ac052007a

[15] Wang, J, Zhou, N, Zhu, Z, Huang, J, and Li, G.Detection of flavonoids and assay for their antioxidant activity based on enlargement of gold nanoparticles. *Analytical and bioanalytical chemistry* 2007; 388(1199-1205. https://doi.org/https://doi.org/10.1007/s00216-007-1295-y

[16] Roy, N, Laskar, RA, Sk, I, Kumari, D, Ghosh, T, and Begum, NA.A detailed study on the antioxidant activity of the stem bark of Dalbergia sissoo Roxb., an Indian medicinal plant. *Food chemistry* 2011; 126(3):1115-1121. https://doi.org/https://doi.org/10.1016/j.foodchem.2010.11.143

[17] Cardoso-Avila, PE, Patakfalvi, R, Rodríguez-Pedroza, C, Aparicio-Fernández, X, Loza-Cornejo, S, Villa-Cruz, V, and Martínez-Cano, E.One-pot green synthesis of gold and silver nanoparticles using Rosa canina L. extract. *RSC advances* 2021; 11(24):14624-14631. https://doi.org/10.1039/d1ra01448j

[18] McNamara, K, and Tofail, SA.Nanoparticles in biomedical applications. *Advances in Physics: X* 2017; 2(1):54-88. https://doi.org/https://doi.org/10.1080/23746149.2016.1254570

[19] Mustapha, T, Misni, N, Ithnin, NR, Daskum, AM, and Unyah, NZ.A review on plants and microorganisms mediated synthesis of silver nanoparticles, role of plants metabolites and applications. *International Journal of Environmental Research and Public Health* 2022; 19(2):674. https://doi.org/10.3390/ijerph19020674

[20] Chopade, B, Ghosh, S, Patil, S, Ahire, M, Kitture, R, Jabgunde, A, Kale, S, Pardesi, K, Cameotra, SS, and Bellare, J.Synthesis of silver nanoparticles using Dioscorea bulbifera tuber extract and evaluation of its synergistic potential in combination with antimicrobial agents. *International Journal of Nanomedicine* 2012; 483.

[21] Nagaich, U, Gulati, N, and Chauhan, S.Antioxidant and antibacterial potential of silver nanoparticles: biogenic synthesis utilizing apple extract. *Journal of pharmaceutics* 2016; 2016(7141523. https://doi.org/10.1155/2016/7141523

[22] Ge, L, Li, Q, Wang, M, Ouyang, J, Li, X, and Xing, MM.Nanosilver particles in medical applications: synthesis, performance, and toxicity. *International journal of nanomedicine* 2014; 9(2399-2407. https://doi.org/10.2147/IJN.S55015

[23] Nabikhan, A, Kandasamy, K, Raj, A, and Alikunhi, NM.Synthesis of antimicrobial silver nanoparticles by callus and leaf extracts from saltmarsh plant, Sesuvium portulacastrum L. *Colloids and surfaces B: Biointerfaces* 2010; 79(2):488-493. https://doi.org/10.1016/j.colsurfb.2010.05.018

[24] Ahmad, N, and Sharma, S.Green synthesis of silver nanoparticles using extracts of Ananas comosus. 2012; https://doi.org/http://dx.doi.org/10.4236/gsc.2012.24020

[25] Ibraheem, SA, Audu, EA, Atabat, AJ, Tanimu, BF, Yahaya, JY, and Barminas, JT.Pectin-stabilized silver nanoparticles: synthesis, optical and antimicrobial activity against E. Coli. *Inorganic Chemistry Communications* 2023; 158(111500. https://doi.org/https://doi.org/10.1016/j.inoche.2023.111500

[26] Raj, S, Jose, S, Sumod, U, and Sabitha, M.Nanotechnology in cosmetics: Opportunities and challenges. *Journal of Pharmacy and Bioallied Sciences* 2012; 4(3):186-193. https://doi.org/10.4103/0975-7406.99016

[27] Miguel, MG.Chemical and biological properties of propolis from the western countries of the Mediterranean basin and Portugal. *Int. J. Pharm. Pharm. Sci* 2013; 5(3):403-409.

[28] Shubharani, R, Mahesh, M, and Yogananda Murthy, V.Biosynthesis and characterization, antioxidant and antimicrobial activities of selenium nanoparticles from ethanol extract of Bee Propolis. *J. Nanomed. Nanotechnol* 2019; 10(2):https://doi.org/http://dx.doi.org/10.4172/2157-7439.1000522

[29] Ramnath, S.Synthesis of copper nanoparticles from bee propolis extract and its antibacterial property. *European J. Pharm. Med. Res* 2017; 4(684-688.

[30] Padil, VVT, and Černík, M.Green synthesis of copper oxide nanoparticles using gum karaya as a biotemplate and their antibacterial application. *International journal of nanomedicine* 2013; 889-898. https://doi.org/https://doi.org/10.2147/IJN.S40599

[31] Heidarieh, M, Chakoli, AN, Shahbazi, S, Shawrang, P, and Zhang, B.Effect of gamma irradiation processing on total phenol and antioxidant capacities of the Iranian extract of propolis. *Radiochimica Acta* 2021; 109(8):635-641.

[32] Matsuda, AH, Machado, LB, and del Mastro, NL.Thermal analysis applied to irradiated propolis. *Radiation Physics and Chemistry* 2002; 63(3-6):353-355. https://doi.org/https://doi.org/10.1016/S0969-806X(01)00524-2

[33] Bosio, K, Avanzini, C, D'avolio, A, Ozino, O, and Savoia, D.In vitro activity of propolis against Streptococcus pyogenes. *Letters in applied microbiology* 2000; 31(2):174-177. https://doi.org/10.1046/j.1365-2672.2000.00785.x

[34] Emrani, S, Zhiani, R, and Dafe Jafari, M.The biosynthesis of silver nanoparticles using plants of Glycyrrhiza glabra and Mentha piperata and its antimicrobial effect on some bacterias that cause tooth decay. *Journal of Rafsanjan University of Medical Sciences* 2018; 16(10):953-968.

[35] Kischkel, B, Castilho, PFd, de Oliveira, KM, Rezende, PS, Bruschi, ML, Svidzinski, TI, and Negri, M.Silver nanoparticles stabilized with propolis show reduced toxicity and potential activity against fungal infections. *Future Microbiology* 2020; 15(7):521-539. https://doi.org/https://doi.org/10.2217/fmb-2019-0173

[36] Singleton, VL.Lamuela-Raventos: Analysis of total phenoles and other oxidation substartes and antioxidants by means of folin-ciocalteu reagent. *Methods in enzymology* 1999; 299(152.

[37] Singleton, VL, and Rossi, JA.Colorimetry of total phenolics with phosphomolybdic-phosphotungstic acid reagents. *American journal of Enology and Viticulture* 1965; 16(3):144-158.

[38] Sharma, VK, Yngard, RA, and Lin, Y.Silver nanoparticles: green synthesis and their antimicrobial activities. *Advances in colloid and interface science* 2009; 145(1-2):83-96. https://doi.org/https://doi.org/10.1016/j.cis.2008.09.002

[39] Badi'Ah, H, Seedeh, F, Supriyanto, G, and Zaidan, A, "Synthesis of silver nanoparticles and the development in analysis method." p. 012005.

[40] Lara, HH, Ayala-Núñez, NV, Ixtepan Turrent, LdC, and Rodríguez Padilla, C.Bactericidal effect of silver nanoparticles against multidrug-resistant bacteria. *World Journal of Microbiology and Biotechnology* 2010; 26(615-621. https://doi.org/https://doi.org/10.1007/s11274-009-0211-3

[41] Gubernat, M, Tomala, J, Frohs, W, Fraczek-Szczypta, A, and Blazewicz, S.De-agglomeration and homogenisation of nanoparticles in coal tar pitch-based carbon materials. *Journal of Nanoparticle Research* 2016; 18(1-13. https://doi.org/10.1007/s11051-016-3362-9

Design and assessment of the ARC fusion reactor integrated energy system to produce power and hydrogen

Sulenur Asal
Gazi University, Faculty of Technology, Department of Energy Systems Engineering, Ankara, Turkiye, sulenurasal@gazi.edu.tr
ORCID: 0000-0003-2711-9290

Adem Acir
Gazi University, Faculty of Technology, Department of Energy Systems Engineering, Ankara, Turkiye, adema@gazi.edu.tr
ORCID: 0000-0002-9856-3623

Cite this paper as: *Asal, S, Acir A. Design and assessment of the ARC fusion reactor integrated energy system to produce power and hydrogen. 7th Int. Conf. on Nucl. and Ren. Energy Res. 27-30 October 2024, Ankara, Turkiye.*

Abstract: This proposed study presents two cases of a novel integrated system for the production of power and hydrogen. In Case 1, the Rankine cycle is modeled for the affordable, robust, compact (ARC) fusion reactor. A hydrogen production cycle is added as a subsystem to the integrated system in Case 2. While the 3-step magnesium-chlorine cycle is employed as a hydrogen production method, the Rankine cycle is modeled depending on the given energy efficiency of the Rankine cycle in the literature and employed in both cases. Analyses of the integrated system are carried out with the approach of energy and exergy by using the first and second laws of thermodynamics. Moreover, the hydrogen production amount of the 3-step magnesium chlorine cycle is calculated using the Shomate heat capacity equation. The Rankine cycle is designed with 39.86% energy and 59.65% exergy efficiencies. The integrated system produces about 1.49 kg of hydrogen per second, while the energy and exergy efficiencies of the overall system are 28.38% and 40.91%, respectively.

Keywords: *ARC fusion reactor, nuclear hydrogen production, 3-step Mg-Cl cycle.*

1. INTRODUCTION

Today, energy demand continually increases depending on the rising population. Moreover, fossil fuels which are employed to meet the majority of the energy demand (about 80% [1]) are used up and there are lots of known drawbacks of them. Researchers tend to find cleaner options to meet the energy demand instead of fossil fuels. As a clean energy medium hydrogen can be a promising solution. Hydrogen production methods require high temperature heat, electrical power or both to run. To meet the demand for high temperature of the hydrogen production methods, nuclear can be a good option.

In the open literature, there are many nuclear-based hydrogen production systems. Pinsky et al. [2] reviewed different nuclear-driven hydrogen production systems. Solid oxide electrolysis cells (SOEC), proton exchange membrane (PEM) electrolysis, alkaline electrolysis, copper-chlorine (Cu-Cl), hybrid sulfur (HyS), and sulfur-iodine (S-I) cycles were employed in the hybrid energy systems. Ryland et al. [3] integrated Canada Deuterium-Uranium Reactor (CANDU) and high temperature electrolysis (HTE). |They investigated the proposed system from the perspective of technical and economical. Kowalczyk et al. [4] designed and assessed a cogeneration system, integrating high temperature reactor (HTR) and SOEC. The energy and exergy analysis of the integrated system were calculated using the first and second laws of thermodynamics. As a result of the study, the integrated system produced about 4392 kg/h of hydrogen with 55.83% energy efficiency. Asal and Acir [5] modified the Laser Inertial Fission Fusion Engine (LIFE) and the S-I, Cu-Cl, and cobalt-chlorine (Co-Cl) cycles were integrated into the system. Norouzi et al. [6] modeled a cogeneration system of power and hydrogen. The HTR and steam methane reforming (SMR) system were employed. Consequently, approximately 144.50 tons/h of hydrogen was produced with 78.50% exergy efficiency. Genc [7] examined the hydrogen production potential of the Advanced Power Extraction (APEX) fusion reactor using SMR, S-I, and HTE hydrogen production methods, and the methods were compared depending on the production rates.

Differing from the above studies, a Rankine cycle is modeled depending on the literature information about the efficiency and the produced power of the ARC fusion reactor's Rankine cycle. Moreover, a hydrogen production cycle is integrated into the ARC fusion reactor for the first time. Within the scope of this study, two different cases are modeled, a Rankine cycle is designed for the ARC fusion reactor in Case 1. The 3-step magnesium-chlorine (Mg-Cl) cycle is integrated into the system for Case 2. The energy and exergy analyses are carried out and assessed for two cases, employing the first and second laws of thermodynamics.

2. SYSTEM DESCRIPTION

Within the scope of this study, the ARC fusion reactor is employed as a heat source. The original design of the ARC fusion reactor is modeled to decrease the plant capital cost of the large-scale designs. The purpose of the ARC design is to minimize reactor size. The fusion power of the reactor is 525 MW and the total of thermal power is 630 MW. The coolant of the reactor is FLiBe and its inlet and outlet temperatures are 800 K and 900 K, respectively [8–10].

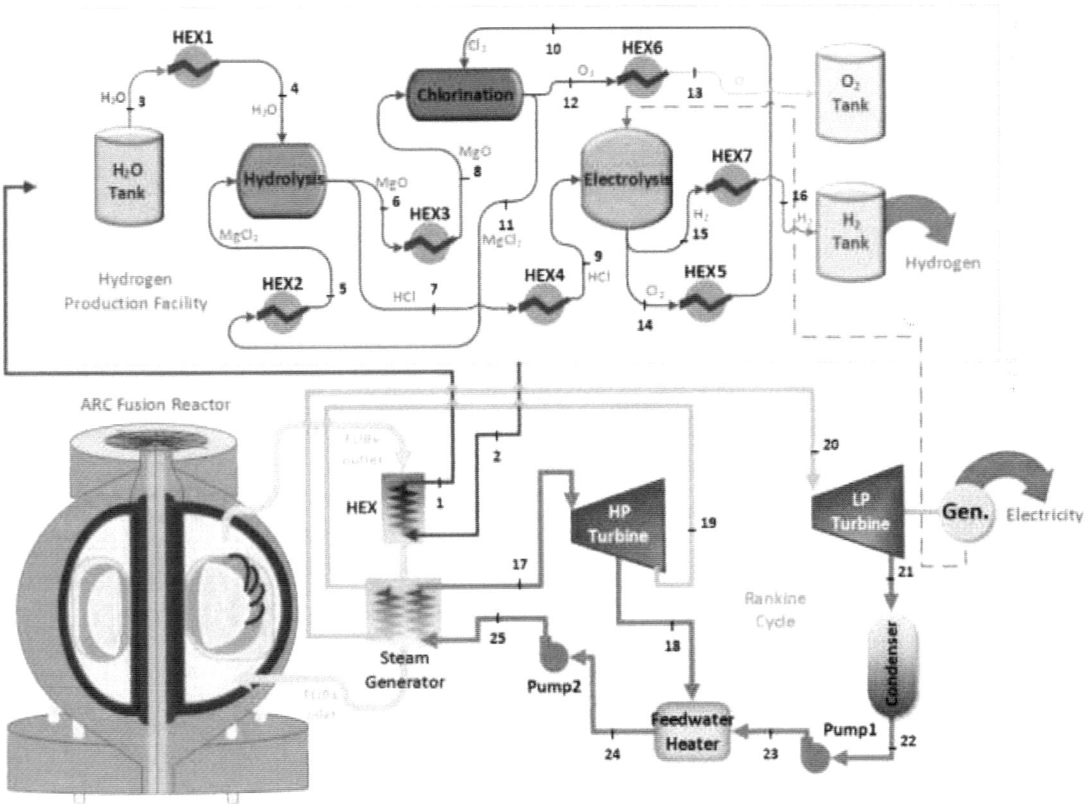

Figure 1. The layout of the proposed integrated systems (Case 2)

Two different cases are modeled for this proposed study. Firstly, a Rankine cycle is modeled for the ARC fusion reactor with 40% energy efficiency depending on the Ref. [8]. In Case 1, the ARC fusion reactor driven Rankine cycle is designed and assessed with the approaches of energy and exergy. While Case 1 only produces power, a hydrogen production cycle is added to the system in Case 2. The 3-step Mg-Cl cycle is chosen since it has high energy and exergy efficiencies as well as lower reaction temperatures. The thermal power that comes from the ARC fusion reactor is conducted via heat exchangers (HEXs). In both cases, the Rankine cycles have the same energy and exergy efficiencies and also the same thermophysical properties for each state of the Rankine cycle. In Case 2, while the thermal power is sent to both the hydrogen production cycle and the Rankine cycle, all produced heat in the ARC fusion reactor is employed in the subsystems to produce hydrogen. Figure 1 illustrates the system layout of Case 2.

3. ANALYSIS

The analyses of the proposed system are made with the energy and exergy approaches using the first and second laws of thermodynamics. While the modeling and assessment of the Rankine cycle depends on the general energy, entropy, and exergy equations, the hydrogen production cycle's assessment is completed using the Shomate heat capacity equation. This study is conducted with the following assumptions:

- All processes are steady-state and steady-flow.
- The kinetic and potential energy changes are accepted as negligible.
- The dead state temperature and pressure values are 25°C and 101.325 kPa, respectively.
- All heat that comes from nuclear fusion reactor is employed in the subsystems.

The general mass, energy, entropy, and exergy equations which is used in the calculations are given with Eqs. (1-4) [11].

$$\sum \dot{m}_{in} = \sum \dot{m}_{out} \tag{1}$$

$$\sum \dot{Q}_{in} + \sum \dot{W}_{in} + \sum \dot{m}_{in}\left(h_{in} + \frac{V_{in}^2}{2} + gZ_{in}\right) = \sum \dot{Q}_{out} + \sum \dot{W}_{out} + \sum \dot{m}_{out}(h_{out} + \frac{V_{out}^2}{2} + gZ_{out}) \tag{2}$$

$$\sum \dot{S}_{in} + \sum \frac{\dot{Q}_{in}}{T_s} + \dot{S}_{gen} = \sum \dot{S}_{out} + \sum \frac{\dot{Q}_{out}}{T_s} \tag{3}$$

$$\sum \dot{Ex}_{in} + \sum \dot{Ex}_{in}^{Q} + \sum \dot{W}_{in} = \sum \dot{Ex}_{out} + \sum \dot{Ex}_{out}^{Q} + \sum \dot{W}_{out} + \dot{Ex}_{d} \tag{4}$$

where, $\dot{S} = \dot{m}\, s$, $\dot{Ex} = \dot{m}\, ex$ and $ex = (h - h_0) - T_0(s - s_0)$.

The Shomate equation is used to calculate the energy requirements of the 3-step Mg-Cl cycle to produce 1 mole of hydrogen. The Shomate heat capacity equation is a method dependent to calculate thermochemistry data using polynomial equations. Eqs. (5) and (6) give the enthalpy and entropy equations of the Shomate [12].

$$\bar{h}(T) - \bar{h}_0 = AT + B\frac{T^2}{2} + C\frac{T^3}{3} + D\frac{T^4}{4} - \frac{E}{T} + F - H \tag{5}$$

$$s = Aln(T) + BT + C\frac{T^2}{2} + D\frac{T^3}{3} - \frac{E}{(2T^2)} + G \tag{6}$$

where, $\bar{h}_f^o$ is entalphy of formation, $\bar{h}(T)$ is sensible entalphy of a given state (kJ/mol), $\bar{h}_0$ is standard sensible enthalpy (kJ/mol) at 25°C and 1 atm, T is 1/1000 of the specified temperature in K, s is the standard entropy (J/molK), and A, B, C, D, E, F and H are constants that are given in Table 1.

Table 1. The required parameters for the Shomate equation [12]

Compound	$\bar{h}_f^o$ (kJ/mol)	A	B	C	D	E	F	G	H
$MgCl_{2(s)}$	-641.62	78.307330	2.435888	6.858873	-1.728967	-0.729911	-667.5823	179.2639	-641.6164
$H_2O_{(g)}$	-241.83	30.092	6.832514	6.793435	-2.534480	0.082139	-250.8810	223.3967	-241.8264
$MgO_{(s)}$	-601.24	47.25995	5.68162	-0.872665	0.104300	-1.053955	-619.1316	76.4618	-601.2408
$HCl_{(g)}$	-92.31	32.12392	-13.45805	19.86852	-6.853936	-0.049672	-101.6206	228.6866	-92.31201
$MgO_{(s)}$	-601.24	47.259950	5.681621	-0.872665	0.104300	-1.053955	-619.1316	76.461760	-601.2408
$Cl_{2(g)}$	0	33.0506	12.2294	-12.0651	4.38533	-0.159494	-10.8348	259.029	0
$O_{2(g)}$	0	30.0235	8.772972	-3.988133	0.788313	-0.741599	-11.32468	236.16630	0
$H_{2(g)}$	0	33.066178	-11.363417	11.432816	-2.772874	-0.158558	-9.980797	172.707974	0

The 3-step Mg-Cl cycle is one of the water splitting methods. The cycle includes three different reactions; two of them are thermochemical and one of them is electrochemical. The first reaction is hydrolysis reaction and occurs endothermically at 500°C. Solid-phased $MgCl_2$ and H_2O react, and MgO and HCl are produced. While solid-phased

MgO is sent to the second reaction to react with Cl_2, HCl enters the electrolysis cell. The second reaction is called as chlorination reaction, where $MgCl_2$ and O_2 occur at 550°C temperature. $MgCl_2$ goes to the first reactor as a reactant, whilst O_2 is obtained as a product. The third reaction is the electrolysis reaction, which occurs at 90°C. In the electrolysis cell, HCl decomposes to H_2 and Cl_2.

$$MgCl_{2\,(s)} + H_2O_{(g)} \rightarrow MgO_{(s)} + 2\,HCl_{(g)} \quad (7)$$

$$MgO_{(s)} + Cl_{2\,(g)} \rightarrow MgCl_{2\,(s)} + 0.5\,O_{2\,(g)} \quad (8)$$

$$2\,HCl_{(g)} \rightarrow Cl_{2\,(g)} + H_{2\,(g)} \quad (9)$$

The energy and exergy efficiencies of the overall system (case 2) is assessed by followed equations:

$$\eta_{overall} = \frac{\dot{m}_{H_2}\,LHV_{H_2}}{\dot{Q}_{ARC}} \quad (10)$$

$$\psi_{overall} = \frac{\dot{m}_{H_2}\,ex_{H_2}}{\dot{Ex}^{Q}_{ARC}} \quad (11)$$

4. RESULTS AND DISCUSSION

The performance of the proposed systems is investigated with the energy and exergy approaches. Figure 2 illustrates the exergy destruction rates of each component in the cases. In Case 1, the steam generator of the Rankine cycle has the highest exergy destruction rate with 102.35 MW. For Case 2, the major contributor to the exergy destruction is the 3-step Mg-Cl cycle. The electrolysis cell's exergy destruction rate is calculated as 184.92 MW. The second contribution belongs to the steam generator of the Rankine cycle. For both cases, pump 1 and 2 have the lowest exergy destruction rates, about 0 MW.

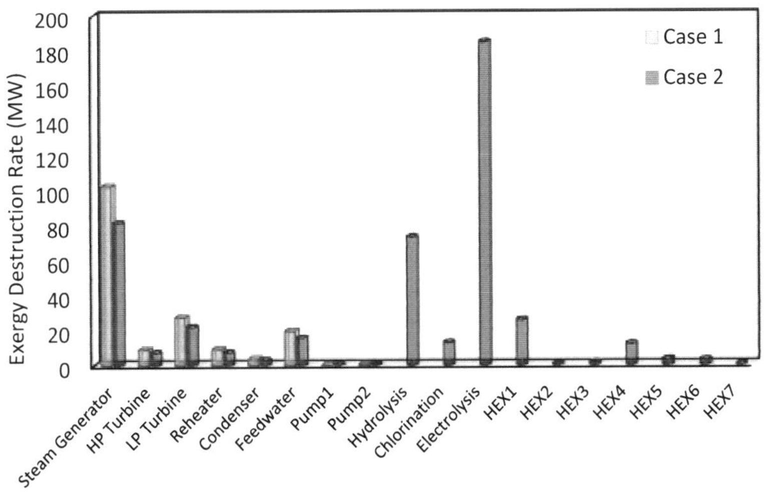

Figure 2. The exergy destruction rates of each component for both cases 1 and 2

Case 1 produces about 251.09 MW of power, while it uses all thermal power that comes from the ARC fusion reactor. The energy and exergy efficiency values of the designed Rankine cycle are calculated as 39.86% and 59.65%, respectively. Case 2 produces about 1.49 kg/s of hydrogen via the 3-step Mg-Cl cycle, while the cycle uses all thermal heat and electrical power that comes from the ARC fusion reactor and the Rankine cycle.

Figure 3 gives the energy and exergy efficiencies of the Rankine cycle, 3-step Mg-Cl cycle, and overall system for Case 2. The 3-step Mg-Cl cycle's energy efficiency is 54.51%, whilst the exergy efficiency is assessed as 60.81%. Ultimately, the energy and exergy efficiencies of the overall cycle for Case 2 are calculated as 28.38% and 40.91% in turn.

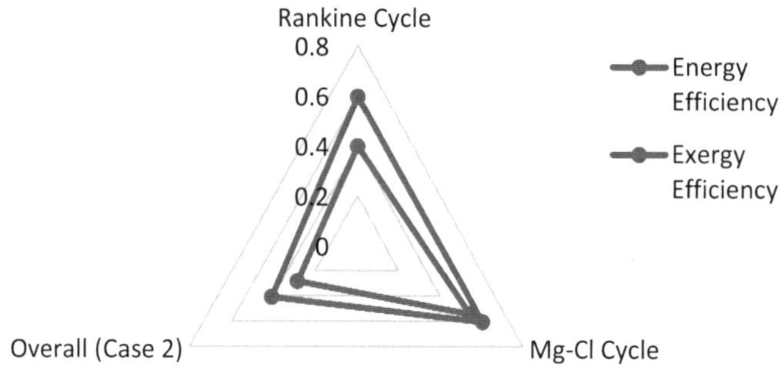

Figure 3. The energy and exergy efficiencies of each cycle

5. CONCLUSION

In this study, a Rankine cycle is designed for the ARC fusion reactor, and also a hydrogen production cycle is integrated into the nuclear reactor to assess the hydrogen production potential. Consequently, the followings are obtained:

- The energy and exergy efficiencies of the Rankine cycle are 39.86% and 59.65% in turn, while it produces about 251.09 MW of power.
- The 3-step Mg-Cl cycle integrated ARC fusion reactor can produce nearly 1.49 kg/s of hydrogen.
- The overall system's energy efficiency is calculated as 28.38%, while the exergy efficiency is 40.91%.

REFERENCES

[1] Wright VP. World Energy Outlook. 1986:23–8.

[2] Pinsky R, Sabharwall P, Hartvigsen J, O'Brien J. Comparative review of hydrogen production technologies for nuclear hybrid energy systems. Progress in Nuclear Energy 2020;123:103317.

[3] Ryland DK, Li H, Sadhankar RR. Electrolytic hydrogen generation using CANDU nuclear reactors. Int J Energy Res 2007;31:1142–55.

[4] Kowalczyk T, Badur J, Bryk M. Energy and exergy analysis of hydrogen production combined with electric energy generation in a nuclear cogeneration cycle. Energy Convers Manag 2019;198:111805.

[5] Asal S, Acir A. Evaluation and comparison of hydrogen production potential of the LIFE fusion reactor by using copper-chlorine (Cu-Cl), cobalt-chlorine (Co-Cl) and sulfur-iodine (S-I) cycles. Int J Hydrogen Energy 2023;48:22791–805.

[6] Norouzi N, Talebi S, Fani M, Khajehpour H. Exergy and exergoeconomic analysis of hydrogen and power cogeneration using an HTR plant. Nuclear Engineering and Technology 2021;53:2753–60.

[7] Genç G. Hydrogen production potential of APEX fusion transmuter fueled minor actinide fluoride. Int J Hydrogen Energy 2010;35:10190–201.

[8] Sorbom BN, Ball J, Palmer TR, Mangiarotti FJ, Sierchio JM, Bonoli P, et al. ARC: A compact, high-field, fusion nuclear science facility and demonstration power plant with demountable magnets. Fusion Engineering and Design 2015;100:378–405.

[9] Meschini S, Testoni R, Segantin S, Zucchetti M. ARC reactor: A preliminary tritium environmental impact study. Fusion Engineering and Design 2021;167:112340.

[10] Kuang AQ, Cao NM, Creely AJ, Dennett CA, Hecla J, LaBombard B, et al. Conceptual design study for heat exhaust management in the ARC fusion pilot plant. Fusion Engineering and Design 2018;137:221–42.

[11] Dincer I (2021) Thermoydnamics: A Smart Approach, 1st ed., John Wiley & Sons, London.

[12] NIST Chemistry WebBook. http://webbook.nist.gov/chemistry/; 2018.

Long-term Planning of Power Generation Volumes for the Akkuyu Nuclear Power Plant

Saltuk Buğrahan Yeşilgöl

[1]AKKUYU NUCLEAR JSC, Mersin, Türkiye, saltukenerji@gmail.com

ORCID: 0009-0006-4761-9566

Fethi Mustafa Çimen

[1]Gazi University, Ankara, Türkiye, [2]University of South Carolina, Columbia, United States, fethimustafacimen@gazi.edu.tr

ORCID: 0000-0002-1290-1469

Cite this paper as: *Yeşilgöl, SB, Çimen FM. Long-term Planning of Power Generation Volumes for the Akkuyu Nuclear Power Plant. 7th Int. Conf. on Nucl. and Ren. Energy Res. 27-30 October 2024, Antalya, Turkey*

Abstract: This report outlines the long-term planning of power generation volumes for the Akkuyu Nuclear Power Plant (NPP). It covers a time horizon from one day to 15 years and aims to provide critical data for calculating both guaranteed power purchase volumes, per the agreement between the governments of Russia and Turkey, and electricity sales on the wholesale market. The report also addresses the contributions to the reserve fund for decommissioning the NPP units. Power generation planning is currently limited by the lack of operational data, so project data is utilized in the interim. Various factors influencing power generation include technical and economic indicators, repair schedules, and a disruption coefficient (k_{dis}), which accounts for unplanned shutdowns. Seasonal restrictions, particularly the temperature of cooling water, also affect power generation efficiency. An example of power generation calculations for Unit 1 shows the significant role of both the disruption coefficient (k_{dis}) and seasonal temperature changes in reducing power output. For 2029, four operational scenarios are evaluated: without seasonal and disruption coefficient, with seasonal restrictions, with the disruption coefficient, and with both factors considered. The analysis demonstrates that while seasonal limitations are uncontrollable, the disruption coefficient (k_{dis}) can potentially be reduced, increasing overall power generation. Future optimization will focus on reducing operational disruptions and enhancing the accuracy of power generation planning through the collection of real-time operational data.

Keywords: *Akkuyu Nuclear Power Plant, Power Generation Optimization, Comparative Analysis, Disruption coefficient, Seasonal restrictions*

1. INTRODUCTION

The planning of power generation volumes at the Akkuyu Nuclear Power Plant (NPP) is carried out over a time horizon ranging from one day to 15 years. The goal is to provide the initial data required to calculate the guaranteed power purchase volume in accordance with the agreement between the Government of the Russian Federation and the Government of the Republic of Turkey on cooperation in the construction and operation of the nuclear power plant at the Akkuyu site in the Republic of Turkey. Additionally, it calculates the volume for sale on the wholesale electricity market, with the possibility of planning over the entire operational lifetime of the NPP. This planning serves to provide the necessary data for calculating contributions to the reserve fund for decommissioning the Akkuyu NPP units. At present, the primary issue hindering more accurate planning is the lack of accumulated statistical data on the operating modes of the Akkuyu NPP equipment under various conditions.

2. INITIAL DATA FOR PLANNING POWER GENERATION

Due to the current lack of actual operational data for the Akkuyu NPP units, power generation calculations can be performed based on project data.

Power generation calculations for the grid during the industrial operation phase of the Akkuyu NPP units are carried out based on:

- Approved technical and economic indicators;
- The approved prospective repair schedule for the units;
- The integrated construction schedule for the Akkuyu NPP Project (Units No. 1, 2, 3, 4);
- The Regulation for the formation of power generation plans at the Akkuyu NPP according to which, in the first year of operation at the nominal power level of the commissioned unit, an increased disruption coefficient k_{dis} = 5% is applied, accounting for unplanned shutdowns and reductions of the startup unit. In the second and subsequent years, the disruption coefficient (k_{dis}), accounting for unplanned shutdowns and reductions of the startup unit, is k_{dis} = 2% [1].
- Dependence of Generator Terminal Power on Cooling Water Temperature" to the Technical Specification "Steam Turbine ARABELLE 1200". The average monthly cooling water temperature is determined based on "Water Temperature" of the Safety Justification Report, Characteristics of the Site and Area [2].

Additionally, to prevent simultaneous repairs on two or more units, the calendar durations of the first three fuel cycles on Unit 1 are considered with respect to reactivity power effects. Consequently, for Unit 1, power reduction is accounted for to extend the reactor's fuel campaign, aiming to bring the unit to repair in optimal terms while working under power effect conditions.

The calculation of power generation volumes during pre-commisioning operation phase is carried out in accordance with the "Phase Program for pre-commisioning operation of Unit 1 of the Akkuyu NPP. Additionally, the calculation follows the recommendations of the Regulation for the Formation of Power Generation Plans at the Akkuyu NPP, and further corrective factors are used:

- A reduction factor accounting for decreased efficiency at partial power levels $k_{\text{efficeny},start\ up}$;
- A reduction factor for conducting dynamic tests at all stages of power ramp-up $k_{\text{dinamic},start\ up}$

3. EXAMPLE OF POWER GENERATION CALCULATION

The calculation is based on the following indicators for Unit 1 of the Akkuyu NPP. The calculation for the other units is performed in a similar manner.

Table 1 Initial Data for Calculation [2]

No.	Parameter	Unit	Value
1.	Nominal power of one unit	MW	1232.5
2.	Disruption coefficient in the first year of operation	%	5
3.	Disruption coefficient in subsequent years of operation	%	2
4.	Operating power in the first year of industrial operation	MW	1170.875
5.	Operating power in the second and subsequent years of industrial operation	MW	1207.85

Table 2 Seasonal Power Reduction [2]

	June	July	August	September	October	November
Power Loss, %	0.6	1.6	2.2	1.8	1.1	0.25
Share	0.994	0.984	0.978	0.982	0.989	0.9975
Power Loss, MW	7.395	19.72	27.115	22.185	13.5575	3.08125
Cooling Water Temp., °C	23.5	26.7	27.9	26.8	25	22.1

Table 2 lists the months in which water temperature affects the reduction in power output. As the temperature of the cooling water increases, the generator's power decreases. This indicates that changes in the cooling water temperature are an important factor in the efficiency and performance of the generator.

Table 3 Power Generation of Unit 1 of the Akkuyu NPP Without Considering k_{dis} and Seasonal Restrictions

Power Generation, MWh	**2025**	**2026**	**2027**	**2028**	**2029**	**2030**	**2031**	**2032**
January	0	916980	0	916980	916980	0	916980	916980
February	0	828240	88740	857820	828240	769080	828240	857820
March	0	916980	916980	916980	916980	916980	916980	916980
April	0	887400	887400	887400	887400	887400	887400	887400
May	0	916980	916980	916980	916980	916980	916980	916980
June	0	887400	887400	887400	887400	887400	857820	887400
July	0	916980	916980	236640	916980	916980	0	916980
August	0	916980	916980	562020	916980	916980	828240	916980
September	0	887400	887400	887400	887400	887400	887400	887400
October	0	916980	916980	916980	916980	916980	916980	916980
November	0	887400	887400	887400	887400	887400	887400	887400
December	88740	798660	916980	916980	857820	916980	916980	621180
TOTAL for the year	88740	10678380	9140220	9790980	10737540	9820560	9761400	10530480

To compare calculations under different conditions, Figure 1 presents the planned values of power generation for Unit 1 of the Akkuyu NPP in 2029. Specifically:

- **Yellow line** – without consideration of seasonal restrictions and without consideration of the disruption coefficient (Ideal condition).
- **Gray line** – with consideration of seasonal restrictions;
- **Orange line** – with consideration of the disruption coefficient;
- **Blue line** – with consideration of both seasonal restrictions and the disruption coefficient.

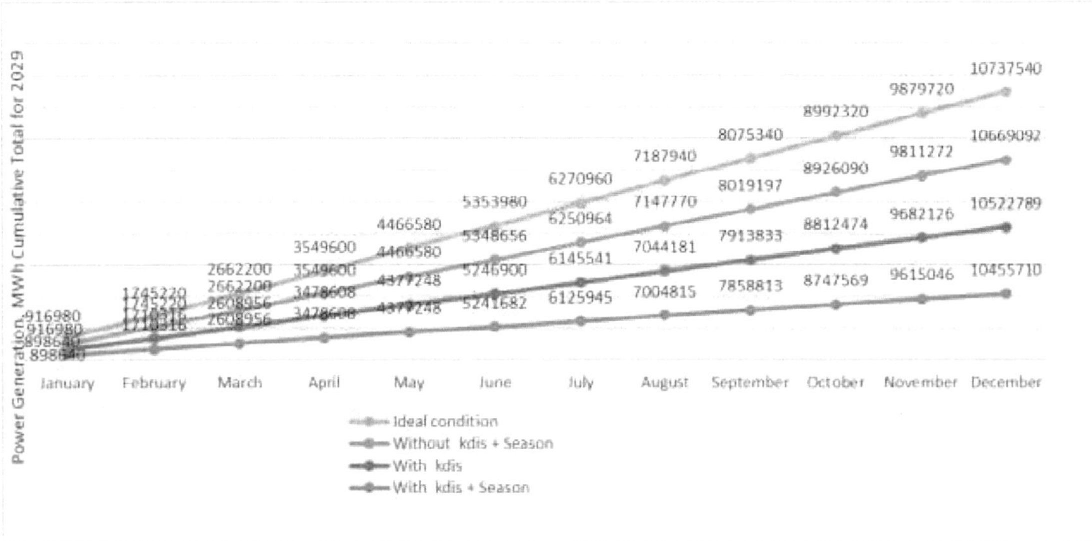

Figure 1 Power Generation for Unit 1 in 2029 Cumulative Total

4. CONCLUSION

Thus, for planning the power generation of the Akkuyu NPP units, project data were used.

To determine the importance of optimizing power generation planning, an analysis was conducted for 4 different operational scenarios for Unit 1 in 2029.

The analysis of the power generation data for Unit 1 of the Akkuyu NPP clearly demonstrates that seasonal restrictions lead to a reduction in the amount of electricity generated. There is no possibility for the station personnel to influence the planned reduction in power generation caused by seasonal restrictions.

During planning, a disruption coefficient (k_{dis}) was applied, which characterizes deviations in the operation of the unit's equipment. These deviations can potentially be prevented, thereby increasing the actual volume of electricity generated. In future planning, it may be possible to reduce this coefficient from 2% to the minimum possible.

For clarity in the performed analysis of electricity production planning for 2029, the following values for different planning scenarios are considered:

1- Excluding seasonal restrictions and the disruption coefficient (Yellow line in Figure 1 – ideal conditions): The generation would be 10,737,540 MWh.
2- Considering seasonal restrictions (Gray line in Figure 1): The generation would be 10,669,091 MWh.
3- Considering the disruption coefficient (Orange line in Figure 1): The generation would be 10,522,789 MWh.

4- Considering both seasonal restrictions and the disruption coefficient (Blue line in Figure 1): The generation would be 10,455,710 MWh.

Expressing these volumes as a percentage of ideal conditions:

With the disruption coefficient, the planned reduction in power generation is about 2%.

With both the disruption coefficient and seasonal restrictions, the planned reduction in power generation is about 2.62%.

The impact of seasonal restrictions is considered an uncontrollable factor in the operation of the NPP and cannot lead to an increase in power generation, unlike the disruption coefficient, which accounts for losses due to operational issues. Without such disruptions, the actual power generation of the NPP increases, leading to an additional 2% of the planned electricity generation.

During the pre-commisioning and industrial operation phases of the Akkuyu NPP, technical directorate departments will organize data collection to determine technical and economic indicators for optimizing the electricity production planning process during operation. Currently, the Company has developed a document for the formation of electricity production plans for the Akkuyu NPP, and the procedures specified in this document will be optimized based on the analysis of the obtained data.

ACKNOWLEDGMENT

The authors would like to thank AKKUYU NUCLEAR JSC for permission to utilise the data used in this study.

REFERENCES

1.Operation Documentation of Akkuyu Nuclear Power Plant.

2.Project Documentation of Akkuyu Nuclear Power Plant.

Heat Pump System of Renewable Energy Sources Utilization As A Basis For The Development Of A Digital Twin

Bauyrzhan Amirkhanov
Al-Farabi Kazakh National University, Almaty, Kazakhstan, amirkhanov.b@gmail.com,
ORCID: 0000-0002-4915-0347

Murat Kunelbayev
Al-Farabi Kazakh National University, Almaty, Kazakhstan,
Institute Information and Computational Technologies CS MES RK, murat7508@yandex.kz,
ORCID: 0000-0002-5648-4476

Gulnur Tuylepberdinova
Al-Farabi Kazakh National University, Almaty, Kazakhstan, tyulepberdinova@gmail.com,
ORCID: 0000-0002-4322-8983

Madina Mansurova
Al-Farabi Kazakh National University, Almaty, Kazakhstan, mansurova.madina@gmail.com,
ORCID: 0000-0002-9680-2758

Gulshat Amirkhanova
Al-Farabi Kazakh National University, Almaty, Kazakhstan, gulshat.aa@gmail.com,
ORCID: 0000-0003-3933-5476

Cite this paper as: *Amirkhanov B., Kunelbayev M., Tuylepberdinova G., Mansurova M., Amirkhanova G. Heat pump system of renewable energy sources utilization as a basis for the development of a digital twin. 7th Int. Conf. on Nucl. and Ren. Energy Res. 27-30 October 2024, Ankara, Turkey*

Abstract: The use of renewable energy sources (RES) in agricultural processes, reducing the consumption of traditional energy carriers, and energy saving is a relevant technological and economic task. The proposed heat pump system allows utilizing solar energy, recycled heat from animals, milk, and the ground. The system is connected to and closely interacts with technological processes, regulating thermal processes for the microclimate, cooling, and product storage, while simultaneously utilizing the excess heat generated by these processes. The system serves as the foundation for building a digital twin of the farm. The article describes the system's structure, operating principle, and key patterns. An assessment of the system's functional characteristics has been conducted during its operation in the microclimate of a cowshed, its compliance with technical requirements, and the quality and stability of the technological process over a specified period.

***Keywords**:* *Livestock farm, heating and cooling supply, heat pump, solar energy, digital twin.*

Nomenclature	
GH	Ground heat
HP	Heat pump
LPS	Low-potential heat sources
MC	Milk cooler
MH	Milk heat
MZS	Multi-zonal system
RES	Renewable energy sources
STH	Secondary technological heat
SC	Solar collector
SEH	Solar energy heat
TEA	Thermal energy accumulator
TA	Thermal accumulator
UNFCCC	United Nations Framework Conventions on Climate Change

1. INTRODUCTION

Modern agriculture faces many challenges related to increasing energy consumption and the need to switch to renewable sources. With increasing climate instability and global economic difficulties, it is critical to reduce carbon emissions and optimize energy resources on farms. In line with international standards such as the Kyoto Protocol and the Paris Agreement, attention is being paid to technologies that can significantly reduce carbon footprints.

One promising solution to these challenges is heat pumps that utilize renewable sources such as solar energy and heat generated by animals. These systems not only reduce the consumption of traditional energy carriers, but also provide efficient indoor climate control, which is particularly important for livestock farms [1-3].

2. MATERIALS AND RESEARCH METHODS

Comprehensive theoretical and experimental studies have been carried out to develop a heat pump based heat supply system utilizing renewable energy sources. The key element of the system is the integration of a digital twin that allows real-time modelling and process control based on data from different sensors. This approach provides not only monitoring of the system state but also active control, which is particularly important for optimizing energy consumption and reducing excessive heat losses.

Literature analysis was conducted using both domestic and foreign sources [4-7]. These studies allowed the identification of design and technological solutions for a heat supply system based on the utilization of several heat energy sources such as ground, milk, animal heat and solar energy. Theoretical studies were based on the methods of thermodynamics, focusing on the development of new approaches to the control of thermal processes.

3. PROBLEM STATE AND RESEARCH

In the proposed system (Figures 1 and 2), the low-potential heat sources (LPS), which will be converted and utilized, include: ground heat (GH), milk heat (MH), secondary technological heat (STH), and solar energy heat (SEH). For these purposes, a system has been developed consisting of: a heat pump (HP) 1, ground heat exchanger 2, milk cooler (MC) 3, microclimate device 4, solar collector (SC) 5, and thermal energy accumulator (TEA) 6. Functionally, the system is divided into four subsystems: "HP-solar collector", "HP-microclimate", "HP-milk cooler" and "HP-ground heat exchanger". The coordinated operation of the multi-zonal system (MZS) is regulated by a microprocessor control unit 33 [3, 8-10].

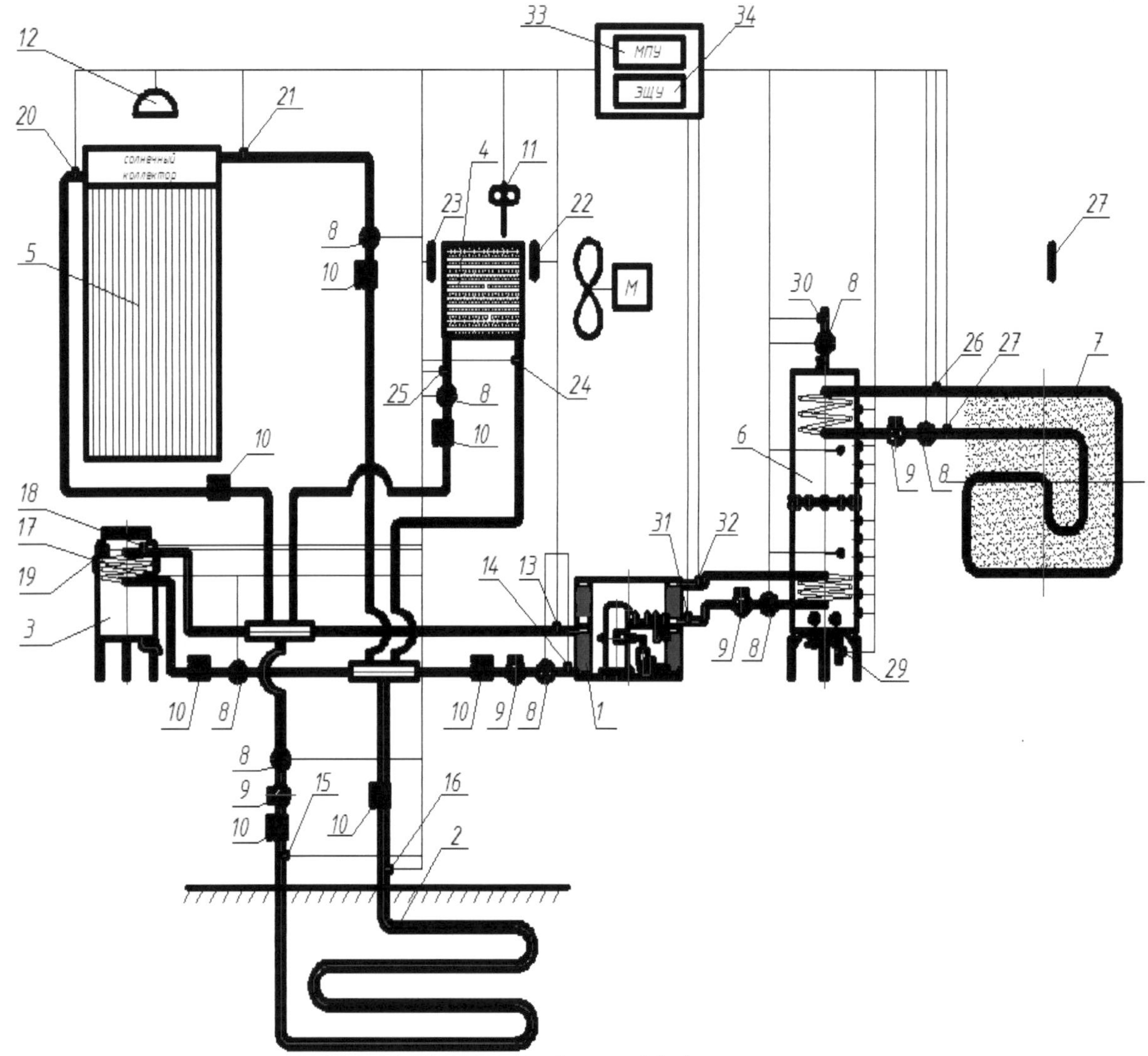

Figure 1. The schematic diagram of the heat pump system

The system operates as follows.

When starting up, the microprocessor control unit 33 connects one of the four sources to the heat pump (HP) based on priority tasks. This means that one of the subsystems is functioning, and heat from the source is supplied to the evaporator of the HP via the circulation pump 9. The heat is then absorbed by the refrigerant circulating in the internal circuit of the HP. The refrigerant, through the compressor, transfers the heat to the condenser. The heat is extracted from the condenser by the heat carrier and stored in the thermal energy accumulator (TEA). For example, at night, when the animals are in the cowshed, the system extracts excess heat from the air in the room using the heat exchanger 4. At the same time, while utilizing the excess heat, the system maintains the air parameters inside the room at the required level.

In the morning, after milking, the HP automatically connects to the milk cooler and extracts heat from it. As a result, the milk is cooled to the required temperature within a specified time.

Figure 2. Overall view of the laboratory setup

During the day, the switching valves connect the evaporator of the solar collectors (5). In the absence of solar radiation, the heat pump (HP) connects to the ground heat exchanger (2) and extracts heat from the soil layer surrounding the heat exchanger until a specific ground temperature is reached. The accumulated heat in the thermal accumulator (TA) can be supplied to the underfloor heating (7) or for hot water supply (30) using a circulation pump. If necessary, the temperature of the hot water can be adjusted to the required technological value using an electric heater (heating element) integrated into the lower part of the TA. The supply of cold water is provided from the water supply system through pipeline (29).
The diagram and general view of the laboratory stand for researching the characteristics of the subsystem "HP-solar collector" are presented in Figure 3. Research is conducted in both winter and summer periods.
During the experiments, the following parameters are recorded: t_1, t_2, t_3, t_4, G_1, G_2. The frequency of measurements is chosen depending on the dynamics of the process. Simultaneously, the readings of the pyranometer (E) and the ambient temperature (t_m) are recorded.
The characteristics of the solar collector with and without a transparent covering are studied. The powers developed by the solar collectors are calculated using the Eq. [1]:

$$P_1 = c_p G_1 (t_2 - t_1)\frac{1}{S}, \qquad (1)$$

where t_1 and t_2 are the temperatures of the heat carrier at the inlet and outlet of the solar collector, respectively, in °C; S is the area of the solar collector in m^2.

Experiments with solar collectors with and without a transparent covering are conducted under the same conditions. To achieve this, a stable solar radiation interval was chosen, from 12:00 to 15:00 hours.
The results of the experimental measurements are shown in Figure 4. On the graphs, the left vertical axis indicates the change in temperature over time, while the right vertical axis shows the change in heat production. The experiments were conducted under winter conditions in both cloudy and sunny weather.

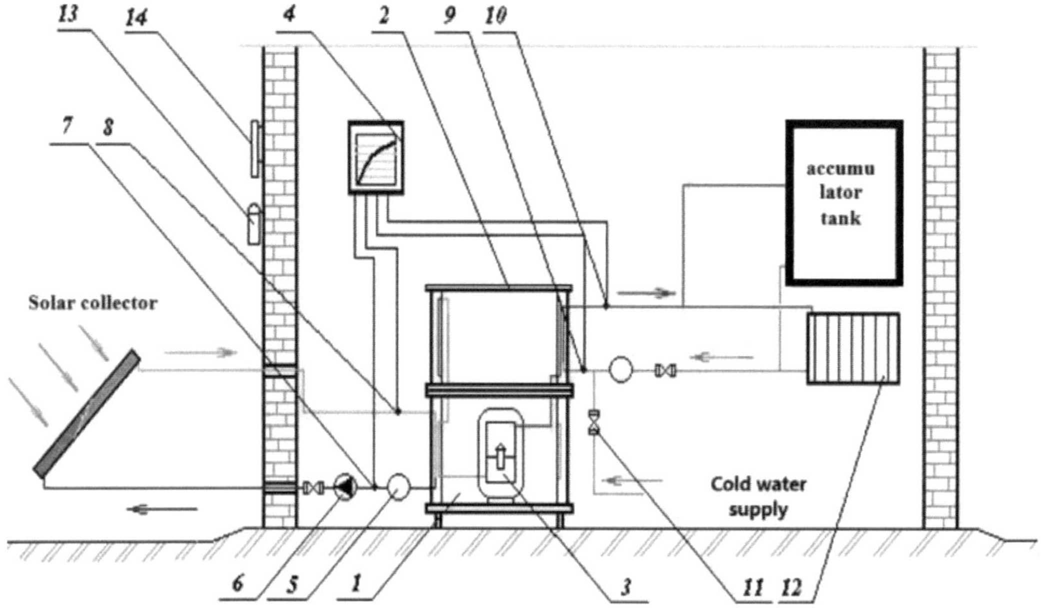

Figure 3. The experimental measurements

1, 2, and 3, respectively: evaporator, condenser, and compressor of the heat pump; 4 – measuring unit; 5 – flow meter; 6 – circulation pump; 7, 8, 9, 10 – temperature sensors; 11 – valve; 12 – heating load; 13 – pyranometer; 14 – thermometer

Figure 4. Diagram and overall view of the stand for researching the subsystem "HP-solar collector"

Figure 4 shows the schematic and general view of the HP-solar collector subsystem research bench. The analysis shows that at an ambient temperature of 10°C and dense cloud cover, the temperature of the antifreeze supplied from the heat pump to the solar collector was 18°C. As it flowed through the solar collector, its temperature increased by 4°C. At the same time, the calculated heat output of the solar collector ranged from 1.1 to 1.5 kWh, and the heat output of the heat pump was: $q^{э}_{T\ I}$ =1,5…2,0 kWh. The average calculated value of the conversion coefficient was $\varphi=1,9...2,5$.
On sunny days, the average intensity of total solar radiation was e_T=0,1…0,3 kW/m^2 (Figure 5).
From the graphs, it can be seen that at an ambient temperature of 8 to 10°C, the temperature of the antifreeze supplied from the heat pump to the solar collector ranged from 5 to 10°C, with a heat carrier flow rate of 350 to 400 L/h. As it flowed through the solar collector, its temperature increased on average by 5 to 8°C.

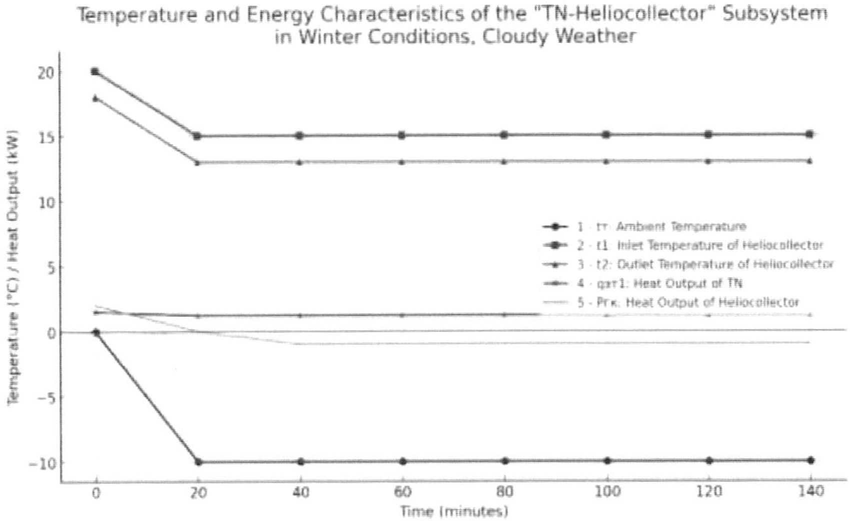

Figure 5. Temperature and energy characteristics of the subsystem "TH - solar collector" in winter conditions, in cloudy weather

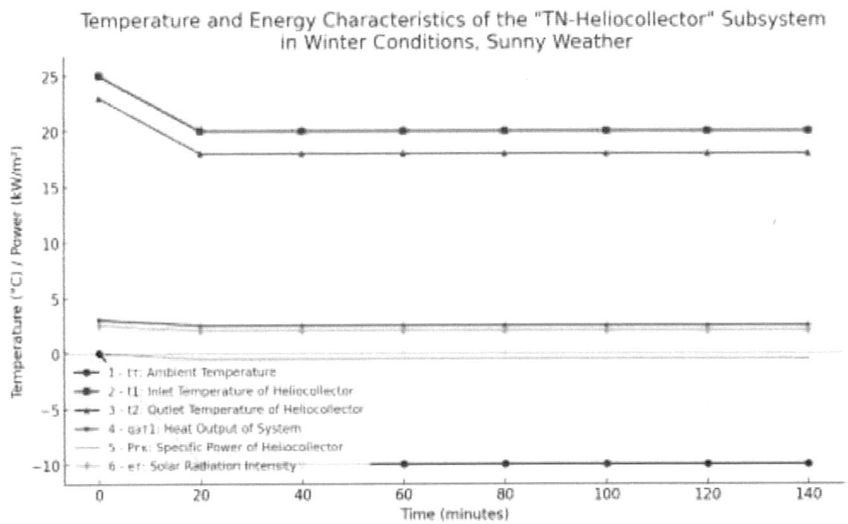

Figure 6. Temperature and energy characteristics of the subsystem "TH-heliocollector" in winter conditions, in sunny weather

The results of the calculations for the average daily heat output of the subsystem "HP-solar collector" for the conditions of the Almaty region are presented in Figure 6.

Six heating months (from October to March) were taken as the calculation period, with an average daily operation of 8 hours. The lower graph (dependency 1) shows the energy consumption for driving the heat pump compressor. The upper heat outputs are for different areas of the solar collector (S_{sc} = 1, 2, 4 and 10 m^2).

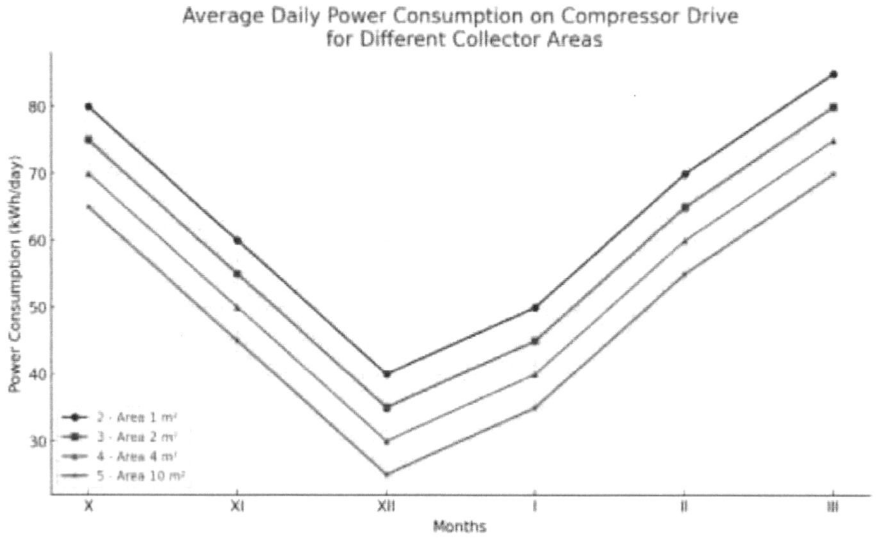

Figure 7. Average daily drive power consumption for different collector zones

Figure 7 shows the average daily power consumption of the drive for the different collector zones.

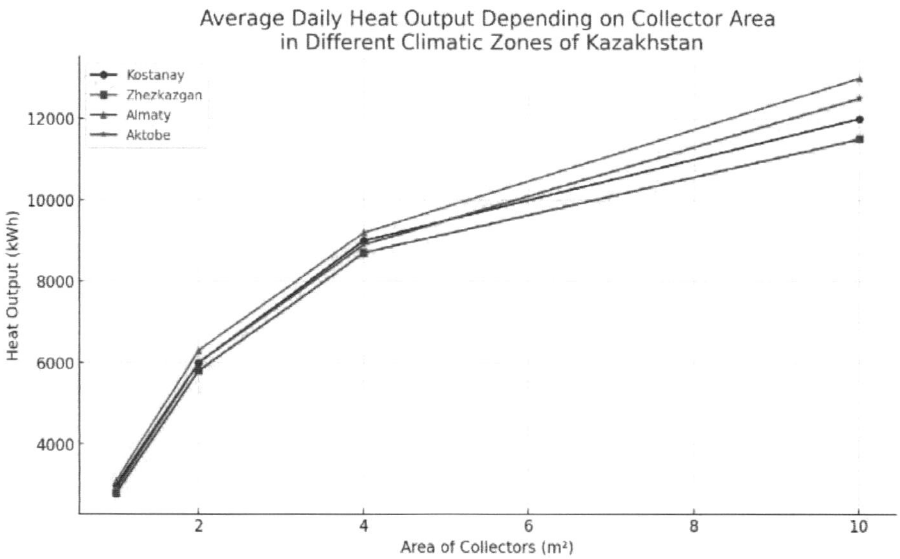

Figure 8. Patterns of average daily (a) and annual (b) heat output of the "HP-solar collector" subsystem as a function of the area of the solar collector for the climatic zones of Kazakhstan

Figure 8 shows the regularities of average daily (a) and annual (b) thermal capacity of the subsystem "HP-solar collector" depending on the area of the solar collector for climatic zones of Kazakhstan. For different zones of Kazakhstan, the area and thermal capacity of the system varies due to weather conditions.

4. ANALYSIS OF RESEARCH RESULTS

During the heating period, the system is efficient because the useful energy produced is greater than the energy consumed to drive the heat pump compressor. The patterns are parabolic, with a minimum heat output occurring during December and January. In October and March, the heat output is highest due to increased ambient air temperature and solar radiation intensity. The rise in the heat carrier temperature in the solar collector leads to an increase in the heat pump's coefficient of performance k_{tr}.

In a conventional system for heating and cooling, traditional energy carriers (electricity, gas, liquid fuel, or coal) are used, along with solar energy and electrical energy to drive the compressor of the refrigeration unit. When using the solar-assisted system, the structure of energy carriers changes (see Table 1).

The sources of exergy in the new system include heat from the surrounding atmospheric air, which is absorbed in addition to solar energy. The "HP-microclimate" subsystem not only regulates the thermal regime but also utilizes heat accumulated in the room from animals through ventilated air. The "HP-milk cooler" subsystem cools the milk while simultaneously recovering the heat from the cooled milk.

Table 1. Structure of energy carriers in the traditional and new systems

Processes	Sources of energy		Sources of exergy generated from the use of solar-assisted systems
	Known, basic system	SAS	
Hot water supply, heating	Traditional sources, solar energy	Traditional sources, solar energy, heat from the ambient air	Heat from the ambient air
Maintaining the microclimate	Traditional sources, solar energy	Traditional sources, solar energy, heat from animals	Heat from animals, heat from the ambient air
Cooling of perishable products.	Electric energy (for driving the compressor of the refrigeration unit)	Electric energy (for driving the compressor of the heat pump)	Heat from milk

These types of energy can be classified as exergy, as they were potentially contained in the known system but not utilized. The new system effectively realizes them. It has been established that their potential is significant.

The "HP-solar collector" subsystem provides heat output at sub-zero air temperatures. The structure of heat output consists of three components: absorbed direct solar radiation, heat from the ambient air, and heat from the operation of the compressor. The ratio of energy shares depends on the area of the solar collector and changes in the ambient air temperature. The majority of energy comes from solar energy and heat from the ambient air, with their share remaining above 50% throughout the heating season. With an area of 4 m^2, their share is higher than with other areas (1, 2, and 10 m^2). The share of heat from the surrounding ambient air during the heating period (October to March) significantly exceeds the share from solar energy.

The "HP-microclimate" subsystem operates within specific, defined temperature limits. It has been established that a heat pump with a compressor power of 1 kW can maintain the temperature in a room for animals between +5 to +25°C for 6 cows. It was calculated that, on average, one cow emits up to 0.7 kWh of heat, the heat pump heats water from 14 to 40°C, the coefficient of performance of the heat pump is 3 to 5, the cooling capacity is 2 to 4 kWh, and the heat output is 3 to 5 kWh. It has been established that the annual energy production, with an average daily animal occupancy of 15 hours in the room, will amount to 14,000 kWh/kW. The economic effect is $630/kW.

The "HP-milk cooler" subsystem, with a heat pump compressor power of 1 kW, under the condition of providing technological cooling parameters for milk from +35 to +5°C within a period of up to 3 hours, will have an average coefficient of performance of 3 to 4. The volume of cooled milk will be 200 liters, and the volume of heated water will be 250 liters, with a temperature range from 14 to 40°C. The annual savings in electric energy will amount to up to 10,800 kWh.

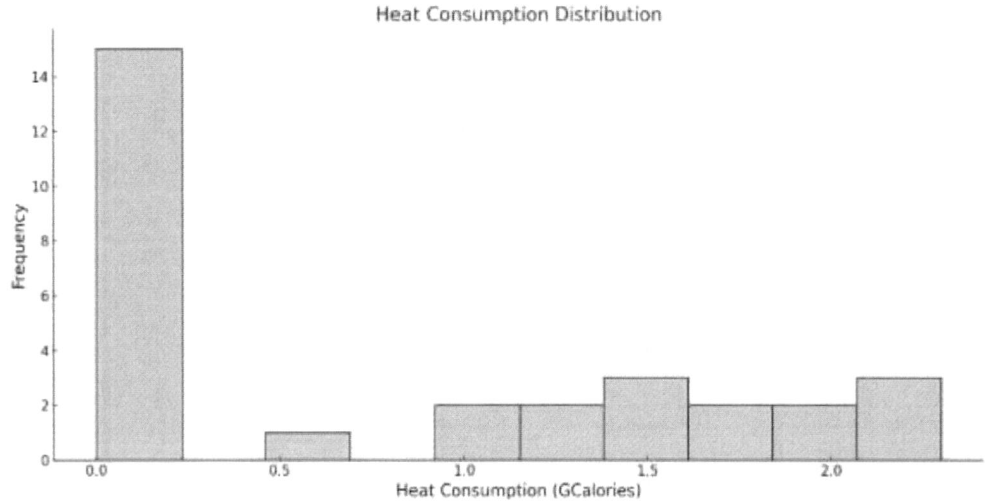

a)

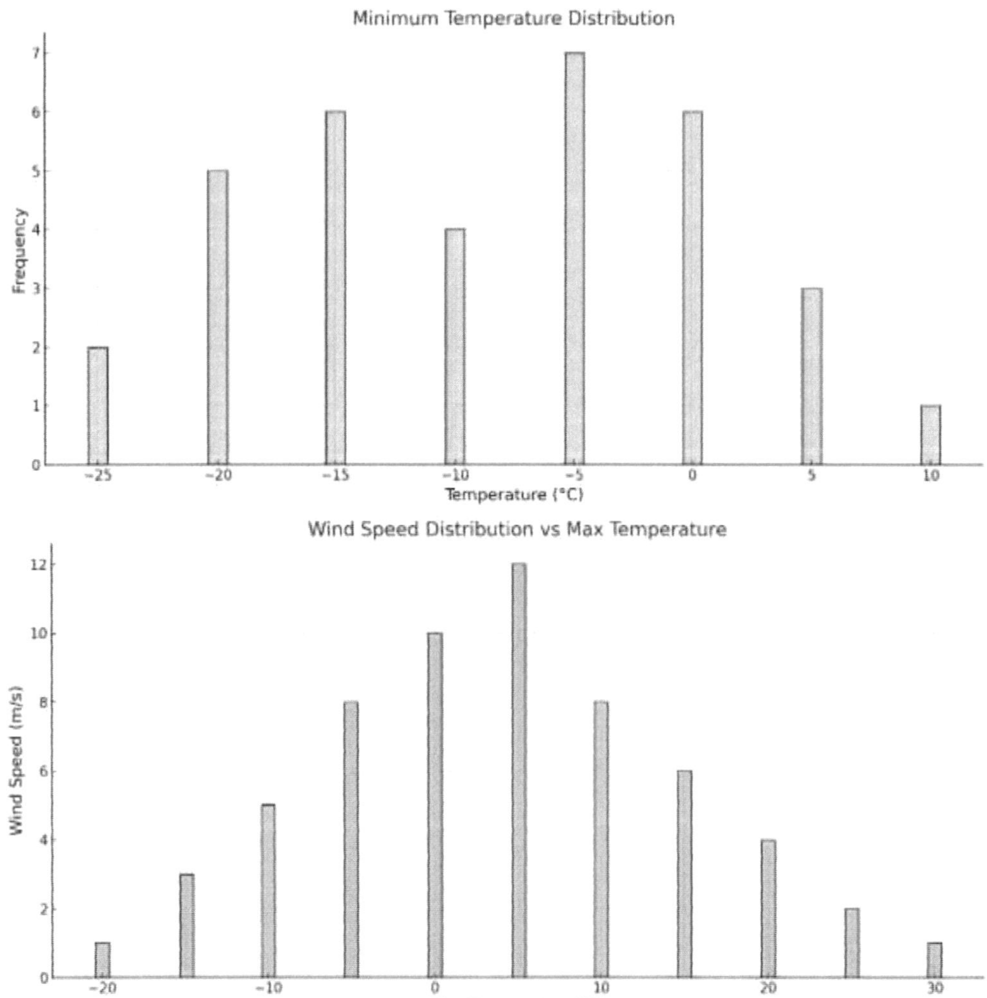

b)

Figure 10. a),b) Temperature distribution: maximum and minimum outdoor temperatures in Celsius and heat usage in GCal

To predict the key variable, we should use daily temperature (maximum) and wind speed (maximum) as inputs by applying one of the models such as linear regression, k-nearest neighbour regression, or random forest regression. We then compared the different models to determine the most accurate one. Figure 10 a, b shows the visual difference between them.

For better understanding, thermal energy was calculated based on the minimum temperature for the day and wind speed, as shown in Figure 11. This allowed us to analyse the effect of these factors on energy consumption. Thirdly, Figure 12 shows the data where wind speed and maximum and minimum temperatures were used as input values. For additional validation, the results were averaged over the largest dataset, which was processed and validated.

A comparison of the data from the three models for the training and test sets is presented in Table 1. The data set was divided into training set (80%) and test set (20%) to model the target variable. With or without the wind component, the maximum prediction coefficient value for linear regression was less than 83% accuracy. However, when using k-nearest neighbour or random forest regression models, the expected values improved to 89-90% (green squares), which shows the effectiveness of these machine learning methods in predicting energy consumption trends.

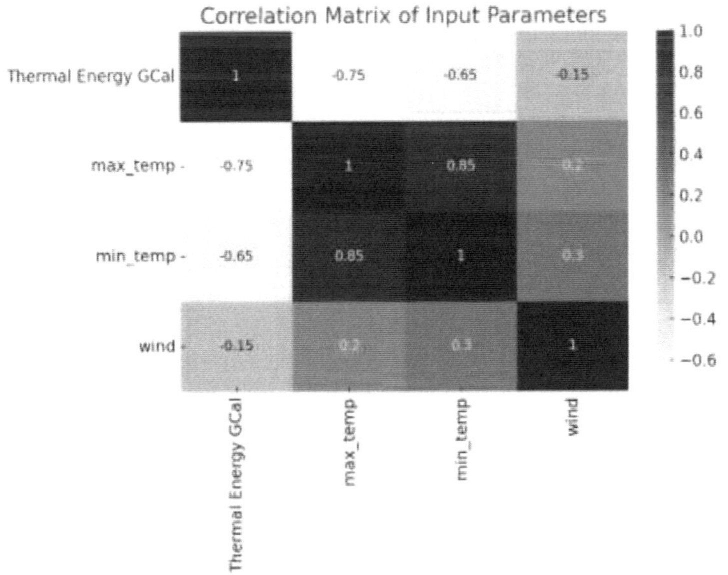

Figure 11. Correlations among the input parameters (heat consumption, temperatures, and wind speed)

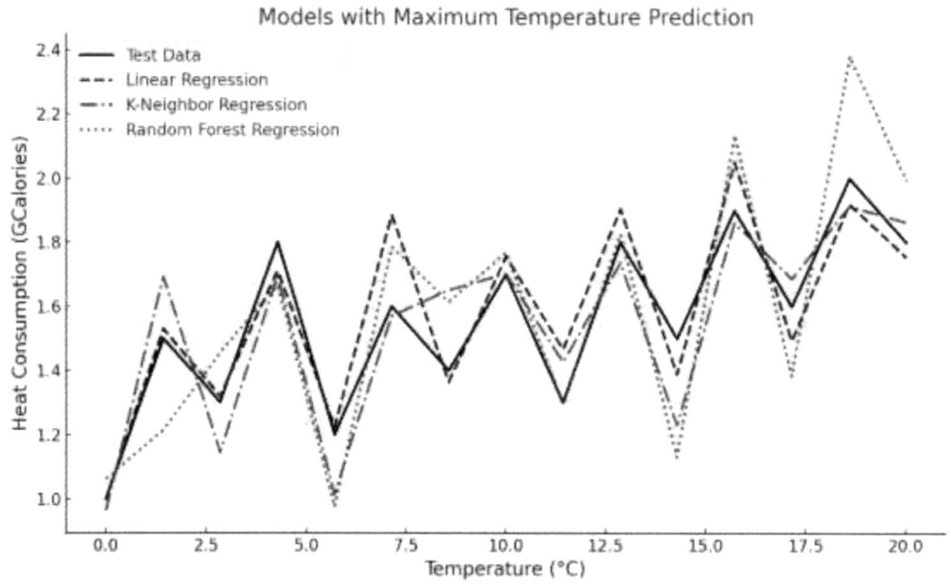

Figure 12. Models with maximum temperature prediction

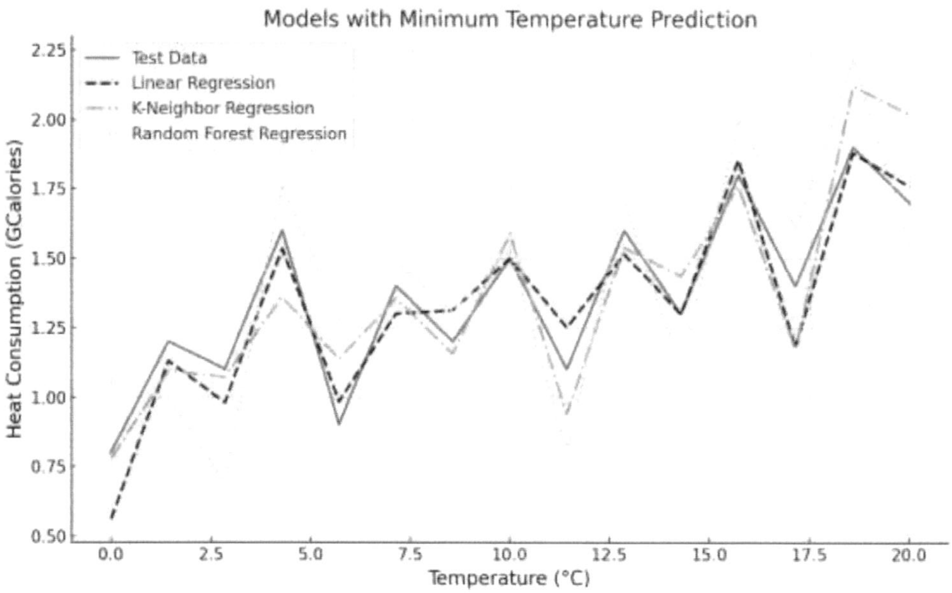

Figure 13. Models with minimum temperature prediction

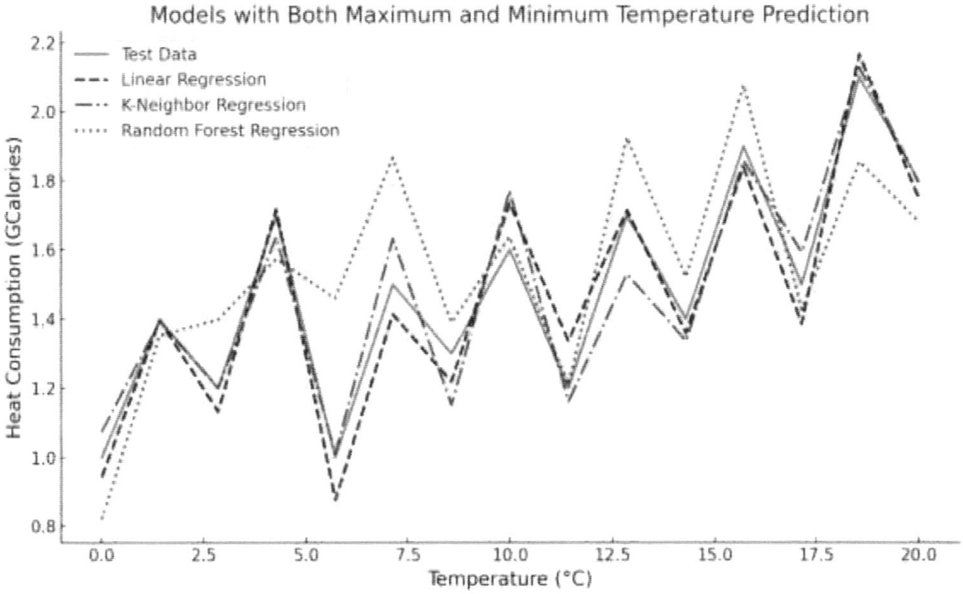

Figure 14. Models with both maximum and minimum temperature prediction

Figure 13, 14 shows the models with both maximum and minimum temperature predictions.

DISCUSSION

This research is the initial stage for developing a digital twin system for the farm. The temperature sensors mentioned in the article allow for the collection of data on seasonal and daily fluctuations in soil and milk temperatures, enabling forecasts for more efficient utilization of the energy consumption system described herein [6, 11-13].
The diagram (see Figure 15) will include the main components of the system and their interconnections.

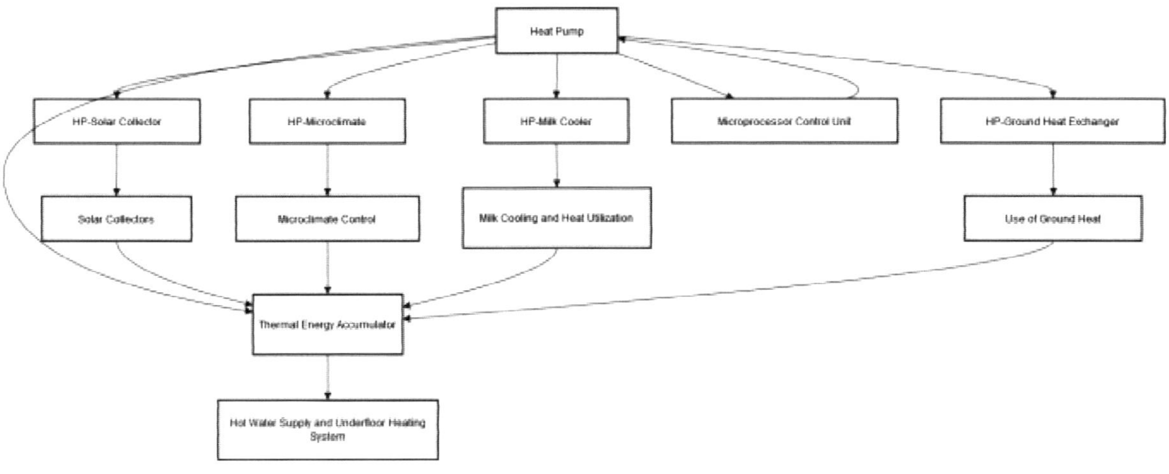

Figure 15..Diagram for the Digital Twin of the Heat Pump System Utilizing Renewable Energy Sources

Key elements included in the diagram:
1. Heat Pump (HP) - the central element that operates with various heat sources.
2. Subsystems: HP-Solar Collector: utilizes solar energy for heating. HP-Microclimate: regulates the microclimate in the premises using heat from animals. HP-Milk Cooler: cools milk and utilizes the released heat. HP-Ground Heat Exchanger: uses ground heat for heating.
3. Thermal Energy Accumulator (TEA) - stores heat for later use.

4. Microclimate Control - achieved through a microprocessor unit for automatic regulation.
5. Energy Flows: Solar Collectors: connected through switching valves to utilize solar heat. Ground Heat Exchanger: used in the absence of solar radiation.

The system operates depending on the time of day, including the use of milk heat, solar energy, and ground heat. The use of a microprocessor control unit allows for the automation and optimization of the system's operation.

CONCLUSIONS

Based on the literature analysis and patent research, a constructive and technological solution for a heat pump system using renewable energy sources has been selected. The system includes subsystems: "HP-Solar Collector", "HP-Microclimate", "HP-Milk Cooler" and "HP-Ground Heat Exchanger". It absorbs and sums up renewable energy sources, secondary technological heat, cools products, conditions air, and effectively utilizes their excess thermal energy. As a result, a significant energy-saving and environmental effect is achieved.

The system is designed for livestock farms. Therefore, it differs from known systems in its structure, functions, and parameters. The results obtained confirm the overall energy and exergy efficiency and operational capability of the new system in the sharply continental climate typical of the territory of the Republic of Kazakhstan. A microprocessor control system has been developed for the automatic operation of the device.

The development of a digital twin for farm heat networks, combined with mathematical modeling and big data analysis, has significant potential to improve the planning and efficiency of heat supply systems. With real-time monitoring, analysis and forecasting capabilities, the digital twin can significantly improve the distribution and optimization of heat networks.

These results have implications for smart farming concepts, such as predicting the optimal pump pressure for each distribution network using machine learning. Although this research area has not received much attention in Kazakhstan, its importance will grow given the global energy consumption problems and the desire of states to coordinate energy-intensive industries, especially in construction. The machine learning algorithms developed in this research can help solve many problems related to heat consumption in large metropolitan areas around the world. The use of digital twin and intelligent control based on heat pumps significantly improves the energy and economic efficiency of the heat supply system. The system meets modern sustainability requirements and provides the ability to adapt to changing operating conditions, reducing carbon footprint and energy costs.

ACKNOWLEDGMENT

This work was supported by the project of the Ministry of Science and Higher Education of the Republic of Kazakhstan "BR24992975 Development of a digital twin of a food processing enterprise using artificial intelligence and IIoT technologies".

REFERENCES

[1] Teng, F., Aunedi, M., Strbac, G. Benefits of flexibility from smart electrified transportation and heating in the future UK electricity system. Applied Energy 2016;167:420–431.
[2] Strbac, G., Aunedi, M., Pudjianto, D., Teng, F., Djapic, P., Druce, R., Carmel, A., Borkowski, K. Value of flexibility in a decarbonised grid and system externalities of low-carbon generation technologies. Imperial College London, NERA Economic Consulting, 2015.
[3] Zhang, X., Strbac, G., Teng, F., Djapic, P. Economic assessment of alternative heat decarbonisation strategies through coordinated operation with electricity system—UK case study. Applied Energy 2018;222:79–91.
[4] Chua, K., Chou, S., Yang, W. Advances in heat pump systems: A review. Applied Energy 2010;87:3611–3624.
[5] Heinen, S., Burke, D., O'Malley, M. Electricity, gas, heat integration via residential hybrid heating technologies—an investment model assessment. Applied Energy 2016;109:906–919.
[6] Carter, E., Lancaster, O., Chanda, F. Early results from the freedom project: fully-optimised hybrid heat pumps providing demand flexibility. In: Proceedings of the 12th IEA Heat Pump Conference, 2017, Rotterdam.
[7] Klein, K., Huchtemann, K., Müller, D. Numerical study on hybrid heat pump systems in existing buildings. Energy and Buildings 2014;69:193–201.
[8] Vuillecard, C., Hubert, C. E., Contreau, R., Stabat, P., Adnot, J., et al. Small scale impact of gas technologies on electric load management—μchp & hybrid heat pump. Energy 2011;36:2912–2923.
[9] Calder, C. Taking a hybrid approach to decarbonise domestic heating. Energy World 2018;30–31.
[10] Chow, T. T., Chan, A., Fong, K., Lin, Z., He, W., Ji, J. Annual performance of building-integrated photovoltaic/water-heating system for warm climate application. Applied Energy 2009;86:689–696.

[11] Arteconi, A., Polonara, F. Assessing the demand side management potential and the energy flexibility of heat pumps in buildings. Energies 2018;11:1846.

[12] D'hulst, R., Labeeuw, W., Beusen, B., Claessens, S., Deconinck, G., Vanthournout, K. Demand response flexibility and flexibility potential of residential smart appliances: Experiences from large pilot test in Belgium. Applied Energy 2015;155:79–90.

Ultra-low NO_x emissions and increased power in a combustor-SOFC System under CDC conditions

Berre Kumuk
Iskenderun Technical University, Hatay, Turkiye, berre.kumuk@iste.edu.tr,
ORCID: 0000-0001-7953-0167

Cite this paper as:	*Kumuk, B. Ultra-Low NOx emissions and increased power in a combustor-SOFC System under CDC conditions. 7th Int. Conf. on Nucl. and Ren. Energy Res. 27-30 October 2024, Antalya, Turkey*

Abstract: This study investigates the performance of a novel hybrid system combining a combustor and a Solid Oxide Fuel Cell (SOFC) model, operating under Colorless Distributed Combustion (CDC) conditions to achieve varied temperature distributions within the combustion chamber. This innovative configuration enables the assessment of SOFCs in an unconventional high-temperature setting. The study examines how changes in oxygen concentration affect temperature distributions and SOFC power output. Under standard conditions, a significant flame is observed at 21% O_2 concentration. As oxygen levels decrease, thermal regions become more uniform, achieving a colorless distributed state at 15% O_2. This transition results in ultra-low NOx emissions and a stable thermal environment, conducive to SOFC initiation and maintenance. The widest power output range is recorded at 21% O_2, with approximately a tenfold increase in maximum power compared to conventional fuel cells. However, the maximum power range contracts as oxygen levels decrease, indicating the potential for additional SOFCs within the chamber. The findings highlight that optimizing temperature distribution could significantly enhance SOFC efficiency and scalability within the same operational regions.

Keywords: *SOFC; Colorless Distributed Combustion; Nitrogen Dilution; Methane Combustion.*

NURER 2024
Nuclear and Renewable Energy Resources

Introduction to Accelerator-Driven Systems: the power and promise of ADS

Jean-Pierre Revol
iThEC

Cite this paper as:	*Revol J.P. Introduction to Accelerator-Driven Systems: the power and promise of ADS, relevance of nuclear power & the thorium fuel. 7th Int. Conf. on Nucl. and Ren. Energy Res. 27-30 October 2024, Antaya, Turkey*

Abstract: My presentation will begin with a historical overview of accelerator-driven systems (ADS), tracing their development and underlying motivations. I will compare the fundamental physical characteristics of ADS to those of critical nuclear reactors, demonstrating how ADS represent a practical solution to addressing the issues of safety, nuclear waste management, proliferation, and sustainability. In this context, I will highlight how these systems open the way to a thorium-based fuel cycle, bringing us closer to a more sustainable and safer future for nuclear fission. The presentation will also include a brief overview of current ADS projects in the world.

Keywords: *Energy, sustainability*

Transmutex START Project: A Thorium-Based Fuel Cycle For Sustainability And Proliferation Safety

Jean-Christophe de Mestral

Cite this paper as: *Revol J.P. Introduction to Accelerator-Driven Systems: the power and promise of ADS, relevance of nuclear power & the thorium fuel. 7th Int. Conf. on Nucl. and Ren. Energy Res. 27-30 October 2024, Antaya, Turkey*

Abstract: The START project, developed by the Swiss company Transmutex, is a commercially focused Accelerator-Driven System (ADS) initiative in nuclear energy. It combines proven and emerging technologies to enable energy production and long-lived nuclear waste transmutation, while closing the thorium-based fuel cycle. The facility will incorporate an on-site pyroprocessing unit to reprocess spent fuel and manufacture new fuel, enhancing proliferation safety. Key features of START include subcriticality and passive safety, with an 800 MeV, 5 mA proton cyclotron, an advanced iteration of the Paul Scherrer Institute cyclotron and a lead-cooled subcritical reactor based on the Italian ALFRED design. An innovative spallation target serves as the neutron source..

***Keywords**:* *Energy, sustainability*

NURER 2024
Nuclear and Renewable Energy Resources

Advanced and Small Modular Reactors' Supply Chain: Current Status and Potential for Global Cooperation

Farooq Ahmad
Nuclear Engineering and Radiation Science Dept., Missouri University of Science and Technology, Rolla, MO, 65409, fahmad@mst.edu

Shoaib Usman
Nuclear Engineering and Radiation Science Dept., Missouri University of Science and Technology, Rolla, MO, 65409, usmans@mst.edu

Cite this paper as: *Ahmad, F, Usman S. Advanced and Small Modular Reactors' Supply Chain: Current Status and Potential for Global Cooperation. 7th Int. Conf. on Nucl. and Ren. Energy Res. 27-30 October 2024, Ankara, Turkiye.*

Abstract: A well-developed and fully coordinated supply chain not only brings down the direct EPC cost but also provides cheaper design options. Most importantly, a local well-established supply chain reduces uncertainty and risks associated with construction time, project cost overrun and hence limits the contingencies which are about 15 % of total investment costs. It is therefore important to fully understand the status of one's supply chain before embarking on a major engineering project. In recent past, the manufacturing and procurement issues along with immaturity of design and licensing, have caused delays and cost overruns in Gen-III FOAK construction projects in many countries This study includes the attributes of Small Modular Reactors (SMRs), their current status and potential capability in the Midwest of the United States for its active participation in nuclear supply chain. Moreover, Applicable codes and standards pertinent to nuclear industry are discussed along with the potential challenges for globalization of nuclear supply chain. Among those, adoptability and acceptance of nuclear codes/standards are major issues. The harmonization of various codes and standards would result in lowering the costs during construction and operation.

1. INTRODUCTION

In the wake of global disruptions[1] in energy security and International Energy Agency's (IEA) Net Zero Emission (NZE) initiative by 2050[2], many developed countries have started to rethink and reshape their energy strategies. Traditional nuclear power plants (NPPs) have been a reliable source of green base-load energy for more than half a century. Further advancements in nuclear technology have made Small Modular Reactors (SMRs) and Advance Reactors economically attractive, inherently safe, compact and above all efficient. SMRs being frontier in new era, can provide several benefits like lower capital cost, enhanced safety with passive systems, low system cost, and off-grid deployment options with beyond electric power (process heat etc.) applications. Cost reduction via modularity, short construction time and potential for extended core cycles are some of the other attractive features of SMRs. In the United States, SMRs have gained government support and public acceptance leading to licensing of NuScale's SMR design by Nuclear Regulatory Commission (NRC) [3,4]. This progress attracted investors and as a part of the Carbon Free Power Project (CFPP) Utah Associated Municipal Power System (UAMPS) started combined licensing process, which was terminated on November 10, 2023[5]. While the effort did not reach the desiredlogical conclusion, there are many milestones achieved and lesson-learned in the process. Recently, many other countries like Canada, UK and France have provided substantial grants for SMR development and demonstration and deployment[1,6].

Typically, for a nuclear project, the Engineering, Procurement and Construction (EPC)[7] costs cover about 60% of the total investment or 80% of the Overnight Construction Costs (OCC)[8]. The capital cost of a nuclear project includes OCC, cost escalation and financing charges. The breakdown of the capital cost is shown in Figure 1. The EPC can be classified into direct costs encompassing civil works, heavy equipmentand components, piping, trained labor whereas indirect costs include design, engineering, procurement, project management, quality assurance, testing, commissioning etc. In terms of equipment, reactor, turbine and electrical equipment have the substantial portion of the cost. On the other hand, major components ofmaterial expenses are piping, steel structures and concrete/civil works adding to the direct EPC costs. Another significant factor is highly skilled manpower required for both direct and indirect cost domains. Abreakdown of EPC direct cost is presented in Figure 2. As EPC has a direct impact on the total plant cost and so its linkage and dependability upon the supply system and project management needs to be assessed.

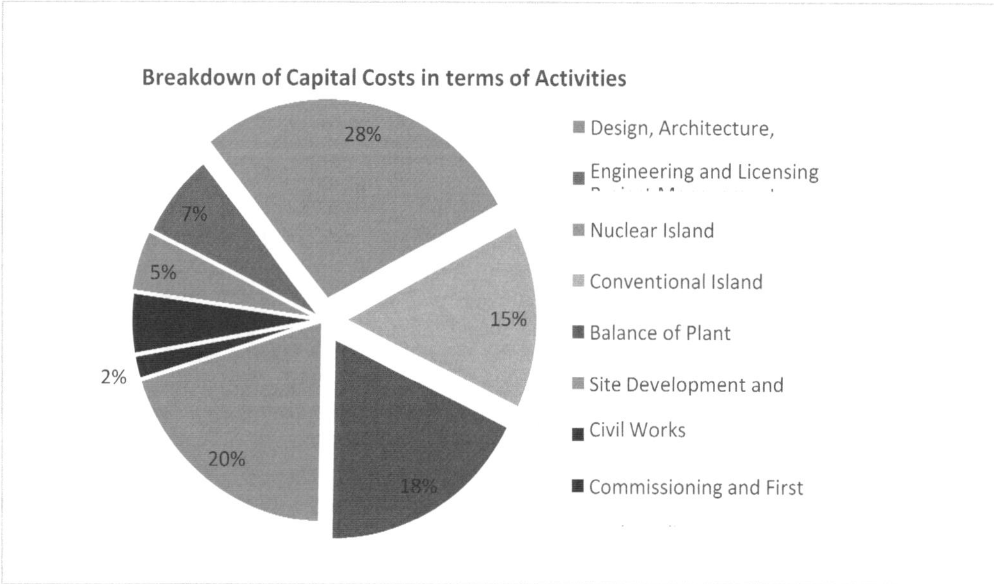

Figure 1: Breakdown of Capital Costs in terms of Activities[8]

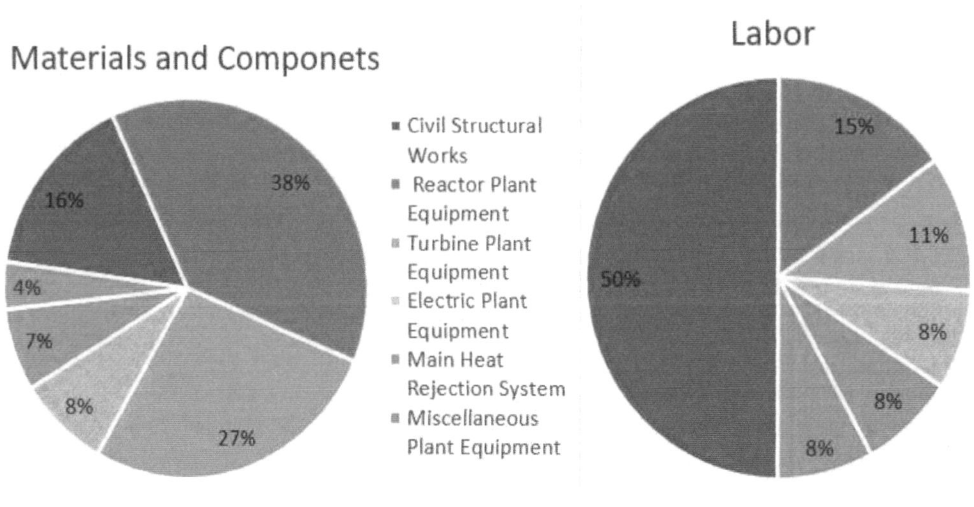

Figure 2: Direct EPC Cost Breakdown in Terms of Material/ Components and Labor[7]

Maturity of nuclear supply chain in terms of **First of a kind** (FOAK)[7] projects and their subsequent series construction is of prime concern because it is closely related to total construction time and cost. In recent past, the manufacturing and procurement issues along with immaturity of design and licensing, have caused delays and cost overruns in Gen-III FOAK construction projects in Europe and North America[7]. The main reason is the dormant period of about three decades which has eroded the competence and capabilities in supply chain across the regions. For innovative reactor designs, the situation is further complicated due to other important factors like significant shortage of trained manpower, new designs and safety regulations. One major concern for new design is the array of **change orders** during construction which adversely impact all aspects of the project[9]. Similar challenges were faced during construction of AP1000 in US (Vogtle 3 & 4) which resulted in cost escalation due to limited supply chain consideration. Weakness and lack of readiness of the supply chain surfaced during this very first nuclear project in the United States afterdecades of inactivity. For example, out of a total of 146 structural modules, 46 were delayed by more than 6 to 18 months[9]. While some of these delays can be attributed to COVID-19, many were avoidable with better planning and understanding the limitations of the supply chain. In the same manner, these issues caused the cost estimates for NuScale projects to escalate from USD 1200 per kWe to USD 5100 per kWe[7].Therefore, the revival and up-gradation of sustainable nuclear supply chain is inevitable regarding new designs and construction for modular reactors.

Multi-aspects SMR technology has several additional challenges related to design, construction, equipmentand components, cost economics, licensing, prototype demonstration etc., yet the prime concern is its cost effectiveness which is directly related to supply chain. Studies[2] have shown that the nuclear build rate wouldbe around USD 5000 per kWe for FOAK nuclear project in advance economies like US, UK, France etc. Through modularization[10], standardization of plant design and construction processes and deployment of multiple units through series production may lower the cost to around USD 2000-3000 per kWe. Dependingon financing costs and interest rates, this would lead to a Levelized Cost of Electricity (LCOE) in the region at rates of USD 40-80 per MWh[2]. Many countries (China, Korea and India) have delivered cost effective nuclear projects with lower LCOE in stipulated time due to presence of their fully developed supply chainand experienced management. Therefore, the challenges for FOAK projects are not solely nuclear technology itself but it strongly depends on conditions and interaction between the stakeholders during design and construction.

A FOAK nuclear power project has several optimization challenges with respect to cost, public support, proliferation concerns, sustainability of local resources and environmental concerns. Over the time, these factors have been given different weightages due to their strong dependance on country's policies, political leadership, society, locality and even individuals' perception. The life cycle of a nuclear facility involves design, construction, commissioning and installation, operation and decommissioning: all steps are to conform certain codes' and regulatory compliances.

This study highlights few aspects of supply chain for SMRs and advanced reactors, their status and their potential impacts in terms of sustainable energy and economy. It includes list of codes and standards in the US and around the world mainly applicable to SMRs and nuclear industry, there would be more workin this area. This work provides useful data for the policy makers to better evaluate the multilateral impactof SMR supply chain on economic development of the region.

2. Status of SMRs

SMRs generally fall in the category of nuclear reactors with electric capacity less than 300 MW per unit[11]. There are some other micro-modular reactors (MMRs) with capacity less than 10 MWe, are also included in this category. Currently, more than 70 SMR designs[11] are in the developing phase in 19 countries aroundthe world employing different technologies. These SMR designs include water, gas, molten metal and saltcooled reactors and use different types of fuel and heat transfer/coolant technologies. About 40 of them are at conceptual design stage, 25 are at Basic and Detailed Design Stage, 2 are licensed, 7 of them are under construction and 4 are in operation. In US alone, there are about 20 designs at various stages of developmentand many of them are also looking at High Assay Low Enriched Uranium (HALEU) with enrichment from6 to 20%. Several of these designs are considering TRISO (TRi-Structural ISOtropic) fuel for its enhanced safety. A few FOAK SMR projects are in the demonstration stage but none of them have reached commercialization stage. Table 1 summarizes SMRs around the globe which are licensed, under construction and operational.

Most of the innovative reactor designs (> 30) are Water-Cooled Reactors (WCRs) which can be deployed at land or marine environment and their outlet temperature lies in the range of 300~330 °C. These reactorscan be used for electricity production and/or heating purposes. The remaining type of reactors can providehigher outlet temperatures which can go up to 600 °C in metal cooled reactors, 800 °C in molten salt reactorsand about 1000 °C in high temperature gas cooled reactors[11]. These higher temperatures make them efficientand favorable to be used for electricity generation and other industrial applications (Pulp and paper manufacturing, petroleum refining, methanol production, hydrogen production and blast furnaces for steelmanufacturing).

SMRs, by design, are to be deployed in series through a sustainable global supply chain as in the case of naval and aircraft manufacturing. They can be deployed as a single unit or multiple units at the same site making it equally feasible for large and small grids. Furthermore, these reactor designs are more favorable due to their small core, inherent safety feature including passive system and reduced Emergency Planning Zones (EPZ)[12,13], they can be installed near industrial or residential areas resulting in lower line losses, less infrastructure, heating for industrial applications. In addition, already existing sites for fossil-fuel-based power plants can be utilized for their deployment. SMRs can also complement other Variable Renewable Energy (VRE) sources (solar, wind, tidal etc.) due to their load following attribute and provide flexible cogeneration options for hydrogen production and thermal storage.

To meet NZE goals and sustainable and secure energy system, SMRs have provided opportunity for the private investors and governments in terms of short construction time and low-risk smaller upfront cost. Although the LCOE may not be lower as compared to other energy sources but SMRs can provide high flexibility, affordability, scalability, reliability and dispatchability. According to one estimate, design development and engineering cost of each SMR technology would be about USD 1.0 B[14]. While adopting a 10% learning rate based on the experience of Navy shipbuilding i.e. the costs can be reduced by 10 % for every doubling in number of modules/plants till a stable cost of **n**th **of** a **k**ind (NOAK) is achieved. The enhanced capacity factors of nuclear plants around the world, especially in US from 66% to 90%[2] will further reduce the LCOE significantly. In addition, lower interest rate for this developing technology may help in lowering LCOE. As per the study, the LCOE from nuclear side is lowest in almost all countries witha 3% discounted rate as compared to other VRE sources[2].

Regardless of the associated uncertainties[15], efforts are being made worldwide for the development and demonstration of SMRs mostly in the developed economies. China being the leader in advance nuclear reactors has connected its first SMR, HTR-PM (2 module system), in 2021[16] while Russia has already connected world's first floating SMR in May, 2020[11]. Various SMRs with different technologies are under near-term demonstration stage in several other countries. For example, ACR-100, Micro-Modular Reactor Program and SSR-W in Canada, CAREM in Argentina, NUWARD in France, RITM-200 in Russia and UK SMR in United Kingdom are near demonstrations. In the US, the DOE has awarded USD 30 M to fiveSMR design teams to meet the technical challenges till deployment[17]. These designs include BWXT Advanced Nuclear Reactor (BANR)[18], eVinci, Kairos Power FHR, Holtec SMR-160 and TerraPower's MCFR all aiming for demonstration by 2030. Moreover, a USD 20 M grant will be awarded to less maturebut novel advanced reactor designs[17].

Table 1: List of SMRs in the World[11, 17,19]

Name	Power (MWe)	Type	Producer	Country	Status
Water Cooled Reactors (WCRs)					
NuScale	50-300	PWR	NuScale Power	United States	Licensed
SMART	100	PWR	KAERI	South Korea	Licensed
ACP100	100	PWR	CNNC	China	Under construction
CAREM	30	PWR	CNEA	Argentina	Under construction
KLT-40S	70	PWR	OKBM Afrikantov	Russia	Operational (floating plant)
RITM-200M	100	PWR	OKBM Afrikantov	Russia	Under construction
RITM-200N	55	PWR	OKBM Afrikantov	Russia	Under construction
RITM-200S	106	PWR	OKBM Afrikantov	Russia	Under construction
High Temperature Gas Cooled Reactor (HTGR)					
HTR-PM	210	HTGR	China Huaneng	China	Operational[20]
HTR-10	2.5	HTGR	INET, Tsinghua University	China	Operational
HTR-30	30 MWt	HTGR	JAEA	Japan	Operational
Liquid Metal Cooled Fast Reactor (LMFR)					
BREST-OD-300	300	LFR	NIKIET	Russia	Under construction
Molten Salt Cooled Reactors (MSR)					
TMSR-LF1	10	MSR	China National Nuclear Corporation	China	Under construction

3. Codes/ Standards Requirement in US

Due to the nature of nuclear reactors, component suppliers are required to provide and/or use materials thatare highly resistant to deterioration caused by radiation damage and other harsh environment. Globally, IAEA provides the framework in developing the evolutionary and innovative reactor designs through a series of safety requirements and guides[21]. In the United States, the USNRC, DOE and various Standard Development Organizations (SDOs) like ASME[22], IEEE[23], AISC[24], ACI[25], ANS[26] etc. have been active in developing nuclear standards for existing and advance reactor designs to provide safe, reliable and well- coordinated nuclear industry. The vendors must meet the standard codes and regulations for construction permits, operational license, design certifications, combined licenses and early site permits which are stipulated by USNRC. The "Safety related" items must meet "Nuclear-Grade standard" while the components categorized as "Supplemental Quality Products" need to meet specific quality requirements set by the industry and Standard Quality Products.[27,28]. The nuclear materials must be from the qualified suppliers which possess the nuclear grade certifications for enhanced safety and security. A pathway for the development of nuclear project is shown in Figure 3. There are many tiers in the hierarchy starting from the governmental act creating and passing on regulatory authority to USNRC in the United States. Regulators rely on other professional bodies for codes and regulations (say ASME). Reactor vendors, their contractors and subcontractor are all required to follow these standards and have to be inspected for compliance as shown in Figure 3.

The nuclear codes/standards are critical, and their adoptability is inevitable during the development and complete life cycle of new reactor designs. These standards provide guidelines for the safe operation of facility and are updated with technology advancement and current industrial practices according toregulatory requirements. For example, ASME codes for the Boiler and Pressure Vessel for Nuclear containment specifies codes for materials, operations and maintenance, in service inspection and construction of NPP components have been updated five times between 2015 to 2023 (Table 2). Similarly,there are a number of codes/ standards by different SDOs which are applicable to nuclear projects. Table 2contains a list of codes/ standards which are mainly applicable to SMRs in the US[29],[30]. Nuclear componentsuppliers and technology providers are required to adhere to these standards and codes set by the industry as well as USNRC to fulfill the requirements of domestic licensing.

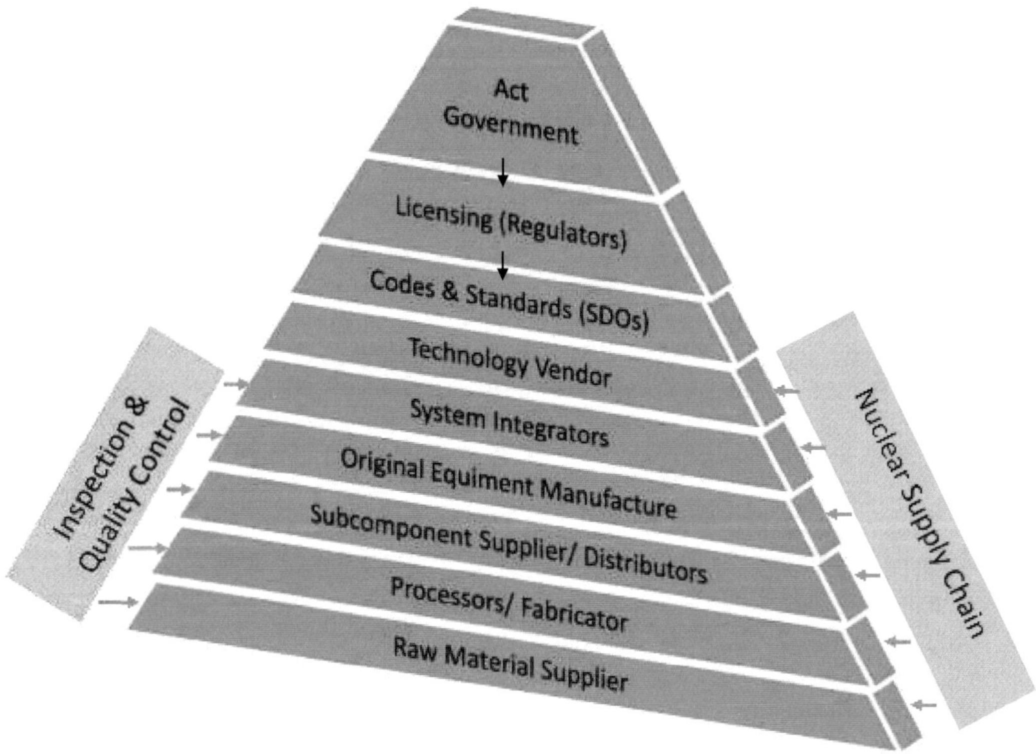

Figure 3: Typical Hierarchy of Nuclear Supply Chain Management[31]

Table 2: List of Codes/ Standards for SMRs and Advance Reactors[22,23,24,25,26,29,30,32,33]

SDO	Codes/ Standards	International Standards
American Institute of Steel	Specification for the Design, Fabricationand Erection of Steel Safety-Related Structures for Nuclear Facilities	• IAEA Safety Standards Series No. NS-G-1.10[34] • EN 1993-1-1 (Eurocode 3)[35]

Construction (AISC)	[ANSI/AISC N690-1984, 1994, 1994(R2024)]	• RCC-MRx (France)[36] • KTA 2201 Standards (Germany)[37] • JSME S NC1 (Japan)[38] • CSA N287 (Canada)[39] • BS 5975 (UK)[40]
American Society of Mechanical Engineers (ASME)	• Boiler Pressure Vessel Code Sections, 2023, 1914, 1956 ○ BPV III - Rules for Construction of Nuclear Facility Components (2015, 2017, 2019, 2021, 2023)	• EN 13445 • D 5500:2018 (Specification for UnfiredFusion Welded Pressure Vessels) • CSA B51 (Canadian Standards Association) • AD 2000-Merkblatt, German Code • JIS B 8265 and JIS B 8243 (Japanese Industrial Standards)
	○ BPV XI - Rules for Inservice Inspection of Nuclear Power Plant Components	• CSA N285.4, "Periodic Inspection of CANDU Nuclear Power Plant Components • IAEA Safety Standards Series No. NS-G- 2.6, "Maintenance, Surveillance and In- service Inspection in Nuclear Power Plants • French RCC-M Code • Japanese Standards (JSME S NA1 Code)
American Concrete Institute (ACI)	Code Requirements for Nuclear Safety-Related Concrete Structures (ACI 349-13)	• Eurocode 2 (EN 1992) and Eurocode 8 (EN 1998) • RCC-CW (Design and Construction Rulesfor Civil Works of PWR Nuclear Islands) • Canadian Standards Association (CSA)N287 Series
American National Standards Institute/American Nuclear Society (ANSI/ANS)	• ANSI/ANS N18.2: Nuclear Safety Criteria for the Design of Stationary Pressurized Water Reactor Plants	• IAEA's "Specific Safety Requirements for the Design of Nuclear Power Plants (SSR- 2/1)" and "NS-G-1.12" • EUR (European Utility Requirements) • RCC-E (Design and Construction Rules for Electrical Components of Nuclear Islands • Canadian Standards Association (CSA) N286 Series
	• ANSI/ANS N18.7: Administrative Controls and Quality Assurance for the Operation Phase of Nuclear PowerPlants	• IAEA Safety Standards GS-R-3 (The Management System for Facilities and Activities) and GSR Part 2 • Canadian Standards Association (CSA) N286 Series • European Utility Requirements (EUR)
	• ANSI/ANS 56.8: Containment System Leakage Testing Requirements	• IAEA Safety Standards Series (NS-G-1.10 and SSG-53) • European Standard EN 1591-1

	• ANSI/ANS 56.2: Containment Isolation Provisions for Fluid Systems • ANSI N170: Standards for Determining Design Basis Flooding at Power Reactor Sites • ANSI/ANS 2.2: Earthquake Instrumentation Criteria for Nuclear Power Plants • ANSI/ANS 2.3: Standard for Estimating Tornado and Extreme Wind Characteristics at Nuclear Power Sites • ANSI/ANS 2.7 Guidelines for Assessing Capability for Surface Faulting at Nuclear Power Sites • ANSI/ANS 2.8 Determining Design Basis Flooding at Power Reactor Sites • ANSI/ANS 2.10 Guidelines for Retrieval, Review, Processing and Evaluation of Records Obtained from Seismic Instrumentation • ANSI/ANS 2.11 Guidelines for Evaluating Site-Related Geotechnical Parameters at Nuclear Power Sites • ANSI/ANS 2.12 Guidelines for Combining Natural and External Man- Made Hazards at Power Reactor Sites • ANSI/ANS 2.26 Categorization of Nuclear Facility Structures, Systems, and Components for Seismic Design • ANSI/ANS 6.4 Nuclear Analysis and Design of Concrete Radiation Shielding for Nuclear Power Plants • ANSI/ANS 6.4.2 Specification forRadiation Shielding Materials • ANSI/ANS 67.02.01-2014 Nuclear Safety-Related Instrument Sensing Line	• French Standard RCC-M (Design and Construction Rules for Mechanical Components of PWR Nuclear Islands) • Canadian Standards Association (CSA) N287.7 IAEA Safety Standards (SSR-2/1 and SSG-53) European Utility Requirements (EUR) French RCC-E (Design and ConstructionRules for Electrical Components) Canadian Standards Association (CSA) N285 Series Japanese Standards (JEAC 4207)

	Piping and Tubing Standard for Use in Nuclear Power Plant • ANSI/ANS 3.1-2014 Selection, Qualification, and Training of Personnel for Nuclear Power Plants • ANS 20.2 Nuclear Safety Design Criteria and Functional Performance Requirements for Liquid-Fuel Molten Salt Reactor Nuclear Power Plants	
American Society of Civil Engineers (ASCE)	• ASCE 4-98 (2000) Seismic Analysis of Safety-Related Nuclear Structures • ASCE 43-05 (2005) Seismic Design Criteria for Structures, Systems, and Components in Nuclear Facilities**Error! Bookmark not defined.**	• IAEA Safety Standards Series (NS-G-1.6and NS-G-1.10) • Eurocode 8 (EN 1998-1 and EN 1998-2) • RCC-G (Design Rules for Civil Engineeringof Nuclear Islands) • Japanese Standards (JEAC 4601 and JEAC 4201) • Canadian Standards (CSA N289 Series) • IAEA Safety Standards (SSG-67 and NS-G-1.6) • Eurocode 8 (EN 1998-1 and EN 1998-2) • Japanese Standards (JEAC 4601 and JEAC 4604) • RCC-G (Design Rules for Civil Engineeringof Nuclear Islands) • Canadian Standards (CSA N289 Series)
National Fire Protection Association (NFPA)	• NFPA 253 Standard Method of Test for Critical Radiant Flux of Floor Covering Systems Using a Radiant Heat Source (also ASTM E119). A new edition of NFPA 253 was released in 2011. • NFPA 259 Standard Test Method for potential Heat of Building Material. • NFPA 251 Standard Methods of Tests ofFire Resistance of Building Construction and Materials. • NFPA 255 Standard Method of Test of Surface Burning Characteristics of Building Materials	•
American Society of Civil Engineers (ASCE)	• Seismic standard • Seismic Design Criteria for Structures, Systems, and Components in Nuclear Facilities	•
American Society for Testing	• ASTM Standards for Steel, Stainless Steel and Related Alloys (A36, A36M, A182, A182M,A240, A240M, A370)	•

and Materials (ASTM)	• ASTM Standards/ Test Methods for Cement (C109, C109M, C114, C150M,C150M, C191) • ASTM Standards/ Test for concrete (C31, C31M, C33, C33M, C39, C39M, C231) • ASTM Standards/ Test for soil and rock (D422, D698, D2487, D3911-16, D4318, D7491-08) • ASTM Standards/ Practice/ Guidelines for Nuclear Fuel Cycle (C753, C967, C1233, C1682) • ASTM Standards/ Practice/ Guidelines for Protective Coating and Living Work for Power Generating Facilities (D3911,D4082, D5144) • Standard Guide for Establishing Procedures to Monitor the Performance of Safety-Related Coating Service Level III Lining Systems in an Operating Nuclear Power Plant**Error! Bookmarknot defined.** Total Over 1700 standards/ Tests	
Institute of Electrical and Electronics Engineers (IEEE)**Error! Bookmark not defined.**	• Human Action Reliability Analysis for Nuclear Power Generating Stations. • Nuclear power plants -- Instrumentationand control important to safety -- Electrical equipment condition monitoring methods. Part 6: Insulation resistance**Error! Bookmark not defined.**	•
Nuclear Energy Institute (NEI)[41]	• NEI 00-01, Rev. 2(b) Guidance for Post-Fire Safe-Shutdown Circuit Analysis • NEI 04-02, Rev. 2 Guidance for Implementing a Risk-Informed, Performance-Based Fire Protection Program Under 10 CFR 50.48(c) • NEI 08-01, Rev. 5 Industry Guideline for the ITAAC Closure Process under 10 CFR Part 52 • NEI 97-04, Appendix B Guidelines and Examples for Identifying 10 CFR 50.2: Design Bases • NEI 99-03 Control Room Habitability Assessment Guidance**Error! Bookmark not defined.**	•

4. Current Status of Nuclear Supply Chain in US

A well-coordinated nuclear supply chain encompasses all activities related to full life cycle of a nuclear project. It involves sourcing, procurement of fuel, materials and equipment, services and logistic support by collaborating with all stakeholders. It ensures the quality of each component for safe and smooth operation during entire project cycle. A supply chain is managed by the technology vendor for new projectswhereas power generating NPPs deal with equipment manufacturer, supplier or fabricator for operational and maintenance activities (Figure 3). The sustainability of supply chain strongly depends on size of project, contract types, reliability and qualification of suppliers and types of standardized components. Subpar components may impact the performance and safe operation of plant and lead to costly shutdowns or failures[31]. Therefore, no subpar materials, workmanship or equipment is permitted to enter the nuclear supply chain.

A significant decline in construction of NPPs over the past three decades has resulted in somewhat limited information about the availability and costs of nuclear materials. Similarly, the advent of innovative reactordesigns has added more uncertainty regarding efficiency and sustainability of nuclear supply chain[8]. According to the World Nuclear Supply Chain: Outlook 2040[42]; the number of suppliers or companies having nuclear certifications from different SDOs are declining in Europe and US. For example, there were 2489 U-Certified and 1124 S-Certified companies in 2012 in the US which have reduced to 2359 and 1005 respectively in year 2024. A comparison has been provided in Figure 4 for other ASME certified companies for the two reference years. It is clear from the figure that many ASME certified companies have gone out of the nuclear business because of low market demand. This situation may get even worse if the local industry remains reluctant to accept the high level of nuclear standards/qualifications especially in the absence of future projects[7]. Recent activities at the technology vendors level and above (Figure 3) have failed to mobilize the downstream of the supply chain, at lease at the time of this writing. Furthermore, new suppliers need lag-time and order commitment to have a new nuclear quality assurance program established and required certifications acquired for manufacturing of safety related components.

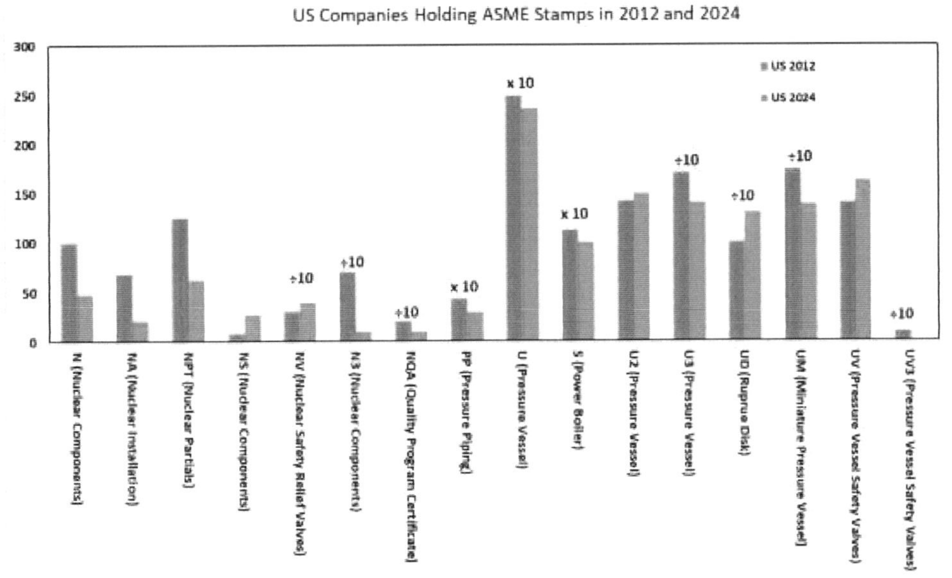

Figure 4: Comparison of US companies Holding ASME stamps in 2012 and 2024.

The most important hindrance in the deployment of the SMRs is their cost competitiveness which can only be achieved through modularization, standardization of plant and series production which is ultimately linked to a sustainable supply chain. As Lyons and Roulstone point out[10], there is a need for well- coordinated effort and task sharing between various layers of supply chain, enabling mutual learning and eventual cost reduction. As many SMRs technology are being developed, the US has the opportunity to establish a large domestic manufacturing based on pre-existing infrastructure and enhance its capability by utilizing naval shipbuilding. Underutilized dormant nuclear industry can be revitalized through incentives. By mobilizing domestic nuclear components and equipment plants and taking initiative in design, manufacturing, supply, and construction activities, a number of jobs and huge revenue can be materializedalong with secure and reliable energy source.

With the demonstration of FOAK SMRs in some parts of the world, many companies have started signing contracts in different countries. Recently, GE Hitachi Nuclear Energy signed a contract for a plant in Ontario in partnership with Ontario Power Generation (OPG) and is expected to create about $1.3 billionsin GDP and 1700 jobs in Canada during manufacturing and construction phase[43]. NuScale[44] is planning to deploy SMRs with different organizations in Eastern Europe, Asia and North America [16]. Similarly, Westinghouse has agreement to build a fleet of AP300 SMRs in Czech Republic with an estimated contribution of $12.7 billion in GDP and 8500 FTE jobs locally[45]. Although the revenues and jobs are high during construction phase but there would be substantial amount of property tax, and other revenue generated during operational phase of nuclear projects.

In the wake of recent technological and commercial development, the Federal Infrastructure Investment and Jobs Act (2021) and Inflation Reduction Act (2022)[7] are positive steps to mobilize private investors and bolstering their confidence in nuclear energy through different provisions including loan guarantees, tax credits for investment and insurance in case of regulatory delays for existing and new NPPs. These stepshave not yet been able to energize the downstream supply chain and only four (UV, UD, NV, NS) out of 16 certification categories have seen a positive trend (see figure 4). In this regard, the Department of Energy (DOE) has also granted funds in order to support Advance Reactor Demonstration Program (ARDP) and FOAK SMR technology through partnership with US industry[1,2]. The revival of local industry may result in less political risks and opposition, fostering good relationship between government and communities, lowering procurement cost, reducing transportation costs and above all flourishing of local economy in terms of employments. So, the need is to revive all levels of the stalled nuclear supply chain and regain market leadership.

5. Non US Codes and Standards

The nuclear supply chain is evolving from traditional local market to international marketplace as the number of countries providing nuclear equipment/ components are increasing. There are more companies holding ASME N-Stamp around the world now as compared to those in North America as reported in nuclear supply chain outlook up to the year 2030. The advent of SMRs and innovative reactor designs haveprovided opportunities for suppliers to locate low-cost labor and cheaper raw materials to delivery high quality and cost competitive products in this expanded global market. There are several challenges for international trade: geopolitics, quality inspection, transportation, tractability of materials and components, communication, cross-border licensing, exchange rates and liabilities. However, international collaborations through IAEA, Nuclear Suppliers Group (NSG), Organization for Economic Co-operation and Development (OECD), Generation IV International Forum (GIF) and other forums can lead to a well-coordinated, sustainable and regulated global supply system[31].

IAEA has taken initiatives to review and evaluate the applicability of safety standards for SMRs and making recommendations for additional guidance. The technological shift from conventional to innovative reactordesigns is pushing Member States to establish new specific design safety requirements. Specific safety requirement related to design of NPPs (i.e. SSR-2/1), IAEA has found that about 10% (10/82)[46] changes and/or interpretations would be required for LW-SMRs applicability from larger reactors' specific safety requirements. This small volume of changes is only due to the similarities between the design characteristicsof water cooled SMRs and large NPPs with same technology. Conversely, by considering the significant differences between design characteristics of the HTG-SMRs and LWRs, 35% (30/82)[46] of the requirementsneed changes and/or interpretation. Most of the developed economies have their own set of codes related to nuclear power plant construction as listed in Table 2. The existence of such a huge number of indigenouscodes have led to certain gaps or inconsistencies. Development of well-defined equivalency between nuclear codes in general and SMR related codes in particular from various part of the world is a major undertaking. So, the need of the hour is to harmonize the regulatory requirements by unanimous consensuswhich will further reduce the cost of nuclear power.

6. Nuclear Supply Chain in US Midwest

Midwest region in the United States has been a hub for heavy machinery production. Steel production coupled with downstream industrial activities has been major portion of the regional economy. Most of the automobile industry is situated in this region. Illinois is the state with maximum number of NPPs. For this reason, it is important to take a close look at this region for potential participation in the future nuclear supply chain.

An effort has been made to assess industrial potential and economic impacts in the Midwest region (AR, IA, IL, KS and MO) of the US if SMR industry was to take off in this region. The study has also been carried out to determine the number of companies with ASME certifications[22] stamps in the Midwest regionas compared to the entire US. The results are summarized in Figure 5. The analysis revealed that there is dormant capacity to provide power boilers that can be converted to operational if favorable market conditions develop. Likewise, out of 1706 AISC[24] certified companies, 145 companies are situated in Midwest. Workforce readiness is as critical as any other element of supply chain. Regarding skilled manpower, there are also about 8000 ACI[25] certified staff available in Midwest. Similarly, there are certifiedcompanies and labor available in the region in other fields as well. So, the overall capacity becomes muchhigher to make this region as manufacturing hub for SMRs. An extensive study is required to collect all relevant data about the certified companies and manpower for any region for its potential to contribute to the nuclear supply chain. Due to enormity of this task, we have only provided Midwest as an example. One needs to not only look at the current certifications but also pay careful attention to the dormant capabilities which can potentially become active partner in the overall nuclear supply chain under favorablemarket conditions.

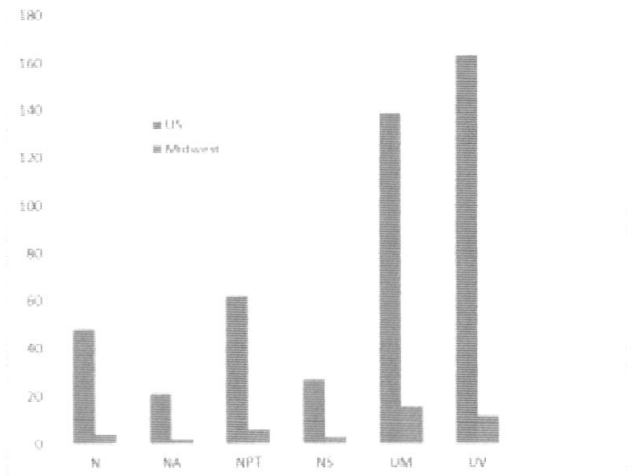

Figure 5: Certification Status for ASME Certified Companies in US and Midwest[22]

7. Nuclear Supply Chain Globalization: Opportunities & Challenges

All SMRs designs are based on the concept of modularity and factory fabrication, they offer lower capital cost due to their short construction time and hence provide greater flexibility and scalability. They are wellsuited not only for developed nations but also for the regions which have budgetary and infrastructural limitations. Their simple, versatile and inherently safe designs make them an attractive choice for energy production and industrial heating applications. SMR deployment in a country would provide a green, base-load alternative to meet future energy demands and climate change commitments. It has the potential for enhancing local manufacturing capabilities resulting in a major economic impact in terms of thousands of highly paid technical/non-technical jobs directly and indirectly related to nuclear supply chain. In addition,a huge amount of revenue is expected for the government through taxes. However, there are many challenges, harmonization of codes and standards being the most significant one. While there may be somelocal expertise holding certification from national or regional SDO, because of the differences between the reactor vendors specific codes and standards requirement, these subcontractors may not be able tocontribute to the supply chain. On the other hand, the project may become cost prohibitive if all componentsand services were to be imported from the country of reactor origin and certification.

For example, if NuScale's SMR was to be constructed in Europe (say, in Türkiye) USNRC License would demand adherence to ANSI/AISC codes for design, fabrication and erection of steel structure. There are alternates available for these quality assurance programs (please see row 1 of Table 2). It might be feasiblefor a project in Türkiye to locate a subcontractor with KTA 2201 (Germany) or EN 1993-1-1 (Eurocode 3) certification within the country or in a neighboring country rather than subcontracting to a US contractor. Without this flexibility, the advantage of modularity in cost saving will be lost and often will impact financial feasibility of the entire project. Looking at Figure 4, the US capacity to sustain nuclear supply chain is limited and is indeed diminishing. To meeting the global energy demand with carbon free optionswithin the ambitious timeline this global cooperation is inevitable.

It is also important that the code compliance alternatives are thoroughly evaluated in the earlier phase of design and analysis and be incorporated in the design certification. This forward thing will enable cost- effective global market to contribute to the supply chain. In cases where the design is already gone throughthe vigorous regulatory scrutiny, equivalency of codes for international deployment must be evaluated andnegotiated in the early phase of the project to manage supply chain risks more effectively. Series construction of reactors with standard design at one site can bring substantial savings. There are some majorchallenges with this globalization which goes beyond workplace cultural differences. For example, in the US nuclear safety related concrete structures must comply with American Concrete Institute code (ACI- 349-13). For a project in Europe, there may be local/national or regional contractor with AFCEN's RCC- SW certification able to contribute to the project cost-effectively yet meeting the overall safety requirementsfor nuclear standards.

As SMR technology is growing rapidly, the challenges with licensing and verification are also rising[13]. Nuclear codes are being modified or developed to be applicable to this innovative technology around the globe resulting in variety of codes for the same system. This challenge of code differences has been recognized and effort are underway to reconcile differences in codes and standards. In this regard, the Cooperation in Reactor Design Evaluation and Licensing (CORDEL)[47] group of World Nuclear Association (WNA)[48] had taken initiative to harmonize codes and standards by involving six SDOs: ASME, AFCEN, KEA , JSME, NIKIET and CSA. Recently, IAEA has launched two interconnected platforms: SMR Platform and Nuclear Harmonization and Standardization Initiative (NHSI) to foster collaboration and knowledge transfer among member states[49]. In this regard, NHSI has established Network for Experiment and Code Validation Sharing (NEXSHARE)[50] with international organizations to facilitate design and licensing through computer codes. The development for harmonized codes and standards would be beneficial to minimize the time for adoption and deployment of innovative reactor technologies. Along withvendors and SDOs, the regulators are required to endorse common codes and standards for innovative features and work together for acceptable international standards[51].

8. Conclusion

Advanced nuclear technology in the form of SMRs has a crucial role in mitigating climate change and hasa potential role in revolutionizing domestic and industrial needs along with secure sustainable energy supply. However, this is closely linked with government policies in implementing robust mechanisms through laws and incentives to promote local industry. SMRs have provided opportunities to the private investors and governments in terms of short construction time with smaller upfront cost at low-risk. Their cost competitiveness can only be materialized through modularization, standardization of plant and series production which is ultimately linked to a sustainable supply chain. Learning-by-doing is regarded as a potentially significant way of reducing capital costs, both through replication at the factory for componentsand at the construction site for installation. Simpler designs with passive safety systems, can also be economical requiring reduced regulatory oversight.

Success of nuclear energy globalization and can only be achieved by cooperation and contribution to the supply chain from all available resources. By avoiding and minimizing risks during project construction through an efficient project management and effective supply system, the commencement of power generation can be made at the earliest to generate revenues. In addition to mature design, a predictable andconsistent licensing process can result in substantial savings. While there is rapid evolution of global marketfor SMRs, there are several challenges associated with full utilization of available supply chain options. Geopolitics, quality inspection, transportation, tractability of materials and components, communication, cross-border licensing, exchange rates and liabilities are few of those challenges requiring close cooperationbetween international players. Among those, adoptability and acceptance of nuclear codes/standards are major issues. Reactor vendors and utilities have been working with national and international regulatory bodies to harmonize regulatory and utility requirements for reactor designs throughout the world. Such harmonization would lower costs for manufacturing, construction, maintenance, and refueling and at the same time help regional economic growth.

9. References:

[1] Nuclear Power and Secure Energy Transitions – From today's challenges to tomorrow's clean energy systems, International Energy Agency (IEA), June 2022.

[2] Nuclear Power Economics and Structuring, World Nuclear Association, 2024.

[3] Standard Design Approval for the NuScale Power Plant Based on the NuScale Standard Plant Design Certification Application, NRC Letter dated: September 11, 2020, https://www.nrc.gov/docs/ML2024/ML20247J564.pdf

[4] Appendix G to Part 52 – Design Certification Rule for NuScale, https://www.nrc.gov/reading-rm/doc-collections/cfr/part052/part052-appn-1.html.

[5] CFPP LLC Request to Withdraw the Limited Work Authorization Application, Topical Reports, and Exemption Request Associated with the Carbon Free Power Project (CFPP) Combined License Application, Letter dated: November 10, 2023. https://www.nrc.gov/docs/ML2331/ML23317A110.pdf.

[6] Economic and Employment Impacts of Small Modular Nuclear Reactor, June 2010

[7] Unlocking Reductions in the Construction Costs of Nuclear: A Practical Guide for Stakeholders, Nuclear TechnologyDevelopment and Economics, NEA No. 7530, OECD 2020.

[8] Economics of Nuclear Power, September 2023

https://world-nuclear.org/information-library/economic-aspects/economics-of-nuclear-power

[9] Stewart, W. R. and Koroush Shirvan, Construction schedule and cost risk for large and small light water reactors,Nuclear Engineering and Design, Vol. 407, June 2023, 112305.

[10] Robbie E. Lyons, Anthony R. M. Roulstone, "Production Learning in a Small Modular Reactor Supply Chain",Conference Paper • April 2018

[11] Advances in Small Modular Reactor Technology Developments, A Supplement to: IAEA Advanced ReactorsInformation System (ARIS), 2020

[12] Small Modular Reactors: Challenges and Opportunities, Nuclear Energy Agency, 2021

[13] IAEA Technology Roadmap for Small Modular Reactor Deployment, 2021

[14] Robert Rosner and Stephen Goldberg, "Small Modular Reactors – Key to Future Nuclear Power Generation in theU.S." Report to DOE from Energy Policy Institute at Chicago, Revision 1 November, 2011.

[15] Elaheh Shobeiri et al, "Small Modular Reactor Deployment and Obstacles to be Overcome", April 2023. https://doi.org/10.3390/en16083468

[16] Casey Crownhart, Small modular reactors could be quicker and cheaper to build, Feb, 2023. https://www.technologyreview.com/2023/02/08/1067992/smaller-nuclear-reactors

[17] Office of Nuclear Energy, "5 Advanced Reactor Designs to Watch in 2030", March 17, 2021 https://www.energy.gov/ne/articles/5-advanced-reactor-designs-watch-2030

[18] BWXT Advanced Nuclear Reactor (BANR)

https://www.bwxt.com/what-we-do/advanced-technologies/terrestrial-micro-rx

[19] IAEA-TECDOC-1936, "Applicability of Design Safety Requirements to Small Modular Reactor TechnologiesIntended for Near Term Deployment", 2020

[20] A Closer Look at Two Operational Small Modular Reactor Designs, 2024

https://www.powermag.com/a-closer-look-at-two-operational-small-modular-reactor-designs/

[21] Safety Reports Series No. 123, "Applicability of IAEA Safety Standards to Non-Water Cooled Reactors and SmallModular Reactors", 2023

[22] American Society of Mechanical Engineers (ASME)

https://www.asme.org/codes-standards/about-standards/the-future-of-asme-standards/smrs-and-advanced-reactors

[23] Institute of Electrical and Electronics Engineers (IEEE)
https://innovate.ieee.org/ieee-nuclear-engineering-standards/ [24]
American Institute of Steel Construction (AISC)

https://www.aisc.org/publications/historic-standards/nuclear-specifications/

[25] American Concrete Institute (ACI)
https://www.concrete.org/topicsinconcrete/topicdetail.aspx?search=nuclear#standards [26]
American National Standards Institute (ANSI)
https://www.ansi.org/search#q=nuclear&sort=relevancy

[27] U.S. NRC's , REGULATORY GUIDE 1.26, Quality Group Classifications and Standards for Water-, Steam-, andRadioactive-Waste-Containing Components of Nuclear Power Plants, Revision 5, 2017.
https://www.nrc.gov/docs/ML1608/ML16082A501.pdf

[28] U.S. NRC's, 10 CFR50.55a, Codes and standards, NRC Technical Library, Two White Flint North, 11545 RockvillePike, Rockville, Maryland 20852. https://www.nrc.gov/reading-rm/doc-collections/cfr/part050/part050-0055a.html [29] Enclosure 7: List of Standards Identified for Advanced Reactors, USNRC

[30] Advanced Reactor Codes and Standards Needs Assessment, NEI 19-03 [Rev 1], 2020

[31] IAEA Nuclear Energy Series No. NP-T-3.21: "Procurement Engineering and Supply Chain Guidelines in Supportof Operation and Maintenance of Nuclear Facilities", 2015

[32] Concrete Codes and Standards for Nuclear Power Plants: Recommendations for Future Development, June 2011

[33] American Society for Testing and Materials (ASTM)
https://www.astm.org/catalogsearch/result
[34] IAEA Safety Standards Series No. NS-G-1.10, "Design of Reactor Containment System for Nuclear Power Plants",2004
[35] https://eurocodes.jrc.ec.europa.eu/EN-Eurocodes/eurocode-3-design-steel-structures

[36] https://www.afcen.com/en/rcc-cw/190-rcc-cw-etc-c.html
[37] https://www.kta-gs.de/e/standards/2200/2201_1_engl_2011_11.pdf
[38] https://www.jsme.or.jp/english/committees/codes-standards/jsme-standards/
[39] https://www.csagroup.org/store/product/2422200/
[40] https://knowledge.bsigroup.com/products/code-of-practice-for-temporary-works-procedures-and-the-permissible-stress-design-of-falsework?version=tracked

[41] Nuclear Energy Institute (NEI)
https://www.nei.org/search?search_term_string_lg=nuclear%20codes%20and%20standards [42]
World Nuclear News, Panelists address SMR supply chain challenges, 25 September 2020[43] The socio-Economic Impact of GE Hitachi SMRs, 2021

https://www.gevernova.com/content/dam/gepower-nuclear/global/en_US/documents/economic-impact/Transforming%20Canada's%20energy%20future%20%E2%80%93%20The%20socio-economic%20impact%20of%20GE%20Hitachi%20SMRs.pdf

[44] https://www.nuscalepower.com/en/projects

[45] The Economic Impact of Westinghouse's AP300 Small Modular Reactor Fleet Project for Czech RepublicWestinghouse Report Czechia AP300 V6 ENGLISH FINAL (hubspotusercontent-na1.net)

[46] IAEA-TECDOC-1936, "Applicability of Design Safety Requirements to Small Modular Reactor TechnologiesIntended for Near Term Deployment", 2020

[47] Benefits Gained through International Harmonization of Nuclear Safety Standards for Reactor Designs, WNA[48] Harmonisation of Design Codes & Standards in Nuclear Industry, NAFEM, 2012 https://www.nafems.org/events/nafems/2012/standards-in-nuclear-

industry/#:~:text=Such%20standardisation%20will%20help%20nuclear%20power%20to,use%20of%20nuclear%20power%20around%20the%20world

[49] The SMR Platform and Nuclear Harmonization and Standardization Initiative (NHSI) https://www.iaea.org/services/key-programmes/smr-platforms-nhsi

[50] New IAEA NEXSHARE Network Promotes Cooperation on Code Validation for SMR Design and Licensing https://www.iaea.org/newscenter/news/new-iaea-nexshare-network-promotes-cooperation-on-code-validation-for- smr-design-and-licensing

[51] "Harmonising the Nuclear Licensing Process for Emerging Technologies: A Global Path Forward", Nuclear SafetyTechnology and Regulation 2022

An Investigation on Large Scale Green Hydrogen Production Through Electrolysis: Case Study of Akkuyu Nuclear Power Plant

Fethi Mustafa Çimen

[1]Gazi University, Ankara, Türkiye, [2]University of South Carolina, Columbia, United States, fethimustafacimen@gazi.edu.tr

ORCID: 0000-0002-1290-1469

Saltuk Buğrahan Yeşilgöl

[1]AKKUYU NUCLEAR JSC, Mersin, Türkiye, saltukenerji@gmail.com

ORCID: 0009-0006-4761-9566

Cite this paper as:	*Çimen, FM, Yeşilgöl, SB. An Investigation on Large Scale Green Hydrogen Production Through Electrolysis: Case Study of Akkuyu Nuclear Power Plant. 7th Int. Conf. on Nucl. and Ren. Energy Res. 27-30 October 2024, Antalya, Turkey*

Abstract: This paper explores the hydrogen production potential of the Akkuyu Nuclear Power Plant using three different electrolyzer technologies: Alkaline, PEM, and High-Temperature SOECs. The study evaluates the hydrogen production potential over an eight-year period (2025-2032), based on the plant's power output. The analysis demonstrates that the SOEC has the highest efficiency and hydrogen production capacity, making it the most suitable for large-scale, continuous energy sources like Akkuyu NPP. PEM Electrolyzers offer a balanced performance with moderate efficiency, while Alkaline Electrolyzers, despite their cost advantages, lag in production capacity and efficiency. These findings highlight the importance of selecting the appropriate electrolyzer technology to optimize hydrogen production in nuclear power plants. The study concludes that nuclear power offers a promising pathway for large-scale hydrogen production, contributing to global clean energy goals.

Keywords: *Akkuyu Nuclear Power Plant, Green Hydrogen Production, Electrolysis, Electrolyzer Efficiency, Decarbonization*

1. INTRODUCTION

Nuclear power plants offer a promising pathway for hydrogen production that addresses the dual goals of meeting energy needs and mitigating climate change. As a versatile energy carrier, hydrogen has the potential to decarbonize industry, transport and energy systems. The use of nuclear energy for hydrogen production, through processes such as low-temperature electrolysis (LTE) and thermochemical water-splitting cycles, makes hydrogen production both efficient and environmentally friendly by harnessing the high thermal and electrical output of nuclear reactors.

Electrolysis, when powered by nuclear energy, is an attractive method because it produces "green hydrogen" with minimal carbon emissions. Unlike renewable sources such as wind and solar, nuclear reactors provide continuous and reliable power, making hydrogen production consistent. A U.S. Department of Energy study underscores the economic feasibility of hydrogen production using nuclear reactors and predicts a significant reduction in production costs by 2030, making this approach competitive with other methods [1,2].

Thermochemical cycles such as sulfur-iodine and copper-chlorine cycles are also promising. These high-temperature processes split water molecules, making hydrogen production more efficient without the need for direct electricity. Researchers continue to explore these methods to further increase their efficiency and reduce associated costs [1,3].

Countries such as the United States, France and Japan are actively investing in nuclear hydrogen projects to meet their climate targets. For example, France has committed to increase its hydrogen production capacity using nuclear energy, aiming to install 6.5 GW of electrolyser capacity by 2030 [1,4]. The UK's 'Net Zero Hydrogen Fund' is also funding nuclear hydrogen projects to bridge the gap between low-carbon hydrogen and high-carbon fuels [5].

Nuclear-based hydrogen production holds significant promise due to its potential for large-scale, low-emission hydrogen production. The integration of nuclear energy into hydrogen production strategies around the world highlights the importance of this technology in achieving global decarbonisation targets.

This study aims to make a comparative analysis of the hydrogen production potential of Akkuyu Nuclear Power Plant according to energy production data.

2. METHODOLOGY

In this study, three different electrolyzer types were used to evaluate the hydrogen production with the electrical energy generated by the Akkuyu Nuclear Power Plant:

- Alkaline Electrolyzer: It is a widely used and economical electrolyzer type with 60-70% efficiency.
- PEM (Polymer Electrolyte Membrane) Electrolyzer: It has a higher performance with 70-80% efficiency and stands out with its fast response time.
- High Temperature (Solid Oxide) Electrolyzer: It is the electrolyzer type that provides the highest performance with 80-90% efficiency.

The efficiency of electrolyzers and the capacity for hydrogen production are calculated using the following equation:

$$\eta = \frac{\text{Higher Heating Value (HHV) of Produced Hydrogen}}{\text{Input Energy (Electricity + Heat)}}$$

In this equation:

- η represents the efficiency of the electrolyzer.
- HHV (Higher Heating Value) is the higher heating value of hydrogen.
- Input energy includes the electricity and heat energy used by the electrolyzer to convert water into hydrogen.

The calculations used in this study are based on the annual power generation data from the Akkuyu NPP.
The hydrogen production potential for each year was calculated using the following formula:

$$\text{Hydrogen Production (kg)} = \text{Power (MWh)} \times 1000 \times 0.03 \times \text{Efficiency}$$

Where:

- Power (MWh) is the annual electricity provided by the NPP.
- 1 MWh of electricity can produce approximately 30 kg of hydrogen.
- Efficiency depends on the type of electrolyzer used.

Hydrogen production efficiency varies depending on the operating conditions of the electrolyzer and the amount of energy used. The following table summarizes the main factors affecting electrolyzer efficiency.

Table 1 The main factors affecting electrolyzer efficiency

Parameter	Description	Effect on Efficiency
Temperature	The operating temperature of the electrolyzer	Higher temperatures increase efficiency by reducing electricity needs
Electric Current	The electrical power used by the electrolyzer to separate water	Higher current increases efficiency, but too high current may reduce performance
Membrane Quality (for PEM)	The quality of the proton-conducting membrane in PEM electrolyzers	Higher quality membranes reduce energy loss and improve efficiency
Electrolyte Quality (for Alkaline)	The purity and conductivity of the electrolyte	Low-conductivity electrolyte may increase energy loss and reduce efficiency
Pressure	The pressure at which the electrolyzer operates	High pressure may require more energy and negatively affect efficiency

Electrolyzer efficiency varies based on the technology used. The efficiency and hydrogen production capacity of the three different electrolyzer types are shown in the table below:

Table 2 Efficiency Variation by Electrolyzer Type

Electrolyzer Type	Efficiency (%)	Hydrogen Production per 1 MWh (kg)	Advantages
Alkaline Electrolyzer	60-70	18-21	Low cost, widely used in industrial applications
PEM Electrolyzer	70-80	21-24	High efficiency, fast response time, clean and quiet operation
High-Temperature Electrolyzer	80-90	24-27	Highest efficiency, reduced electricity consumption, high-temperature advantage

3. RESULTS AND DISCUSSION

In the study, the hydrogen production potential of Akkuyu NPP through electrolysis was analysed by taking into account the energy production forecast of Akkuyu NPP depending on the years. Table 3 shows the data on the electricity generation of Unit 1 of Akkuyu NPP without taking into account seasonal constraints.

Table 3 The electricity generation of Unit 1 of Akkuyu NPP

Power Generation, MWh	**2025**	**2026**	**2027**	**2028**	**2029**	**2030**	**2031**	**2032**
January	0	916980	0	916980	916980	0	916980	916980

February	0	828240	88740	857820	828240	769080	828240	857820
March	0	916980	916980	916980	916980	916980	916980	916980
April	0	887400	887400	887400	887400	887400	887400	887400
May	0	916980	916980	916980	916980	916980	916980	916980
June	0	887400	887400	887400	887400	887400	857820	887400
July	0	916980	916980	236640	916980	916980	0	916980
August	0	916980	916980	562020	916980	916980	828240	916980
September	0	887400	887400	887400	887400	887400	887400	887400
October	0	916980	916980	916980	916980	916980	916980	916980
November	0	887400	887400	887400	887400	887400	887400	887400
December	88740	798660	916980	916980	857820	916980	916980	621180
TOTAL for the year	88740	10678380	9140220	9790980	10737540	9820560	9761400	10530480

During the study, hydrogen production potentials for three different electrolyser types were compared and the results are shown below.

Table 4 Hydrogen production potentials for three different electrolyzer types

Year	Alkaline Electrolyzer	PEM Electrolyzer	SOEC
2025	1,729 kg	1,996 kg	2,263 kg
2026	2,142,827 kg	2,403,038 kg	2,663,249 kg
2027	1,835,405 kg	2,056,550 kg	2,277,695 kg
2028	1,964,430 kg	2,202,970 kg	2,441,510 kg
2029	2,097,856 kg	2,413,945 kg	2,730,033 kg
2030	1,984,112 kg	2,205,026 kg	2,425,941 kg
2031	1,959,279 kg	2,191,815 kg	2,424,352 kg
2032	2,114,688 kg	2,363,359 kg	2,612,030 kg

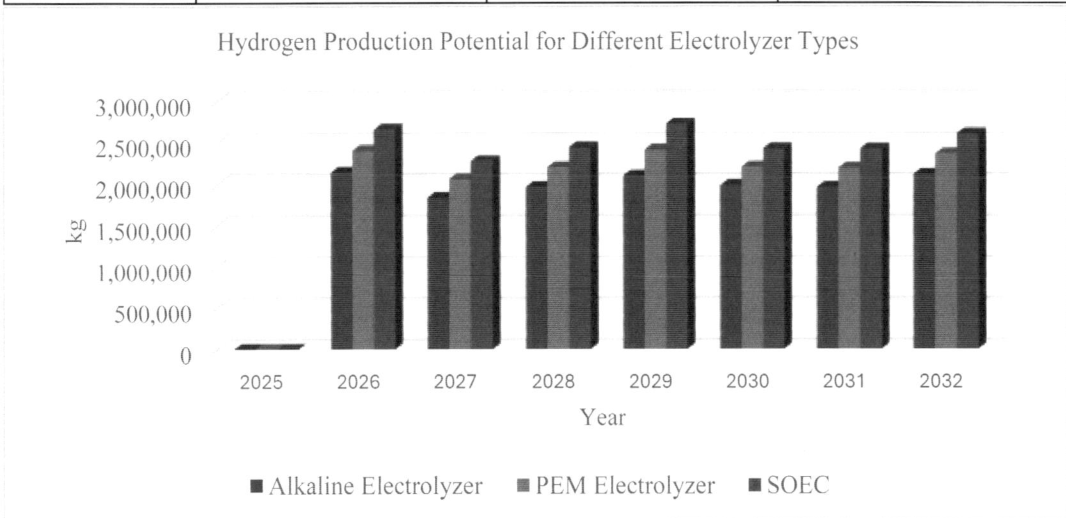

According to the given table, three different types of electrolyzers exhibit various performances in terms of hydrogen production with energy from the Akkuyu NPP. Although the alkaline electrolyzer is a low-cost and widely used technology, it has the lowest hydrogen production capacity in this plant. Especially in a power plant such as Akkuyu NPP, which generates large amounts of energy, this low efficiency is predicted to lead to an underutilization of the hydrogen production potential in the long term. While the PEM electrolyzer offers higher efficiency, it also exhibits a more stable hydrogen production performance in fixed energy producing power plants such as Akkuyu NPP. While its high flexibility allows it to adapt more quickly to variable energy sources, this flexibility may not always be suitable for a continuous power plant like Akkuyu. In contrast, the High Temperature (SOEC) electrolyzer has the highest hydrogen production capacity and works most efficiently with the high-temperature energy produced by Akkuyu NPP. SOEC has achieved maximum efficiency at this power plant, peaking in 2029 with 2,730,033 kg of hydrogen production, and is also advantageous in terms of energy consumption. Therefore, SOEC technology stands out as the most efficient and profitable solution in the long term for large-scale and continuous power supply nuclear power plants such as Akkuyu NPP.

4. CONCLUSION

This study demonstrates the significant potential of hydrogen production through electrolysis using the Akkuyu Nuclear Power Plant's power output. Among the three electrolyzer types analyzed, the High-Temperature SOEC emerged as the most efficient and productive, achieving the highest hydrogen production potential, especially in years of stable and high energy output from the plant. The PEM Electrolyzer, while slightly less efficient, showed a more balanced performance, with stable hydrogen production potential that makes it a reliable option for power plants with consistent energy supply, like Akkuyu. On the other hand, the Alkaline Electrolyzer, despite its cost-effectiveness and wide industrial use, was the least efficient in this study, potentially leading to underutilization of the plant's energy production capacity in the long term. Overall, SOEC technology is recommended as the most suitable solution for large-scale, continuous power generation facilities like Akkuyu NPP, providing both energy efficiency and maximum hydrogen production potential. This analysis confirms that hydrogen production through nuclear power plants represents a promising path for achieving large-scale green hydrogen production, contributing to global decarbonization efforts.

ACKNOWLEDGMENT

The authors would like to thank AKKUYU NUCLEAR JSC for permission to utilise the data used in this study.

REFERENCES

1. Abd El-Hameed, A., Soja, R. J., Ramadhan, H. H., Nandutu, M., & Hyun, J.H. Estimation of the levelized cost of nuclear hydrogen production from light water reactors in the United States. *Processes* 2022:10(8), 1620.
2. International Atomic Energy Agency (IAEA). (2021). Hydrogen production from carbon-free nuclear energy. IAEA Nuclear Energy Series. https://www.iaea.org
3. Dincer, I., Naterer, G. F., & Zamfirescu, C. Hydrogen production from nuclear energy. *Springer*, 2013.
4. International Atomic Energy Agency (IAEA). Hydrogen production using nuclear energy. IAEA Nuclear Energy Series No. NP-T-4.2, 2012.
5. Hydrogen Europe. (2021). Hydrogen production using nuclear energy. https://www.hydrogeneurope.org

Neutronic Study of a Molten Salt Uranium Mixture in a Fusion Reactor Blanket

H. Mehmet Şahin
Gazi University, Ankara, Turkiye, mesahin@gazi.edu.tr,
ORCID: 0000-0001-8538-1399

Güven Tunç
Gazi University, Ankara, Turkiye, guventunc@gazi.edu.tr,
ORCID: 0000-0001-7038-8168

Alper Karakoç
Karabük University, Karabük, Turkiye, @gmail.com,
ORCID: 0000-0002-3692-5895

Cite this paper as:	*Şahin,H.M.,Tunc, G, Karakoç, A. Neutronic Study of a Molten Salt Uranium Mixture in a Fusion Reactor Blanket. 7th Int. Conf. on Nucl. and Ren. Energy Res. 27-30 October 2024, Antaya, Turkey*

Abstract: In this study, a Fusion-Fission Hybrid Reactor (FFHR) for the ITER reference geometry using a uranium molten salt mixture as a dual-purpose medium for coolant and fissile fuel production has been investigated. SS 316 LN-IG was selected as the first wall material of the reactor and LiF-UF4 molten salt fuel mixture was used as the coolant considering the eutectic points of the material. For neutronic analyses, the MCNP6 nuclear code was used together with the ENDF/B-VIII and CLAW-IV nuclear data libraries. The evolution of isotopes in the reactor over time was calculated with the MCNPAS interface code. The study results were evaluated in terms of tritium production rate, energy multiplication factor, radiation damage, fissile fuel production and fuel combustion value. A 4-year study history of the total TBR value was calculated and is always above 1.05 and increases over time. In the reactor, LiF-UF4 molten salt and fuel mixture, the uranium mass decreased from 720.8 tons at the beginning to 639.6 tons, while 239Pu production during this period was 21.3 tons. According to these results, the first wall replacement period was calculated as 3.92 years.

***Keywords**:* *ITER, TOKAMAK, Hybrid, Uranium, Molten salt mixture*

1. INTRODUCTION

The concept of a fission-fusion hybrid reactor was introduced by Edward et al. to reduce the limitations of both fission and fusion reactor technologies [1]. The primary goal of a hybrid reactor is to harness the high-density neutron flux generated by a fusion reactor to sustain a nuclear fission reaction and simultaneously produce fissile fuel from fertile materials such as 232Th and 238U. Additionally, hybrid reactors can utilize waste from conventional fission reactors as fuel. Modeling of a hybrid reactor based on a TOKAMAK design was conducted, using D-T fueled plasma as the neutron source. Unlike standalone fusion reactors with limited power output, the hybrid reactor demonstrated significant potential for high power generation based on the results of these calculations [2]. Wu et al. [3] proposed a conceptual hybrid reactor design called the Fusion Driver System-Energy Multiplier (FDS-EM). This design, based on parameters from the EAST (Experimental Advanced Superconducting TOKAMAK) reactor, was shown to be suitable for sustainable energy production.
In this study, a hybrid reactor design was analyzed using the geometry of ITER and incorporating a D-T fueled plasma neutron source. Stainless steel 316 LN-IG was selected as the material for the first wall and thermal shield, while a eutectic mixture of LiF-UF4 was chosen as the coolant and fuel. The material selection was guided by the eutectic points of these substances. Key performance metrics for the reactor design—such as the tritium breeding ratio, radiation damage, fissile fuel generation, and fuel burn-up—were evaluated using the Monte Carlo N-Particle

Transport Code (MCNP). To address time-dependent parameters, the MCNPAS interface code was utilized. The objective of this study is to optimize the neutronic performance of the hybrid reactor, ensuring efficient energy production and resource utilization.

2. CALCULATION MODEL OF HYBRID REACTOR

Figure 1 shows the blanket structure of the hybrid reactor modeled within the scope of the thesis study. In the modeled reactor, the plasma, the neutron source, is adapted to the D shape geometry, and the blanket structure surrounds the plasma chamber. The reactor layers are listed fallow; first wall, coolant zone, fuel zone, reflector, vacuum chamber, thermal shield, magnetic coil.

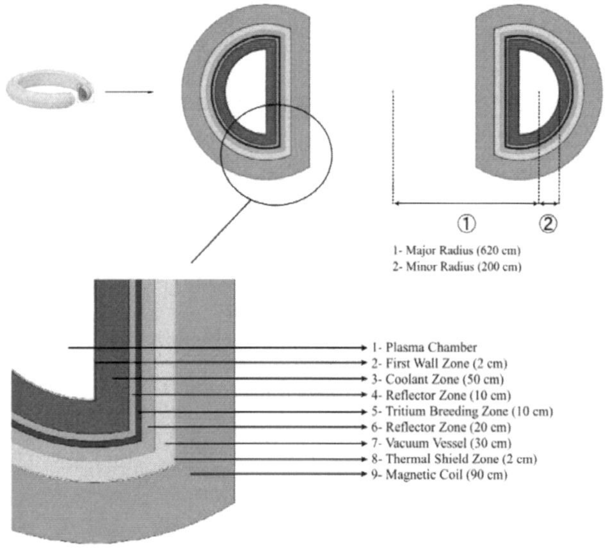

Figure 1. Structure of Hybrid Reactor Blanket.

2.1. Monte Carlo N-Particle Code

MCNP, Monte Carlo N-Particle, is a comprehensive and widely used computer program for the simulation of nuclear processes [4,5]. The neutron energy ranges for all isotopes in ENDF/B-V and ENDF/B-VI are 10-11 MeV to 20 MeV, and for some isotopes it reaches 150 MeV [6]. These libraries are essential for accurate simulations. It uses the built-in continuous energy nuclear and atomic data libraries, such as the Evaluated Nuclear Data File (ENDF) system [7].
In the thesis study, two different codes were used to carry out the simulation study of the hybrid nuclear reactor. As the first code, an input file compatible with the work in MCNP6 was created. In this input file, the required tallies within the scope of reactor geometry, materials and neutronic performance constitute the input file. The resulting output file is introduced as an input file to the created interface programmer. The interface program MCNPAS (MCNP Assessment Code), written in FORTRAN 90, is executed for time-dependent analyses and monitoring of isotope abundance changes. The actinide isotope transformation scheme and the flowchart of the interface code used to process MCNP output via MCNPAS. The atomic densities of the materials in the MCNP input file and the mass and volumetric proportions of this mixture used in the hybrid reactor are given in Table 1. [8].

2.2. Molten Salt-Fuel Mixture

In the reactors, the mixture of molten salt (lithium, beryllium and zirconium fluoride) and fuel (thorium, uranium, plutonium and MA) can be used as both coolant and fuel. A eutectic mixture, characterized by its lowest melting point, was selected to improve the cooling efficiency of the hybrid reactor blankets. Considering their thermophysical properties, the LiF-UF4 fuel mixtures were selected based on their respective ratios at the eutectic point. Furthermore, the Newtonian flow properties of this molten salt-fertile fuel mixture facilitate the straightforward application of thermohydraulic calculations [9]. The phase diagram of LiF-UF4 was shown in Figure 2.

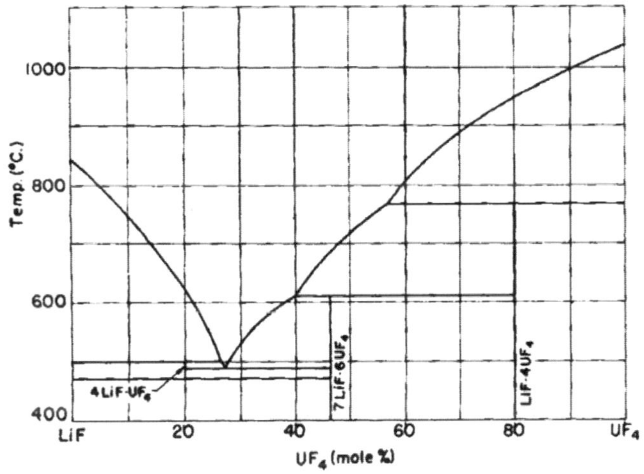

Figure 2. Phase diagram of LiF-UF4 [10].

3. RESULTS AND DISCUSSION

3.1. Tritium Breeding Ratio

The tritium breeding ratio (TBR) in hybrid nuclear reactors is defined as the ratio of tritium generated in the reactor's blanket layer to the amount of tritium consumed in the plasma. To account for tritium decay and other potential losses within the reactor, TBR must be maintained at a minimum value of 1.05 [11,12]. In hybrid reactors, tritium production occurs through interactions between lithium isotopes and both thermal and fast neutrons. These interactions include the exothermic ^{6}Li(n,T) reaction with thermal neutrons and the endothermic ^{7}Li(n,T) reaction with fast neutrons. Among these, the ^{6}Li isotope plays a more significant role in achieving a sufficient TBR. The influence of lithium enrichment on the TBR value, as a function of coolant thickness, is illustrated in Figure 3, highlighting the critical role of ^{6}Li in optimizing tritium production.

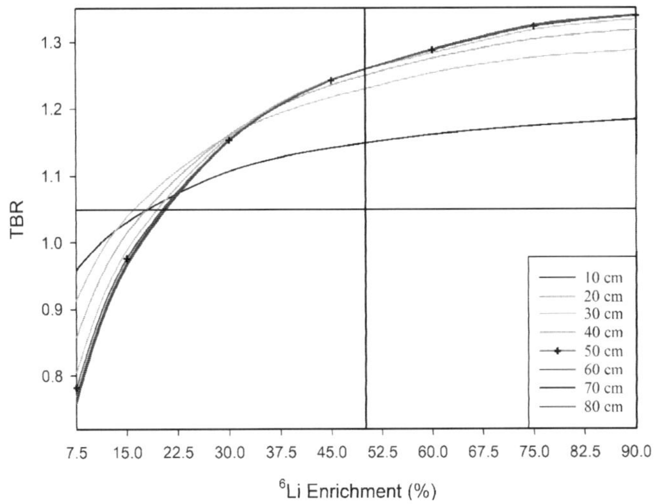

Figure 3. Effect of ^{6}Li enrichment on TBR depending on coolant thickness

3.2. Energy Multiplication Factor

The energy multiplication factor (M) is a measure of the ratio between the energy input and energy output in the reactor. The energy inputs in the reactor can be accounted for in the calculations by considering the energy consumed by the magnets used to control the plasma and the energy consumed by other subcomponents of the reactor. The reactor's energy outputs can be determined by quantifying the electrical energy, waste heat, and radiation emitted as a consequence of fusion and fission reactions. In order to counteract the decrease in reactor power and generate additional power, the energy output of the reactor must surpass the energy input. The minimum requirement for the energy multiplication factor to be reached is 1.5 [13-16].

In hybrid reactors, the amount of fission reaction and fission fuel formation are directly proportional, and the same proportion is valid for the energy multiplication factor. The amount of exothermic $^{6}Li(n,\alpha)T$ reaction included in the equation increases the energy multiplication factor and releases 4.786 MeV of energy per reaction. The endothermic $^{7}Li(n,n\alpha)T$ reaction consumes 2.467 MeV energy per reaction and negatively affects the energy multiplication factor value. The energy multiplication factor values obtained as a result of the calculations are shown in Figure 4.

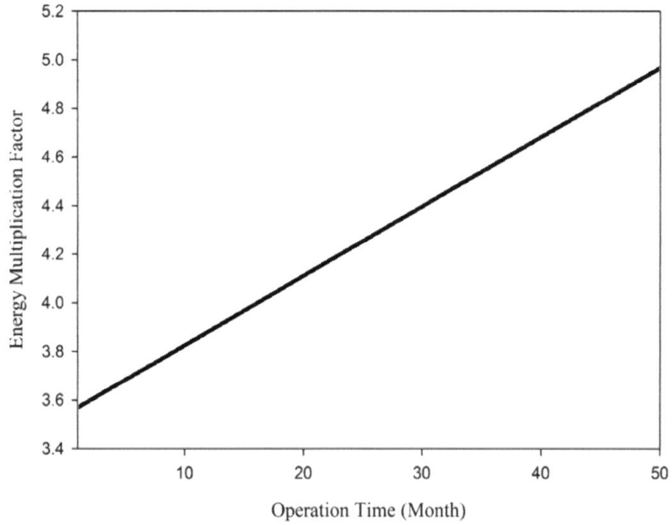

Figure 4. Energy multiplication factor

3.3. Radiation Damage

High energetic neutrons released in fusion reactions react on the first wall of the reactor and cause radiation damage. Helium and hydrogen gas are created in the first wall material as a result of the reactions that have taken place. High working temperatures cause hydrogen gas to escape from the structure of the material, but helium gas stays inside. Helium gas causes the first wall material to lose its qualities over time, which in turn causes the material's lifespan to decline dramatically. Therefore, it is essential to investigate the production of helium gas in the first wall. More specifically, in order to compute the replacement period of the first wall material, the upper production value of helium gas was established as 500 appm, and the calculations were carried out in accordance with one full power year (FPY) [17]. In the first wall material, the term "DPA" refers to the movement of atoms inside the structure of the material as a result of the action of high-energy neutrons and helium isotopes that are produced from the plasma. This parameter, which is used to measure radiation damage on the first wall material of the reactor, should be at as low levels as possible to avoid deterioration and loss of function of the first wall material. Within the scope of the study, the upper limit value of DPA was determined as 100 appm [11, 14,17]. The priority of evaluating damage parameters is determined by the He/DPA ratio. The production of helium gas precedes the displacement of atoms when the ratio between He and DPA (500/100 appm) is greater than 5, the exchange period of the first wall is computed using the production of helium gas. Radiation damage and lifetime of first wall values can be seen in Table 1 [18].

Table 1. Radiation damage

Parameter Mixture	H/FPY (appm)	He/FPY (appm)	DPA/FPY (appm)	He/DPA	First Wall Lifetime (year)
UF_4 - LiF	335.13	127.51	12.43	10.26	3.92

3.4. Fissile Fuel Production

There are two main methods for fissile fuel production, enrichment and breeding respectively. Breeding is the process of creating fissile isotopes from non-fissile isotopes. This is done by exposing a non-fissile isotope to neutrons in a nuclear reactor. In the thermonuclear fusion reactor's core, neutrons are used as a high-intensity external neutron source

for hybrid nuclear reactor. Neutrons released from fusion reactions in hybrid reactors are used to produce fissile fuels from fertile materials in addition to tritium and energy production.
The reserves of fissile fuels used in today's nuclear energy applications are quite limited. On the contrary, the amount of thorium and non-fissile uranium is quite high compared to these reserves. The isotopes in question, upon absorbing a neutron, form ^{233}U and ^{239}Pu isotopes, respectively. The fuel cycle is valid for uranium. The ^{238}U isotope captures neutrons to form the ^{239}U isotope, which turns into the ^{239}Np isotope with beta decay. ^{239}Np, which decays to fissile ^{239}Pu via beta decay with a half-life of 2.35 days. In this study, fuel composition using molten salt including uranium halides is used to produce fissile fuel. The results obtained are shown in Figure 5 [18]..

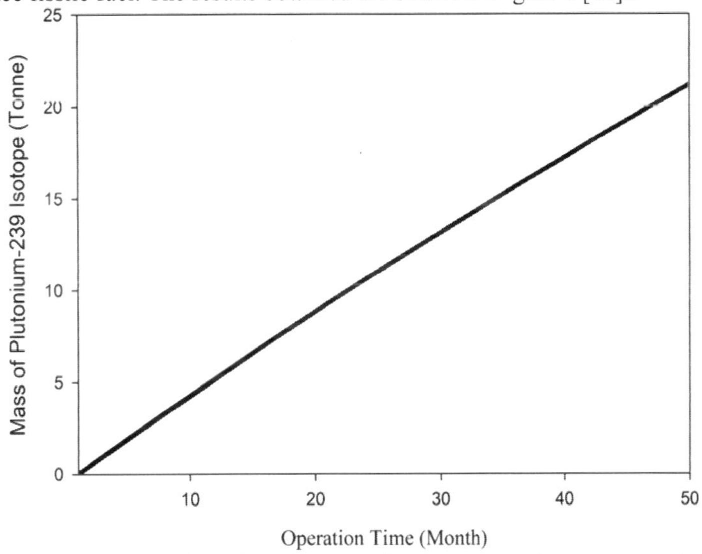

Figure 5. Pu-239 isotope through nuclear transformations over fission reaction in coolant

3.5. Fuel Burnup

The amount of energy extracted from nuclear fuel is measured by fuel burnup (also known as fuel utilization), which is also used as a measure of fuel depletion in nuclear engineering. Burnup is an important factor in determining the efficiency and lifetime of nuclear fuel. How much fuel is consumed affects the energy production capacity. High burnup values indicate that the fuel has the capacity to produce more energy, while low burnup values indicate that the fuel has the capacity to produce less energy. Burnup is expressed in gigawatt days per tonne of fuel (GW.day/tonne) and the burnup values obtained in the study can be seen in Figure 6 [18]..

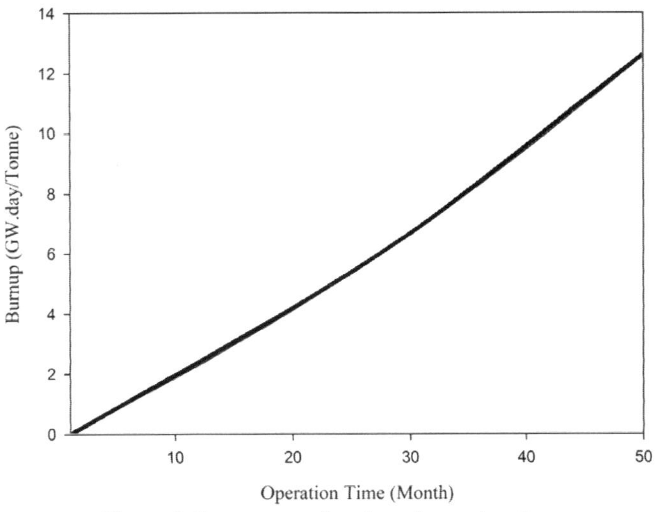

Figure 6. Burn-up as a function of operating time

4. CONCLUSIONS

In this study, a hybrid reactor design incorporating a D-T fueled plasma neutron source based on the ITER geometry was developed. Stainless steel 316 LN-IG was selected as the material for the first wall, while a eutectic mixture of LiF-UF4 was chosen as the coolant and fuel, leveraging the material's favorable eutectic points. Neutron calculations were conducted using the MCNP6 nuclear code, utilizing the ENDF/B-VIII and CLAW-IV nuclear data libraries. The time evolution of isotopic changes within the reactor was analyzed using the MCNPAS interface code.
Key results for the LiF-UF4 molten salt-fuel mixture are summarized as follows [18]:

- Tritium Breeding Ratio (TBR): Over the reactor's 4-year operation, the total TBR value consistently exceeded 1.05 and increased over time.
- Lithium Isotope Changes: The mass of ^{6}Li decreased from 18.68 tonnes to 18.25 tonnes, while ^{7}Li declined from 102.55 tonnes to 102.46 tonnes.
- Energy Multiplication Factor (M): The energy multiplication factor ranged from 3.57 to 5.2 during operation, consistently remaining above the minimum required value.
- Fuel Burnup: The reactor achieved a burnup value of approximately 13 GW·days/tonne by the end of 50 months.
- Isotopic Changes in Uranium and Plutonium: The ^{238}U inventory decreased from 720.79 tonnes to 639.61 tonnes, while ^{239}Pu production reached 21.31 tonnes during the same period.
- First Wall Replacement: The first wall's replacement period was calculated to be 3.92 years.

REFERENCES

[1] Bethe, H. The fusion hybrid. ***Phys. Today*** 32, 44–51, 1979.

[2] Reed, M., et al. A Fission-Fusion Hybrid Reactor in Steady-State L-Mode Tokamak Configuration with Natural Uranium, ***MIT Department of Nuclear Science & Engineering***, January 2011.

[3] Wu, Y., Chen, H., Jiang, J., Liu, S., Bai, Y., Chen, Y., Jin, M., Liu, Y., Wang, M., Hu, Y. and Team, F.D.S. (2008, October). The fusion–fission hybrid reactor for energy production: a practical path to fusion application. In 22nd ***Int. Conf. on Fusion Energy (FEC-22)*** Geneva, Switzerland.

[4] Tunç, G., The neutronic analysis of the performances of thorium and various nuclear fuels in a fusion-fission reactor system. Ph.D. Dissertation. In: ***Department of Energy Systems Engineering. Gazi Univer*****sity** (in Turkish), 2017.

[5] Baxi, C. B., & Wong, C. P. C. Review of helium cooling for fusion reactor applications. ***Fusion Engineering and Design***, 51-52, 319–324, (2000).

[6] X-5 Monte Carlo Team, 2005. MCNP5 – A General Monte Carlo, N-Particle Transport Code, Version 5, LA-CP-03-0245, ***Los Alamos National Laboratory***

[7] Rose, P.F., Compiler and Editor. 1991. ENDF-201, ENDF/B-VI Summary Documentation, BNL-NCS-17541, ***Brookhaven National Laboratory***

[8] Capelli, Elisa, O. Beneš, and R. J. M. Konings. Thermodynamic assessment of the LiF–ThF4 –PuF3 –UF4 system. ***Journal of Nuclear Materials*** 462 (2015): 43-53.

[9] Thoma, R. E., Phase Diagrams of Nuclear Reactor Materials. ORNL-2548, ***Oak Ridge National Laboratory,*** 1959.

[10] Moir, Ralph W., Manheimer, Wallace, Chapter 14. 4"Fusion fission hybrid reactors. In: Dolan, T.J. (Ed.), Magnetic Fusion Technology. ***Springer*** Verlag, London, 2014.

[11] Cadwallader, L., Qualitative Reliability Issues For In-Vessel Solid And Liquid Wall Fusion Designs, ***Fusion Technology***, Vol.39, No.2, S.991-995, 2001.

[12] H.M. Sahin, G. Tunc,, N. Sahin, Investigation of tritium breeding ratio using different coolant material in a fusion-fission hybrid reactor. ***Int. J. Hydrogen Energy*** 41, 7069–7075, 2016.

[13] Sawan M., Abdou M. Physics and technology conditions for attaining tritium self-sufficiency for the DT fuel cycle, ***Fusion Engineering and Design***, 81, 1131-1144, 2006.

[14] Şahin, H. M. Monte Carlo calculation of radiation damage in first wall of an experimental hybrid reactor. ***Annals of Nuclear Energy***, 34(11), 861-870, 2007.

[15] Tunç, G., Şahin, H. M., & Şahin, S. Evaluation of the radiation damage parameters of ODS steel alloys in the first wall of deuterium-tritium fusion-fission (hybrid) reactors. ***International Journal of Energy Research***, 42(1), 198-206, 2017.

[16] Şahin, H.M., Tunç, G., Karakoç, A. et al. Neutronic study on the effect of first wall material thickness on tritium production and material damage in a fusion reactor. ***NUCL SCI TECH*** 33, 43 (2022).

[17] Duderstadt, J.J., Moses, G.A., 1982. Inertial Confinement Fusion. ***John Wiley and Sons***, New York.

[18] Karakoç, A. Neutronic investigation of fissile material production from molten salt fuel mixture in a fusion reactor blanket, Ph.D. Dissertation. In: ***Department of Energy Systems Engineering. Karabük University*, 2024**

Association for Scientific and Academic Research (ASAR)
www.asar.net.in

ISBN 979-8-3313-2087-4

9 798331 320874

CURRAN ASSOCIATES INC.
proceedings.com